Site Planning and Design Handbook

Site Planning and Design Handbook

Thomas H. Russ

Illustrations by Stephen M. Russ

McGraw Hill

Boston, Massachusetts Burr Ridge, Illinois
Dubuque, Iowa Madison, Wisconsin New York, New York
San Francisco, California St. Louis, Missouri

Cataloging-in-Publication Data is on file with the Library of Congress.

McGraw-Hill

A Division of The ***McGraw-Hill*** *Companies*

3 4 5 6 7 8 9 BKM BKM 0 9 8 7 6 5

The sponsoring editor for this book was Cary Sullivan, the editing supervisor was Caroline Levine, and the production supervisor was Sherri Souffrance. This book was set in Century Schoolbook per the MHT design by Paul Scozzari of McGraw-Hill Professional's Hightstown, N.J., composition unit.

For Joseph and Marjorie Russ

Contents

Preface

The character of site design is such that the profession requires a combination of applying one's experience while always learning new methods and materials, and then educating peers, clients, and the public. Site design professionals more so than most other professionals are always faced with having their work reviewed and often modified by nonprofessionals. I recently sat in a meeting with representatives of a state department of education, all former teachers and not a planner or site professional among them. They were discussing the need to review plans prepared by professionals to be sure the plans were correct! Their primary concern was the environmental impact of school construction and seeing that the plan addressed this concern.

Land use planning in the United States takes place in the public arena with citizen-planners the final arbiters of how land is used and developed. While some citizen planners may be professionals themselves or have the benefit of a staff of professionals many more have no specific training or education. They bring only experience, an interest in public service, and a point of view to the process. Among the most difficult challenges of site design is the necessity to educate planning officials about their project and the elements of design synthesis.

The purpose and objective of this book has been to bring together the traditional practices and methods of site planning and design with those emerging practices that figure into the new sustainable site design paradigm. In addition, I have taken the liberty of making an argument for these practices from time to time throughout the book. Site design is a profession with both a technical and artistic character. Creativity is found in both realms. We have long recognized the esthetic beauty of particular natural and manmade landscapes without appreciating the beauty of the underlying form. Sustainability requires us to appreciate the underlying beauty of landscape as a process, not merely a picture.

Throughout the following chapters I have tried to provide a link to consensus standards and guidelines that will support the professional's design choices and provide a basis for educating clients and the public. When it is necessary to defend or argue for a new design principle or element, such consensus standards may provide the support and strength needed to persuade. In most cases

I have relied on the standards developed by the American Society for Testing and Materials (ASTM).

Chapter 1 is, in essence, the argument for a move toward a deeper design ethic, one that recognizes the environmental functions of landscape, considers the impact of design choices, and incorporates that knowledge into the practice of design. The pressures of population growth and anticipated changes in global climate have local implications that should be understood and considered by site design professionals. Chapter 1 includes some environmental data and projections of impacts for various states or regions of the United States. These projections are used as the basis for suggesting that local site design practices, and perhaps regulations, should anticipate these changes. Additionally, Chap. 1 briefly discusses the evaluation of the impact of design choices and suggests some principles for designers to consider in their practice.

The scope of site analysis changes with the scope of professional practice. Chapter 2 provides an overview of the site analysis in contemporary professional practice. In addition to traditional elements of site analysis, such as the study of the physiographic features of a site and issues relating to site access and utilities, the chapter discusses geologic hazards and environmental issues. Standards for the acquisition and use of aerial photogrammetry and historic aerial photography are included as well as resources available over the Internet. Site grading is discussed in Chap. 3. Particular attention is paid to minimizing the environmental impacts of site grading and to the creation of a site that respects the local topography to the extent practicable. The character and limitations of soils and slopes are also addressed as are the principles of erosion and sediment control.

Chapter 4 is concerned with the design of space for people, such as open space, community facilities, and walkways. In light of the anticipated aging of our population, the issues of accessibility and design for older people are addressed in several places in the chapter and throughout the book. Younger folks are also considered in the section on playground design concerns and the development of play fields. An analysis of community interests and open space requirements is also discussed in Chap. 4. Throughout the chapter, construction details and tables of design criteria are included.

The automobile is a pervasive influence on site design and planning. Chapter 5 is concerned with the design of local streets and parking facilities. Larger volume highway design is a specialty and is not addressed in this book. Local streets are considered from the viewpoint of driver and pedestrian. Specific performance criteria for local streets are presented as a basis for design. Street design is discussed in terms of its impacts on communities and pedestrian behavior. Alternatives to typical street design and traffic-calming techniques are presented as are considerations for reducing the environmental impacts of streets and parking lots. Paving materials are also discussed. Many of the concepts and methods are illustrated with construction details and photographs.

Chapter 6, the chapter on infrastructure, is concerned primarily with storm water management. Most utility design is prescribed by the local authority and so is not addressed. Storm water impacts are among the most significant impacts of site development. In general the chapter addresses storm water as a resource to be managed on the site to the extent possible. A variety of methods of infiltration and retention are profiled. Concerns with water quality are also addressed in the design standards to include required retention and contact time for water quality purposes. The chapter includes many construction details and photographs.

Landscape restoration has become a major area of practice and interest for landscape architects and engineers, and it is the topic of Chap. 7. The chapter provides an overview of methods and issues involved in stream restoration, site stabilization, wetland protection and restoration, and brownfield redevelopment. The chapter is illustrated with many construction details and photographs. Keeping in mind that the development that has the least impact is redevelopment, the focus in the chapter is the restoration of environmental function to impacted sites as part of the redevelopment process.

Chapter 8 addresses site layout and design of residential and local commercial property. The focus of the chapter is the design and construction of livable communities that are sensitive to people, the environment, and the marketplace. Community and homeowner preferences and Smart Growth objectives are used as the basis for design criteria. A variety of alternative layout strategies are illustrated using graphics and photographs.

Vegetation is often used primarily as window dressing or eyewash in the developed landscape. Chapter 9 is concerned with the use of vegetation as a functional element within the site design. While the esthetic contribution of plants is important and undeniable, there are many excellent references on planting design for that purpose. This chapter focuses instead on the functional aspects of vegetation. The use of vegetation to reduce energy loads and the design of wind and noise screens are addressed. In addition, the issues of native and invasive exotic plants as well as plant toxicity are discussed. Lists of invasive and toxic plants are included.

Design has been described as the process of avoiding failure. If this is true it is project management as much as any element of the process that is directly engaged in project success. Although project management is a large subject, it is an integral element of site planning and design. Chapter 10 addresses the key elements of project management that are directly related to the planning and design process. Subjects include communication, contracting, professional liability, and getting paid, among others. The approach of the discussion is practical and is addressed to the design professional cum project manager.

Chapter 11 is concerned with historic landscapes and land preservation. The preservation of important sites for historic, cultural, or other reasons has become an area of practice for some professionals and is a growing area of interest. This chapter deals with design concerns of historic and cultural sites and land preservation as distinct subjects. Methods of evaluation and planning are reviewed, and a brief overview of the historical context for various landscapes is included. Land preservation is addressed in terms of planning concerns and opportunities such as the transfer of development rights.

Throughout this book there is an awareness that the professional practice of site design is in a period of flux. We are in the process of sorting through our historic practices and methods with an increasing awareness of the need to incorporate sustainability as a critical element in all of our considerations. Much of the new design paradigm will be written in the next 25 years. It is an exciting time.

Tom Russ

Acknowledgments

While many have provided assistance and contributed in many ways, any shortcomings or errors must be attributed to the author. My thanks to the many people who provided assistance in the preparation of this book.

I am grateful to a number of people and organizations who allowed their work to be used in this book, especially Tom Scheuler of The Center for Watershed Protection, Diane Carstens of John Wiley and Sons, the American Association of State Highway and Transportation Officials, The McGraw-Hill Companies, the American Society of Civil Engineers, Faddis Concrete Products, and Charles Miller, P.E., of Roofscapes, Inc. Thanks also to the agencies of the United States who provided public domain information: the Environmental Protection Agency, Department of Commerce, United States Geological Survey, Department of Agriculture, the Natural Resource Conservation Services, and the U.S.E.P.A.

I am especially indebted to Zolna Russell, ASLA, Niall Kirkwood, and George B.L. Gibson for their assistance in developing this project, and I'm grateful to the many professionals at McGraw-Hill for their skill and effort in bringing this project to fruition, especially Cary Sullivan, Carol Levine, and Marci Nugent. As always, I appreciate the education and experience gained from my association with Robert B. Ludgate, Sr., PE, PLS. I thank Wendy Lochner, for her guidance and confidence in this project. The contributions of these and many other colleagues, clients, and teachers cannot be overstated. The best of my work is a reflection of what they have taught me.

I appreciate the indulgence of my colleagues during the time this book was being prepared, especially Chris Fox and Donna Linksz, Ph.D., at CCBC, and Bill Montgomery and Tim Keating at the College of Southern Maryland. Many thanks to my brother, Stephen M. Russ, for his effort at producing the many figures and details used in the book, but, more importantly, for his good humor and friendship. I am indebted to Tom Mudra for his valuable criticism, support, and friendship. Last, but most of all, for Karla, my friend and wife, for her love, patience, and enthusiasm for this and all of my projects.

Chapter

1

Sustainability and Site Design

The activities of human beings have had and will continue to have a significant impact on the earth's environment. It has been said that 60 percent of the earth's land surface is under the management of people, but 100 percent of the earth's surface is impacted by the practices of that management. Paul Erhlich (1994) used the formula $I = PAT$, or impact = population × affluence × technology, to illustrate the relationship of the number of people, the per capita rate of consumption, and the economic efficiency of consumption. Thus, for example, although the United States may have more efficient and cleaner technologies than some nations, its rate of consumption afforded by its relative affluence may offset those efficiencies. In contrast, although China has a high population, its relative low levels of affluence and technology may offset its high population. In both countries, however, the environmental footprint is clearly significant.

In 1987 the Brundtland Commission published *Our Common Future,* which said that to avoid or at least minimize the environmental impact of human behavior, it is necessary for society to adopt a sustainable approach to development. "Sustainability" was defined as "meeting the needs of the present without compromising the ability of future generations to meet their own needs."

In February 1996 the President's Council on Sustainable Development (PCSD) published *Sustainable America—A New Consensus for Prosperity, Opportunity and a Healthy Environment for the Future.* The PCSD identified 10 goals, but the first 3 could be viewed as encompassing them all: health, economic prosperity, and equity. Equity refers to social equity (equal opportunity) and intergenerational equity (equity for future generations).

To meet the challenges of sustainability, we need to change our behaviors and adapt to a paradigm of economic prosperity, social equity, and environmental sustainability. Unfortunately, these goals have traditionally been viewed as antagonistic or mutually exclusive. We tend to think of extremes: the most damaging economic activities affecting the best of the environment or the most restrictive environmental regulations resulting in dire economic consequences.

So we tend to think of economic health and environmental sustainability as mutually exclusive. The challenge we face is to reconcile our economic interests with our environmental interests.

We have learned that gains in some factors may be offset by losses in other factors. Between 1980 and 1995, per capita energy consumption in the United States fell, but total energy consumption increased by 10 percent because of a 14 percent increase in population. Likewise, while cars built in 2001 are 90 percent cleaner than cars built in 1970, there are so many more cars that the efficiency gains have been offset to some degree.

The impacts of development and land use patterns have been well documented during the last half of the twentieth century. Impacts range from a loss of water quality, a loss of wildlife habitat, a decrease in human health, the loss of native plants caused by the spread of invasive exotic plants, the loss of biodiversity, an increase in the cost of infrastructure maintenance, a decrease in groundwater tables, and more. In addition to these local impacts, human activities are having significant impacts on global climate. People around the world have become more aware of general environmental degradation, and they are turning to action.

Generally it takes from 20 to 30 years for technology to move from research and development to implementation in the land development and construction field. Reasons for the lag time vary but include the time it takes to raise public awareness of problems and available technological solutions to those problems. It takes still more time for the public to adopt the new solutions, both funding them and passing the necessary ordinances to implement them. Yet another reason for the lag time is the natural and predictable resistance of people to change. The various parties to a development project all have interests that they bring to the process, and all of them assess the development differently—how will the site fit into the community, will it be a financial success, does the plan meet codes and ordinances, and so on.

It is the job of the designer to synthesize all of these, often adversarial, views. It is also the designer who has the greatest opportunity to innovate and introduce alternatives to the planning and design of sites and landscape. As a professional, with a duty and responsibility for the health and safety of the public, it is the designer that has the burden to make the site "work." With the realization of the environmental impacts of a site's development, the introduction of innovative, more sustainable practices to a site's development can best be done by the site design professionals. While regulatory agencies may create a framework for more sustainable design practices, it is in the final analysis the site design professional that must implement these guidelines. Public officials and reviewers, however, share the responsibility to educate the public and elected officials as to the importance and desirability of change.

Most people's experience with change has been based largely on the introduction of new materials or methods into design and construction and new regulatory or permitting programs. However, the need for change has accelerated. Contemporary site planning and design must take into consideration much of the knowledge and information gained in recent years as our awareness of environmental impacts has improved. The leadership in incorporating this knowledge in site planning is coming from many different places, and it may require many of us to reevaluate our past work and assumptions in new terms and to begin approaching design differently. There can be a great deal of resistance to such change; methods and principles that have been acceptable in the past and that we thought were successful may have to be abandoned for other methods and new ways of thinking. Some of the logic we have used to plan and design sites will be augmented with new and additional considerations (see Table 1.1). In some cases such logic may be replaced entirely. In studying the impacts of past practices, it will be clear that a new paradigm is in order. This period of change is an exciting time for design professionals as we determine principles of land development for a sustainable postindustrial society.

In the United States site design has always been an issue of local control and practices because, in part, the conditions and needs of local communities and landscapes are too diverse to be addressed entirely in any single ordinance or set of regulations. Nonetheless, there have been common, if not universal, practices and methods that have served design professionals and communities well. The increasing awareness of the need for more sustainable land development includes emergent practices that also have broad application and value. In recent years the federal government and many states have passed incentives to encourage green building. Some states offer tax incentives to encourage energy efficiency and the use of green methods and materials. It is a practical certainty that being able to provide such service to clients will be a competitive necessity in only a few short years. It is through the design professionals that these changes to land development, site planning, and design will be introduced to most communities.

Population and Demographics

Trends in population and demographics have important implications for planners. Projections call for there to be an increase of about 130 million people in the U.S. population by 2050. Much of the population growth in the United States is occurring in the southwest and southeastern United States, the Sun Belt. Much of this area has semiarid to arid climate, and water may be in

TABLE 1.1 Environmental Risks As Ranked by Scientists

Relatively high risk problems	Relatively medium risk problems	Relatively low risk problems
Habitat alteration and destruction	Herbicides and pesticides	Oil spills
Species extinction	Toxics, nutrients	Groundwater pollution
Overall loss of biodiversity	Biochemical oxygen demand and turbidity in surface water	Radionuclides
Stratospheric ozone depletion	Acid deposition	Acid runoff to surface water
Global climate change	Airborne toxics	Thermal pollution

Adapted from *The Report of the Science Advisory Board Relative Risk Reduction Strategies Committee to the EPA* (Washington, D.C.: Government Printing Office, September 1990).

short supply. Shifts in populations will put increasing pressure on existing supplies and require more conservation. The use of more Xeriscaping and infiltration of storm water are already becoming standard practice as part of conservation efforts.

The energy issues that arose in California in 2000 and 2001 is an example of the complexity of the problems we face. The consumers are interested in access to affordable power but have been reluctant to authorize the construction of new generating plants. Although conservation is not a significant part of our national energy strategy, designers might anticipate more opportunities for innovation in site design that contribute to energy and water use efficiency as well as conservation. Conservation-related design is viable because it pays for itself and contributes to the bottom line of businesses.

At the time the 2000 U.S. Census was conducted, there were 77 million people in the United States over 50 years old. The midwestern and northeastern states are growing older. In some northern states the number of births per year is less than the replacement level. It is possible some northern states may experience a decline in population even while other parts of the country are expanding rapidly. Florida is already well known as a retirement destination, but other states such as Pennsylvania, West Virginia, Iowa, and North Dakota are seeing growth in their populations of retired people. In part this is because many younger people are moving to the Sun Belt states while older folks tend to remain close to home even in retirement—"aging in place" it has been called. The number of older people is expected to double in Montana, Idaho, Wyoming, Colorado, New Mexico, Arizona, Utah, Nevada, Washington, Oregon, the Carolinas, and Texas by 2025. Another factor contributing to the shifts occurring in the U.S. population is immigration. The number of immigrants to the United States promises to continue to grow, and immigrants tend to concentrate in "gateway" cities like Chicago and New York.

With the anticipated increase in population, the need for water and energy conservation and planned growth becomes even more important. Issues of "smart growth" will become more critical. For communities in some parts of the country, development pressure will only grow. Local government will have the opportunity to deal with growth-related issues including open space and public facilities before the crush. The community consideration of the standards to be used for that future growth should be undertaken as soon as possible, in accordance with the community's vision for its future.

The growing population of older Americans presents opportunities for design firms as well as significant challenges in some states where the majority of population growth is among the oldest people (see Table 1.2). It is expected that the baby-boomers will enjoy a relatively healthy and active retirement that may yet increase the continuing demand for housing and recreation. The nature of these products should be expected to change, however. Some cultural observers anticipate a return to simpler values and even a growing spirituality in the culture as the boomers reach retirement. These trends may indicate a growing philosophical awareness of the boomers or simply a reflection of lower retirement income. Some communities that allow for real estate and school tax

TABLE 1.2 Population Change from 2000 to 2025

	Total population			Population age 65 and older		
State	2000	2025	Change, %	2000	2025	Change, %
Alabama	4,451	5,224	17.4	582	1,069	83.7
Alaska	653	885	35.5	38	92	142.1
Arizona	4,798	6,412	33.6	635	1,368	115.4
Arkansas	2,631	3,065	16.5	377	731	93.9
California	32,521	49,285	51.5	3,387	6,424	89.7
Colorado	4,168	5,188	24.5	452	1,044	131.0
Connecticut	3,284	3,739	13.9	461	671	45.6
Delaware	768	861	12.1	97	92	(5.2)
District of Columbia	523	655	25.2	69	92	33.3
Florida	15,233	20,710	36.0	2,755	5,453	97.9
Georgia	7,875	9,869	25.3	779	1,668	114.1
Hawaii	1,257	1,812	44.2	157	289	84.1
Idaho	1,347	1,739	29.1	157	374	138.2
Illinois	12,051	13,440	11.5	1,484	2,234	50.5
Indiana	6,045	6,215	2.8	763	1,260	65.1
Iowa	2,900	3,040	4.8	442	686	55.2
Kansas	2,668	3,108	16.5	359	605	68.5
Kentucky	3,995	4,314	8.0	509	917	80.2
Louisiana	4,425	5,133	16.0	523	945	18.2
Maine	1,259	1,423	13.0	172	304	76.7
Maryland	5,275	6,274	18.9	589	1,029	74.7
Massachusetts	6,199	6,902	11.3	843	1,252	48.5
Michigan	9,679	10,072	4.1	1,197	1,821	52.1
Minnesota	4,840	5,510	13.8	596	1,099	84.4
Mississippi	2,816	3,142	11.6	344	615	78.8
Missouri	5,540	6,250	12.8	755	1,258	66.6
Montana	950	1,121	18.0	128	274	114.1
Nebraska	1,705	1,930	13.2	239	405	69.5
Nevada	1,871	2,312	23.6	219	486	121.9
New Hampshire	1,224	1,439	17.6	142	273	92.3
New Jersey	8,178	9,558	16.9	1,090	1,654	51.7
New Mexico	1,860	2,612	40.4	206	441	114.1
New York	18,146	19,830	10.9	2,358	3,263	38.4
North Carolina	7,777	9,349	20.2	991	2,004	102.2
North Dakota	662	729	10.1	99	166	67.7

TABLE 1.2 Population Change from 2000 to 2025 (Continued)

	Total population			Population age 65 and older		
State	2000	2025	Change, %	2000	2025	Change, %
Ohio	11,319	11,744	3.8	1,525	2,305	51.1
Oklahoma	3,373	4,057	20.3	472	888	88.1
Oregon	3,397	4,349	28.0	471	1,054	123.8
Pennsylvania	12,202	12,683	3.9	1,899	2,659	40.0
Rhode Island	998	1,141	14.3	148	214	44.6
South Carolina	3,858	4,645	20.4	478	963	101.5
South Dakota	777	866	11.5	110	188	70.9
Tennessee	5,657	6,665	17.8	707	1,355	91.7
Texas	20,119	27,183	35.1	2,101	4,364	107.7
Utah	2,207	2,883	30.6	202	495	145.0
Vermont	617	678	9.9	73	138	89.0
Virginia	6,997	8,466	21.0	788	1,515	92.3
Washington	5,858	7,808	33.3	685	1,580	130.7
West Virginia	1,841	1,845	0.2	287	460	60.3
Wisconsin	5,326	5,867	10.2	705	1,200	70.2
Wyoming	525	694	32.2	62	145	133.9

Adapted from U.S. Bureau of the Census, 2000.

abatement for older taxpayers may experience a shrinkage in local tax income as local population ages in place at the same time as demand for services for older citizens rises.

Anticipated Effects of Global Climate Change

Global climate change models anticipate a broad range of impacts. These impacts are believed to be underway already, and we are to expect that many will begin to manifest significant impacts on the environment within the next 25 to 100 years. Many of these changes and impacts have direct implications for the development of land.

North America has a largely urban population: 75 percent of the population lives in cities or the suburban fringe of metropolitan areas. Moreover, 75 percent of the population lives in what are termed *coastal communities,* that is, communities influenced or situated by large bodies of water. The United States is the world leader in the production of greenhouse gases—the human-caused component in climate change. As governments around the world have recognized the trends indicating that climate change is already occurring, there has been growing international pressure on the United States to change its behavior.

Most climate change models are based on a doubling of carbon dioxide in the atmosphere. Carbon dioxide is a minor constituent in the atmosphere, representing only about 0.03 percent. At the time the industrial revolution began, there were about 280 parts per million (ppm), down from 1600 ppm about 300 million years ago. Much of the carbon dioxide from earlier epochs has been sequestered in deposits of coal and oil, in peat bogs and tundra. In 2001 there was about 360 ppm carbon dioxide, approximately a 30 percent increase. The increase is estimated to be about 2 percent each year, and it is predicted that a doubling of carbon dioxide over preindustrial revolution levels will occur in the second half of the twenty-first century. It is anticipated that there will be important changes in world climate with such a rapid and dramatic increase in carbon dioxide levels.

The models used to predict climate change trends are projections based on complex sets of factors. Different models give different results, but in general there is a valid and significant agreement on the global climate trends (see Table 1.3). There is a great deal of variability in the climate and weather of the United States and Canada, which means that projections based on these models may have limited use on a local level. Nevertheless, it is important to note that observed changes in weather and climate are consistent with the predictions of global climate change. Uncertainty exists in the models partly because of the limitations of data and science's ability to model something as complex as world climate and partly because it is unknown how people and governments will react to the information. If governments and business respond and reduce the emissions or alternatively increase the sequestration of carbon, for example, the impacts and degree of change may be less. All of the models presume a doubling of carbon dioxide by 2100 although more recent data from the International Panel on Climate Change indicate that the doubling may occur faster than originally expected. Recent work has indicated that the global average temperature increased 1°F in the twentieth century, but most of that increase occurred in the last 30 years, indicating that the rate of warming was increasing.

The area of greatest temperature change is expected to be in a zone from northwestern Canada, across southern Canada and the northern United States to southeastern Canada and northeastern United States. Average temperatures are expected to increase as much as 4°F over the next 100 years. This increase in temperature will decrease the area and length of time of annual snow cover and should result in earlier spring melts. The risk of rain-on-snow storms will also increase, and with it the risk of associated floods will increase. Most of the increase in temperature is occurring as warmer nights, that is, our daytimes are not necessarily significantly warmer but our nights are not as cool as they used to be so there is an increase in the average daily temperature. In effect, there is less cooling at night because greenhouse gases tend to hold the heat longer so the rate of nighttime cooling is slowed. Average temperatures are rising a great deal because the lows are not as low as they used to be.

In addition, the world's oceans are warming as well. The temperature of the sea is expected to rise and influence the weather. Thermal expansion of the ocean

TABLE 1.3 Anticipated Impacts of Global Climate Change on Temperature and Precipitation in Individual States

	Temperature change, °F				Precipitation change (−), %			
State	Spring	Summer	Fall	Winter	Spring	Summer	Fall	Winter
Alabama	3	2	4	2	10	15	15	N/C
Alaska	5	5	5	10	15	10	Slight	Slight
Arizona	3–4	5	3–4	5	20	Slight	30	60
Arkansas	3	2	3	2	15	25	15	N/C
California	<5	5	<5	5	20–30	N/C	20–30	>20–30
Colorado	3–4	5–6	3–4	5–6	10	Little change	10	20–70
Connecticut	4	4	4	4	<10–20	<10–20	10–20	>10–20
Delaware	3	4	4	4	<15–40	15–40	<15–40	>15–40
Florida	3–4	3–4	3–4	–4	Little change	Little change	Little change	Little change
Georgia	3	2	4	3	10	15–40	15–40	10
Hawaii	3	3	3	3	Uncertain of changes	Uncertain of changes	Uncertain of changes	Uncertain of changes
Idaho	4	5	4	5	10	Little change	1	20
Illinois	3	2	4	3	10	25–70	15–50	10
Indiana	3	2	4	3	10	10–50	20	10
Iowa	3	2	4	4	10	20	15	10
Kansas	2	3	4	4	15	15	15	Little change
Kentucky	3	<3	>3	3	20	30	20	<10
Louisiana	3	3	>3	<3	Little change	10	10	Little change
Maine	<4	>4	>4	<4	Little change	10	10	30
Maryland	3	4	4	4	<20	20	<20	20
Massachusetts	4	5	5	4	10	10	15	20–60
Michigan	4	4	4	4	5–15	20	5–15	5–15
Minnesota	4	<4	4	4	Little change	15	15	15
Mississippi	3	2	4	2	10	15	15	Little change
Missouri	3	2	3	3	15	20–60	15	Little change
Montana	4	4	5	5	10	10	10	15–40
Nebraska	3	3	4	4	10	10	10	15
Nevada	3–4	5–6	3–4	5–6	15	(10)	30	40

TABLE 1.3 Anticipated Impacts of Global Climate Change on Temperature and Precipitation in Individual States (Continued)

	Temperature change, °F				Precipitation change (−), %			
State	Spring	Summer	Fall	Winter	Spring	Summer	Fall	Winter
New Hampshire	4	5	5	5	Little change	10	10	25–60
New Jersey	4	>4	>4	4	<10–20	10–20	10–20	>10–20
New Mexico	3	5	4	5	15	Slight decrease	Slight increase	30
New York	4	>4	>4	4	<10–20	10–20	10–20	>10–20
North Carolina	3	3	3	3	15	>15	>15	15
North Dakota	4	3	4	4	5	10	20	25
Ohio	3	3	4	3	5–25	25	20	5–25
Oklahoma	2	3	3	4	20	20	Slight increase	Little change
Oregon	4	5	4	5	Slight increase	Light decrease	15	15
Pennsylvania	<4	>4	>4	<4	10	20	50	20
Rhode Island	4	5	5	4	10	10	15	25
South Carolina	3	3	3	3	15	>15	>15	<15
South Dakota	3	3	4	4	10	10	10	20
Tennessee	2–3	<2–3	2–3	2–3	20	30	20	Slight increase
Texas	3	4	4	4	10	10	10	(5–30)
Utah	3–4	5–6	3–4	5–6	10	(10)	30	40
Vermont	4	5	5	5	Little change	10	10	30
Virginia	3	3	4	3	20	20	20	20
Washington	4	5	4	5	Little change	Little change	Little change	10
West Virginia	3	3	4	3	20	20	20	>20
Wisconsin	4	<4	4	4	Little change	15–20	15–20	15–30
Wyoming	4	5	4	6	10	Decrease slightly	10	30

Compiled from USEPA. www.epa.gov/globalwarming/

and increases in runoff from glaciers and ice fields are expected to continue and result in rising ocean levels (see Table 1.4). In some places such as Texas and Louisiana, the effect of rising seas may be made worse by concurrent land subsidence. The world's oceans are expected to rise by nearly 20 in by 2100. Such an increase has significant implications for coastal communities. Will warmer oceans influence hurricane frequency and storm intensity?

TABLE 1.4 Climate and Sea Level Changes in Individual States

State	Temperature change, (−), °F	Precipitation change, %[a]	Sea level change, in[b]	Anticipated sea level change, in, 2000–2100
Alabama (Tuscaloosa)	(0.1)	20	9.0	20.0
Alaska (Anchorage)	3.9	10	—	10.0
Arizona (Tucson)	3.6	20[c]	—	—
Arkansas (Fayetteville)	0.4	20	—	—
California (Fresno)	1.4	20	3.0–8.0	13.0–19.0
Colorado (Fort Collins)	4.1	20	—	—
Connecticut (Storrs)	3.4	20	8.0	22.0
Delaware (Dover)	1.7	10	12.0	23.0
Florida (Ocala)	2.0	—[d]	7.0–9.0	18.0–20.0
Georgia (Albany)	(0.8)	10	13.0	25.0
Hawaii (Honolulu)	4.4	20	6.0–14.0	17.0–25.0
Idaho (Boise)	<1.0	20	—	—
Illinois (Decatur)	(0.2)	20	—	—
Indiana (Bloomington)	1.8	10	—	—
Iowa (Des Moines)	(0.02)	20	—	—
Kansas (Manhattan)	1.3	<20	—	—
Kentucky (Frankfort)	(1.4)	10	—	—
Lousiana (New Orleans)	N/C	5–20	—	—
Maine (Lewiston)	3.4	(20)	3.9	14.0
Maryland (College Park)	2.4	10	7.0	19.0
Massachusetts (Amherst)	2.0	20	11.0	22.0
Michigan (Ann Arbor)	1.1	20	—	—
Minnesota (Minneapolis)	1.0	20	—	—
Mississippi (Jackson)	2.1	20	5.0	15.0
Missouri (Jefferson City)	(0.5)	10	—	—
Montana (Helena)	1.3	(20)	—	—
Nebraska (Lincoln)	(0.2)	10[e]	—	—
Nevada (Elko)	0.6	20	—	—
New Hampshire (Hanover)	2	20	7.0	18.0
New Jersey (New Brunswick)	1.8	5–10	15.0	27.0
New Mexico (Albuquerque)	(0.8)	20	—	—
New York (Albany)	>1.0	20	10.0	22.0
North Carolina (Chapel Hill)	1.2	5	2.0	12.0
North Dakota (Bismarck)	1.3	(10)[f]	—	—
Ohio (Columbus)	0.3	10[g]	—	—

TABLE 1.4 Climate and Sea Level Changes in Individual States (Continued)

State	Temperature change, (−), °F	Precipitation change, %[a]	Sea level change, in[b]	Anticipated sea level change, in, 2000–2100
Oklahoma (Stillwater)	0.6	20	—	—
Oregon (Corvallis)	2.5	20[h]	4.0	6.0
Pennsylvania (Harrisburg)	1.2	20	—	—
Rhode Island (Providence) 12.4		3.3	20	2.0
South Carolina (Columbia)	1.3	20	9.0	19.0
South Dakota (Pierre)	1.6	20[i]	—	—
Tennessee (Nashville)	1.0	10	—	—
Texas (San Antonio)	0.5	(20)	25.0	38.0
Utah (Logan)	1.4	20	—	—
Vermont (Burlington)	0.4	5	—	—
Virginia (Richmond)	0.2[j]	10	12.0	23.3
Washington (Ellensburg)	1.0	20	8.0	19.0
West Virginia (Charleston)	10.0	10	—	—
Wyoming (Laramie)	1.5	(20)	—	—

[a]Change may not address all parts of a given state.

[b]Rate of change historically.

[c]Some parts of Arizona have experienced a 20 percent decline in precipitation.

[d]Precipitation has decreased in the south and the keys and increased in the north and panhandle.

[e]Precipitation has decreased as much as 10 percent in some parts of Idaho.

[f]Except in western Nebraska where precipitation has fallen by 20 percent.

[g]Precipitation has decreased in southern Ohio.

[h]Except leeward side of Cascade mountains where precipitation has decreased by 20 percent.

[i]Except southeastern part of South Dakota where precipitation has risen slightly.

[j]Other parts of Virginia have shown a decrease in temperature.

SOURCE: Compiled from USEPA. www.epa.gov/globalwarming/

Increases in shore and beach erosion should be anticipated along coastlines. Barrier island communities may experience significant losses. Local and state governments will be required to devise strategies for impacted communities that may require significant public expense. Insurance for coastal properties can be expected to rise significantly. Reinsurance companies have predicted catastrophic insurance losses associated with weather to increase to $300 billion worldwide through 2010. Beach replenishment will become an increasingly expensive (Table 1.5), and perhaps futile, effort. Barrier islands should be expected to shift landward in response to deepening oceans. Necessary mitigation methods such as the construction or improvement of existing sea walls

TABLE 1.5 Estimated Cost of Sand Replenishment for a 20-Inch Rise in Sea Level

State	Cumulative costs of shoreline protection, millions of dollars
Alabama	60–200
California	174–3,500
Connecticut	500–3,000
Delaware	34–147
Florida	1,700–8,800
Georgia	154–1,800
Hawaii	340–6,000
Louisiana	2,600–6,800
Maine	200–900
Maryland	35–200
Massachusetts	490–2,600
Mississippi	70–140
New Hampshire	39–104
New Jersey (Long Beach Island only)	100–500 (bulkheads and sea walls)
New York (Manhattan island only)	30–140 (bulkheads and sea walls)
North Carolina	660–3,600
Oregon	60–920
Rhode Island	90–150
South Carolina	1,200–9,400
Texas	4,200–12,800
Virginia	200–1,200
Washington	143–2,300

Compiled from U.S.E.P.A. information.

or bulkheads and the installation of revetments or levees on bayside beaches would add additional costs to the beach replenishment efforts. It is important to note that some of these costs are already being paid. Sea level rise has significant implications for water supply as well. Saltwater encroachment may become a larger problem as coastline communities continue to grow and groundwater use increases. It is expected that as much as 50 percent of the coastal wetlands will be inundated. Louisiana is currently losing 35 mi^2 of wetland each year due to saltwater intrusion.

Rising sea levels will complicate floods of tidal-influenced rivers and streams. Increased storm surges may back up streams and change floodplain characteristics. It has been calculated that a sea level rise of 40 in (1 m) would result in a flood with a frequency of 15 years actually inundating the same area a 100-year flood covered previously. The Federal Emergency Management

Agency (FEMA) estimated that a rise of 12 in and 36 in would increase the area impacted by a 100-year flood from 19,500 mi^2 to 23,000 and 27,000 mi^2, respectively. Damage resulting from these floods would be expected to rise 36 to 58 percent for a 12-in increase and from 102 to 200 percent for a 36-in increase.

Changes in precipitation patterns along the Gulf Coast, central and northern plains, and parts of the midwestern and northeastern United States may experience as much as 10 to 20 percent increase in annual precipitation. The distribution of the precipitation may also change as it arrives in more frequent storms of higher intensity. The more intense storms may result in less infiltration and a greater amount of runoff. The result would be falling groundwater tables, streams, and lakes. The shortened snow season may result in less snowpack in western states and earlier runoff. Reservoirs built to collect runoff for use throughout the year may begin to have a longer service period and experience shortages earlier in more frequent dry years. Earlier runoff may result in lower streams and river flows later in the summer as well. Reduced flows could impact hydroelectric production in some places. More frequent and intense rains in some places will result in increases in storm runoff, erosion, and slope instability. The increase in runoff may require a rethinking of the maximum probable storm (MPS) event in many places. It may require retrofitting of exiting storm water collection and control devices to retain more water and encourage infiltration.

Paradoxically with an increase in precipitation there is expected to be an increase in the number and severity of droughts. Increased temperatures will result in an increase in evaporation and a loss of soil moisture. The loss of soil moisture, and the increased runoff associated with more intense storm events, may result in lower streams and rivers but also warmer streams and rivers. Cold-water fisheries may become endangered in the southern-most ranges. Falling levels in the Great Lakes have already been observed, and it is possible that falling levels could limit commercial traffic in the Saint Lawrence River during certain times of dry years. This may be offset, however, by a longer ice-free season in the Great Lakes.

An increase in carbon dioxide should result in more robust plant growth. Some have observed that this is the “upside” to global climate change and will increase food and fiber production. Other studies have found that as carbon dioxide levels increase, some plants actually reduce the rate of photosynthesis. Still others observe that the increased production of plant mass results in an increase in plant litter, which alters the carbon/nitrogen ratio in the soil, in effect reducing the amount of nitrogen available for plants. The increase in leaf area will also increase the amount of transpiration, which will contribute to the drying of soils.

The implications of climate change may be significant. It is possible that most of the United States will experience an increase in the frequency of precipitation as the amount of rainfall increases and its distribution changes. Increased erosion and perhaps slope destabilization in some places can be expected with an increase in precipitation. Coastal communities may experience an increase in flooding and beach erosion. Flood-prone areas may increase in size as the sea levels rise. Public health officials and communities may become more sensitive to areas of standing water as subtropical and tropical diseases expand their range. Design strategies in impacted coastal communities may provide significant opportunities for innovation and problem solving.

Site planners and designers will have to respond to these climate changes by retrofitting existing facilities and designing new projects. While infiltration will continue to be an important element of site planning, perhaps the wet pond will be less desirable with the spread of the West Nile virus or malaria. Clearly, in their designs and planning, site planners will have to account for the life cycle and habitat preferences of the mosquitoes that transmit such diseases.

Anticipated warming in most places will result in increasing cooling costs for all buildings, including homes. Properly locating a building and plantings on a given site so as to lower energy costs will become even more important. As temperatures increase, plants growing in the extremes of their southern range may be subject to significant heat- and drought-related stresses. Some places may see a shift in species considered to be "native," particularly those living at the margins of their tolerance.

Figure 1.1 Photograph of a traditional street and neighborhood.

Land Use

Since World War II the growth of the suburbs has been the most important development, and possibly environmental, trend (Fig. 1.1). At the beginning of the twenty-first century more people live in suburbs than in the former urban centers. At the same time, awareness is growing that suburbs as they have evolved are unsustainable, but this knowledge has done little to slow the growth in consumer preference to live in suburban areas. There is a general acknowledgment that cities offer a greater cultural experience, but in general, populations have not started to return to urban areas in significant numbers. In fact, as they vote with their feet and checkbooks, people have shown their preference for suburban living over city living. Builders respond to market demand; they do not create it. Thus changing the trend to urban living will require changing public policy, which is politically difficult, if not impossible. Local ordinances tend to favor low-density development and highways, not parks and higher-density development. It is difficult for planners and designers to influence this suburban growth trend on a site-by-site basis. Instead, planners and designers will have to address the impacts of suburban development through design.

Paradoxically many people living in suburbs seem to prefer what might be considered urban values and character. A survey by the National Association of Home Builders (NAHB) found that those surveyed would prefer to live within walking distance of schools, shops, and community facilities. The study also found that in spite of the standard practices of most ordinances, most people would rather live in a place with narrower streets and more public open space. During the time that American families became smaller by nearly half, new houses have ballooned to more than twice the size. As the population has become older, however, there is an increasing interest in smaller homes. In some metropolitan areas most of the homes built and purchased are townhouses and condominiums (Fig. 1.2). Part of this popularity may be due to the cost of housing in some urban areas, but many of these units are higher-end dwellings located near shopping or social and cultural features of the city.

The southwestern parts of the United States are becoming more popular places to live, and designing for those areas presents significant challenges. The influx of people from more humid parts of the country has brought with it an expectation of life and an esthetic that often is simply out of place in the desert. The southeastern part of the United States is already facing problems with water supply. The native people of these dry places long ago found ways to live that recognized the character of their region. Our culture is faced with learning and acting on the lessons already known by so many, while our footprint is so much larger and deeper. These areas of growth are experiencing significant declines in other environmental indicators such as air quality, biodiversity, and human health. It remains to be seen if we can find the ways to live sustainably and successfully in the desert. Figure 1.3 is an example of good design.

Figure 1.2 Photograph of a contemporary urban neighborhood.

The shift of population to the South and the suburbs leaves many northern cities with declining populations and tax bases, underutilized infrastructure, and the remains of an industrial past that lasted only 50 years in many places. The recognition of brownfield redevelopment opportunities has been important in the last few years for cities and for designers. The challenges of redeveloping brownfields, however, require site designers to confront the impacts of industrial contamination; we can no longer assume a site to be clean and healthy. This requires a different mindset and more than a few new skills. Consequently, design professionals find themselves working on more diverse project teams. The roles of professional boundaries often blur within the context of projects looking for innovative solutions to complex problems.

Sustainable Development Principles

Our culture context for sustainability is in its infancy, but the dialogue is well underway. There are important voices encouraging us not to go back but forward, to solve problems through design. No single set of guidelines has emerged, but there is a growing recognition of the principles that lead to sustainable design and development. The views of leaders such as William McDonough and Emory Lovins are moving into boardrooms and legislatures and are beginning to change the expectations of design professionals.

Figure 1.3 Photograph of a southwestern home.

The definitions of "sustainable development" are too numerous to recount. The phase itself is in danger of becoming another meaningless mantra. Design professionals need to recognize the intellectual and professional challenge presented to them in the need to find a workable balance with nature. This may be the most important time for the design professions since they have emerged. Architects have made important advances in designing green buildings, though the practices are hardly mainstream yet. There are excellent but too few examples of sustainable site development practices.

Sustainable site planning must include considerations of the impact of the site development on the local ecosystem, the global ecosystem, and the future. Principles of green site work encourage the designer to consider the nature of the materials and the flows of energy and materials not only to build the project but to maintain it over its useful life and to dismantle and dispose of it eventually if necessary (Table 1.6). More than just modifying the way storm water is handled, for example, the designer should consider the life cycle costs of the materials being used, the ultimate disposition of the site and the materials, and ways in which any negative impacts can be reduced or mitigated.

The longer the useful life of a building or a site, the longer the environment has to "amortize" the impacts. But designing a site with an extended life span requires the designer to consider future and possibly different uses and to incorporate that thinking into the design. The most sustainable development is

TABLE 1.6 Guidelines for Green Site Planning and Design

1. Minimize cooling loads through careful building location and landscaping.
2. Utilize renewable energy resources to meet site energy demand for lighting.
3. Install energy-efficient lighting.
4. Use existing buildings and infrastructure instead of developing in "greenfields."
5. Design should create or contribute to a sense of community.
6. Design to reduce dependence on the automobile.
7. Reduce *material use* or increase the efficiency of material use.
8. Protect and preserve local ecosystem. Maintain the environmental function of the site.
9. Specify low-impact or green materials.
10. Site and buildings should be designed for longevity and to be recycled.
11. Design to minimize the use and runoff of water. Treat stormwater as a resource not a problem.
12. Minimize waste.

redevelopment. Reuse increases density and eliminates the loss of open space. Materials should be selected on the basis of durability and low environmental impact. Recycled materials are low impact and efficient. Better than recycling materials is reusing entire buildings. Many construction materials have significant environmental impacts either in the manufacturing process or in their final disposition as waste material. Others contain ozone-depleting compounds that continue to volatilize and pollute even after installation.

Reducing the impact of development may be possible by reducing the footprint of a building either by modifying the footprint to the most efficient shape or by building multiple stories. Reducing the surface area of a structure will reduce energy requirements as well.

Sites should be designed to treat storm water as a resource and to use water efficiently. This means not only capturing runoff and encouraging infiltration but also using native plants that are suited to the local climate and precipitation and using Xeriscaping techniques where applicable. Site planning should incorporate the existing environmental function of a site to the extent it is possible. Wetlands and important ecosystem elements such as wildlife habitat, tree masses, and stream corridors should be preserved. The ubiquitous lawn has a notoriously high environmental impact because of its requirements for pesticides, fertilizers, irrigation, and continual mowing. Lawns should be minimized in size and replaced with native species of plants selected for their esthetic quality and drought resistance. Buildings and tree masses can be located to help to minimize cooling costs.

Green Building Materials

The choice of building materials is as important as the site design or choice of construction methods. Designers have significant influence over the materials used through the specifications they make in design and planning. Many designers are not aware of the implications beyond cost and specific performance criteria of choosing one material over another. To be sure, site designers have fewer materials to choose from than do architects, but their awareness of the characteristics of site materials is just as important. As a matter of practice, materials should be selected in part because of their durability. The process of manufacturing materials is energy and material intensive, and durable materials usually require less maintenance over a longer service life. Materials that require less maintenance or whose maintenance has a lower environmental impact are also preferred. Materials that are heavily processed or manufactured have a higher *embodied energy*—that is, there are greater energy inputs required to manufacture the product. Locally produced products require less transportation energy and produce less pollution. Designers should seek a durable, locally produced, low-maintenance product with a low embodied energy. For example, local hardwoods are preferable to tropical woods, and local stone to imported stone.

The best choice for materials may be recycled materials. Using recycled materials reduces solid waste, reduces the energy needed for manufacturing, and reduces the impact on natural resources. Using fly ash in concrete, recycled plastic in site furniture, and ground tires in pavement are all possible ways of incorporating recycled materials in site work. Use of materials, such as pressure-treated lumber that contain toxins should be avoided by specifying alternatives such as recycled plastic lumber.

Determining whether a building material is green involves the consideration of the entire life cycle of the material: the manufacture of the material, the impacts of its use, its distribution and service life, and finally its disposal. Every stage of the material's life involves energy use and environmental impacts. There are a variety of different life cycle assessment techniques, including the Building for Economic and Environmental Sustainability (BEES) model developed by the National Institute of Standards and Technology (NIST) with support from the Environmental Protection Agency (EPA) and the Department of Housing and Urban Development (HUD). The BEES model considers 10 potential environmental impacts of building materials:

1. Global warming
2. Acidification
3. Eutrophication
4. Natural resource depletion
5. Indoor air quality
6. Solid waste
7. Smog
8. Ecological toxicity
9. Human toxicity
10. Ozone depletion

Each of the calculations involves converting impacts to a known and given reference point provided in the BEES documentation. The program then calculates the environmental loading of the product to allow designers to compare alternative materials. BEES software is available from the National Institute of Standards and Technology (NIST) along with a manual that describes the use of the software, explains the algorithms, and provides examples of material and product data already evaluated using BEES.

The American Society for Testing and Materials (ASTM) has developed the *Standard Guide for Environmental Life Cycle Assessment of Building Materials/Products,* E 1991–98. The standard guide describes a four-step process for conducting a life cycle assessment (LCA): a definition of goals, an analysis of inventory, an impact assessment, and an interpretation of findings. The LCA is broad based and comprehensive in scope and includes considerations of embodied energy, raw materials acquisition, and environmental impacts from cradle to grave, as well as performance considerations. Other more approachable methods have also emerged. There are public and private green building initiatives throughout the world. Many of these organizations have established standards or thresholds that products must meet to be listed as green. Since site work involves fewer materials as a rule than building construction, most of the work has been done on materials used in buildings. Still materials used in site development are not without their environmental "signature" as it were. The general elements of green building materials are summarized in Table 1.7.

The ASTM Subcommittee on Sustainability has developed the *Standard Practice for Data Collection for Sustainability Assessment of Building Products,* E 2129. This standard includes a checklist to guide the process of evaluating the environmental character of products. Most of the processed or manufactured materials specified in site work are related to paving and utility or storm water pipes. Even with these few categories of materials there is a wide range of choices designers may consider.

TABLE 1.7 Green Building Material Requirements

1. Products made from recycled or salvaged materials
2. Products made from wood harvested from Forest Stewardship Council Certified forests
3. Products made from materials that are renewable in the short term (10 years or less)
4. Products that do not contain toxics or environmentally damaging materials
5. Products (or methods) that reduce the material volume required
6. Products that reduce environmental impacts during the manufacturing process, construction, renovation or demolition
7. Products (or methods) that are energy efficient or that reduce the heating and cooling loads on a building
8. Products that are reusable or recyclable
9. Local products rather than products from far away

Pipe materials

Pipes are selected primarily for channeling storm water, conveying sanitary sewage, or distributing water. While in the past water systems commonly required ductile iron or steel pipe, today there are many choices of material for storm water and sewage collection. Selecting pipe materials might be a matter of complying with local ordinance or preference, but selecting materials should also involve a consideration of the costs and benefits of the possible choices.

Acryolonitrile butadienne styrene (ABS). ABS is used primarily for waste and storm water pipe. ABS is lighter than PVC, but it is more than twice as expensive. There have been reports of instances of off-quality material making it to the marketplace, resulting in failures in the field. ABS has almost twice the thermal expansion capacity of PVC. The resin material from which it is made is expensive to manufacture. ABS manufacturing involves a number of toxic materials, which have environmental impacts.

Cast iron. Many building codes still require cast iron pipe, but these codes tend to be related to political and economic pressures rather than to the value of the material itself. Cast iron is durable, and it has a low thermal expansion coefficient, but its great weight and associated labor costs would seem to offset those values. Cast iron is no more durable than PVC, for example. The energy and environmental impacts of cast iron pipe manufacture are quite high.

Concrete. Although very durable and resistant to wear, concrete pipe is heavy and expensive to install. It is still required in some local and state codes because of its durability. The strength of concrete makes it useful in applications where there is minimal cover or where significant loads are expected.

High-density polyethylene (HDPE). HDPE is the least expensive, lightest, and most flexible of the pipe materials. HDPE is relatively simple to manufacture, and it is the most easily recycled pipe material. It is manufactured in long sections, and it is familiar as the coils of pipe material used often to reline old pipelines and sewers. For all of its positive characteristics, HDPE unfortunately has the greatest expansion coefficient of any of the popular pipe materials; its capacity is more than twice the thermal expansion capacity of PVC, which limits the usefulness of HDPE for many applications.

Polyvinyl Chloride (PVC). PVC has become widely used because it is very strong, durable, lightweight, inexpensive, and easy to work with. It is used in a wide array of products, but in site design, it is used primarily as pipe or site furniture. About 60 percent of the PVC used in the United States is used in the construction industry. Available pipe diameters in PVC range from ⅛ to 36 in. Nearly all wastewater sewers constructed in the United States today are built of PVC pipe.

Manufacturing PVC has some environmental costs. Vinyl chloride is a carcinogen produced from ethylene and chlorine. PVC manufacturing produces about 4.6 million lb of vinyl chloride emissions each year. PVC manufacturing has been associated with the presence of dioxin—one of the most toxic substances known—in the environment; however, research has not established a clear risk associated with the quantities observed. More dioxin is produced when PVC is burned, however.

Some concerns associated with the decomposition of PVC are associated primarily with architectural or electrical uses of plasticized PVC and do not appear to be relevant to the exterior site applications of the material. PVC is difficult to recycle into consumer goods, however, primarily because of the wide range of formulations used in making different PVC products. Incinerating PVC is problematic because it has a low fuel value and it turns into hydrochloric acid as it burns, increasing the wear on incinerators.

Many products made of PVC include formulations that include lead and other toxins, and although these products are not usually associated with site development applications, there is noteworthy concern about the environmental costs and impacts of PVC manufacture, use, and disposal. There have been calls for stopping the manufacture of PVC because of these concerns.

Vitrified clay pipe (VCP). Vitrified clay pipe has been replaced in most applications by PVC, but it is still used for some applications. Many VCP installations are still in use for well over 100 years. It is durable and resistant to chemical corrosion, and it has the lowest thermal expansion coefficient of any pipe material. The weight of VCP (8.9 lb/ft for a 4-in VCP versus 2.0 lb/ft for a 4-in Schedule 40 PVC) leads to more handling and greater installation labor costs. As PVC has replaced VCP as the material of choice, the availability of VCP has dropped in some areas.

Cement and concrete

Concrete is widely used in all types of construction because it can be cast into a desired form and it is durable once it is cured. Cement manufacturing and concrete mixing make up a large business sector involving about 210 cement plants and almost 5000 ready-mix plants in the United States. Most ready-mix concrete for residential purposes is approximately 12 percent cement. The most common cement used is portland cement.

Manufacturing cement involves mixing a source of calcium (usually limestone) with finely ground additives (such as bauxite or iron ore) in a rotary kiln heated to about 2700°F (1480°C). As the kiln mixes the heated materials, a series of chemical reactions occurs: The materials form a molten mass, which is cooled and then ground to a powder, which is mixed with some gypsum to become cement. In turn, cement is mixed with sand, aggregate, and water and possible admixtures as specified to control setting time or plasticity of the final material.

Environmental considerations. The raw materials of cement are common enough. It takes about 3400 lb of raw material to produce 2000 lb of finished concrete. The most significant environmental impacts of cement manufacturing and concrete use are the amount of energy consumed, the energy-associated emissions of carbon dioxide and other greenhouse and acid-forming pollutants, the dust that results from the manufacturing process, and the pollution of surface waters from runoff and "washout water."

Manufacturing cement is an energy-intensive process involving burning fossil fuels to generate the high temperatures of the rotary kiln. Some cement plants have been converted to burn hazardous wastes or other solid waste to extract the energy value. The high temperature of the kiln can provide a fairly complete combustion with low levels of residual air pollution. According to the Portland Cement Association, a single cement kiln can consume more than a million tires each year. Other elements of concrete do not require the substantial energy inputs of cement manufacturing, and the use of fly ash in concrete reduces the energy load even more (EBN 1993).

In addition to the energy costs, there are environmental impacts associated with fugitive dusts. The EPA has estimated that for every ton of cement manufactured, there is about 360 lb of alkaline dust generated. Much of this occurs during the manufacturing process, but some is generated in handling and transporting the cement and in mixing it. At the cement manufacturing plant, much of the dust is captured in baghouses or other pollution control equipment. Ultimately some of the dust is used for agricultural soil amendments, but much is discarded in landfills. Dust generated at ready-mix facilities or construction sites is usually not controlled.

The alkaline character of cement may result in runoff or washout water with a pH as high as 12. High alkalinity is particularly harmful to aquatic life. Runoff from most concrete and ready-mix sites requires a surface water discharge permit. Washout on construction sites should be properly collected and managed on site.

Fly ash concrete. Fly ash is a residual by-product of burning coal that has become a more common substitute for portland cement in concrete. Fly ash is produced in the generation of electricity and industrial processes. In the past fly ash has been used for a variety of purposes but most commonly as landfill. The use of fly ash as a replacement for or in combination with portland cement reduces the need to produce portland cement and offsets the environmental costs to some degree. The advantages to using fly ash are well documented. Fly ash concrete results in stronger concrete, though it may take longer for strength to develop. Fly ash tends to increase the time it takes for concrete to set. While this may be an advantage in the summer because it allows longer working times, it may be a disadvantage in the winter. Concrete mixes can be adjusted for weather conditions. Local ready-mix plants are usually able to provide mixtures that are seasonally adjusted to a given area. The time for strength to develop can be reduced to be comparable to portland cement if a fly ash–portland cement mixture is used (15 to 30 percent fly ash). Fly ash concrete requires less water per unit of volume and so reduces shrinking and cracking. Fly ash concrete may not accept color dies or acid finishes with the same results as portland cement concrete.

Strategies for environmentally safe use of concrete. The key to the wise use of concrete begins with proper specification of materials and estimates of volumes. Alternative designs or products that minimize the amount of material necessary may be possible. Precast products, for example, may use less material than cast-in-place alternatives and may reduce on-site waste. Specifying fly ash concrete or fly ash–portland cement mixtures can improve strength and perhaps reduce the amount of material required. Solid waste that may be produced can be crushed and used as fill. Arrangements should be made to collect washout water and to control runoff from such areas.

Recycling paving involves milling the top surface of a roadway or parking lot and removing as little as ¾ in to as much as 3 or 4 in of the pavement. There are several methods for recycling the removed material. Recycling may be done in place, crushing the milled materials and mixing them with new asphalt emulsion and perhaps new asphalt materials to be used in repaving the surface. This is often done using a "train" of equipment to mill, crush and mix, and repave the road in a continuous "ribbon." In other cases the material is transported off site to a plant where remixing occurs. In some cases the pavement is heated during repaving to soften the material and bond the new surface. Recycling paving has been demonstrated to be a very cost efficient approach. Other recycled materials may be used as road base material.

Treated lumber

Wood is widely used in the landscape, and in most contemporary applications treated wood is specified because it lasts up to 30 times longer than untreated lumber. It could be argued that the extended service life helps to save trees that would otherwise be harvested and that this offsets the environmental problems associated with treated wood. In the past, wood was treated primarily with creosote, essentially a coal tar distillate, but creosote-treated wood is less commonly used today. The remaining wood preservatives fall into two categories: oil based and water based (Table 1.8).

Some concerns with using treated wood include whether the material will come into contact with people or animals or any water body, including groundwater. Alternative materials should be considered if the treated wood is to come into direct contact with food supplies. Treated lumber should not be used where it will come into direct contact with water that is used for drinking. However, federal guidelines allow for incidental uses such as docks and bridges. The type of treated lumber should be carefully considered in constructing playground equipment or picnic facilities; creosote and penta should not be used for these purposes.

TABLE 1.8 Types of Common Wood Preservatives

Preservative	Type	Character
Creosote	Oil	Restricted use only
Pentachlorophenol (penta)	Oil	Teratogenic properties, restricted use only
Chromated copper arsenate (CCA)	Water	Pressure treatment only, contains arsenic and chromium
Ammoniacal copper quaternary compound (ACQ)	Water	Pressure treatment only, does not use toxics, arsenic, and chromium

Disposing of treated wood presents a difficulty; it is, after all, treated to resist decomposition. Ideally waste wood is recycled, but it should not be composted. Some states prohibit burning treated wood. If treated wood is to be used, the best option for the environment is ammoniacal copper quaternary (ACQ) compound; however, some consideration should be given to specifying rot-resistant species from native trees or recycled plastic lumber.

Measuring Sustainability

Sustainability concerns go beyond the selection of materials. The layout of a site, the types and character of ground cover, and the management of the various landscape functions—all are critical issues that have implications in site design. First, what is the role of site development in contributing to these effects, and how might those effects best be mitigated? Next, given that some of the implications will influence the use and function of a site, how can these changes be accounted for in planning and design?

Site planning, design, and development are moving toward including sustainability as a matter of practice. As it has been in the past, it is the planner and designer's responsibility to find the synthesis of all the issues and interests and then educate the parties involved as to the value of considering sustainability in the plan and design. To include issues of sustainability, the planner and designer should become students of those subjects, giving them as much attention as they give any other site planning subject.

Chapter

2

Site Analysis

Site Analysis

In many respects site analysis is the most important step in the successful site design process. The purposes of the preliminary site analysis are to gather data for preliminary planning, evaluate the site for compatibility with the proposed project or use, recognize concerns requiring additional study, and form an understanding of the administrative requirements of the project such as building permits and approvals. The value of an analysis is in its clear and complete identification of issues and the character of the site as they relate to a proposed use. Although it is usually subject to fairly limited resources, it should be as far-reaching and broad in scope as feasible. The nature of the design business is that very often the initial site assessment is part of the proposal effort and is completed "out-of-pocket." Even more troublesome is that the effectiveness of a particular analysis may be difficult to measure until well into the design process or even after site work has actually begun. Corners cut or inaccurate assumptions made in the site analysis for expediency or economy may result in expensive rework and change orders during the design process or worse yet, during construction.

The site designer rarely has the resources or time to complete a comprehensive site investigation on speculation of winning work. Instead, site analyses are usually conducted in two steps: a proposal phase to facilitate winning the work and a postcontract phase. The *proposal phase site analysis* is extremely important because the proposal, sometimes even including preliminary design and costs, will be based on the outcome. Since the in-house resources provided for the assessment are usually limited, it is important that they be carefully used. The costs of collecting physical information at this stage of a project may be problematic so other sources of information must be found.

Site characterization is a more detailed site investigation that is usually undertaken after some degree of preliminary site planning. Site characterization generally includes a geotechnical analysis of subsurface conditions such as depth to bedrock, depth to groundwater, seasonal high water table, and soil makeup. The American Society of Testing and Materials (ASTM) has developed the *Standard Guide to Site Characterization for Engineering, Design and Construction Purposes,* ASTM D- 420. The standard guide provides the site designer with a consensus standard with which to plan and evaluate site characterizations.

Location

The first consideration of the site analysis is to locate the site. Site location entails more than simply locating the site on a map. "Location" in this sense is referring to the site in terms of the project's relationship to the community. Commercial projects will be concerned with visibility, site access, and traffic. Is the traffic past the site adequate or too congested? Is the street infrastructure adequate for the anticipated increase? What sort of improvements might be anticipated? Is the site accessible from the street? What sort of on-site improvements might be expected to facilitate access? Is the interior of the site visible from the street? From how far away will drivers be able to see the site? Can traffic access the site from both directions? Is a left-hand turn possible? Are the neighboring sites commercial or residential? Are off-site improvements required? Are the necessary utilities nearby?

Residential projects raise different concerns. How far away are schools, government services, and shopping? Are local roads and streets adequate to handle increased traffic? Is the character of the area conducive to the proposed project? Will future residents be able to enter and leave the site without traffic congestion? Are adjacent properties developed? If not, what will zoning allow?

Collecting Site Information

There are a number of existing sources of site information for the site designer. In many cases these should be readily available within the office. The development of the Internet has significantly increased the availability of other sources. In many cases this information is available in fairly specific forms that may contribute to the site analysis effort at little cost.

Site analysis is an interpretive process. The site assessment process involves collecting a broad array of information from what are individually fairly limited sets of information and combining the data collected for the purpose of projecting a future use of the land. In general, preliminary site assessments are based on precious little new information—that is, much of the analysis is based on the existing sources of information or first-hand observation. It is how the site information is understood and used that makes the difference. Site analysis of course is not conducted in a vacuum; it is the context of the proposed use that frames the scope and character of the effort. For example, among the most important considerations is the topography of the site. Sites with significant change in elevations are typically difficult and more expensive to develop. Of course, the same steep slopes that are a source of concern for the commercial builder may be the bread and butter of the resort or high-end residential developer.

Topography

The U.S. Geological Survey (USGS) is a common and valuable source of topographic information. A local selection of the 7.5-min quadrangle series of topographic maps is found in every design office. The amount of detail and relative accuracy for the cost is difficult to improve upon. USGS maps are available from a variety of sources, including the Internet (www.usgs.gov). There are commercial sources of topographic information on CD-ROM as well, and some of these are accessible through the USGS Web site. These commercial sites are operated by firms working in partnership with the USGS on a variety of projects.

The most basic element of site analysis is the lay of the land. The topography of a site may dictate the purposes for which the site may be practically used and eventually the layout of the proposed project. The location of buildings and roads, pedestrian circulation, and the arrangement of storm water features are all commonly affected by topography. The analyst must consider how the existing topography affects the proposed use and vice versa. Although the contour intervals are fairly large, the relative accuracy of the quad maps allows for interpolation for general planning purposes although they are not adequate for design.

The analysis of the site in the context of the proposed development program provides an early look into how the proposed development will fit into the site. Will significant earthwork be necessary? Will retaining walls or other appurtenances be required? Can the site be accessed from adjacent roads? Is there visibility into the site from adjacent roads?

The nature of the material making up the slope is also important. Though soils will be discussed in greater detail in another section, it is important to mention that soils surveys may provide important information pertaining to the erodability of soils and the risks associated with cut-and-fill operations. Removing established vegetation from slopes may create unstable conditions requiring additional engineering and construction costs. Many land development and zoning regulations include restrictions on the development of steep slopes.

A slope analysis is done to identify the areas of steep slopes and the possible location for building sites and access. The slope analysis is usually a graphic representation of slope shown in classes or ranges. The ranges are sometimes established by local ordinances that describe the parameters to be observed when conducting a slope analysis and steep slope development restrictions. The slope analysis may identify possible routes for on-site traffic circulation as well as drainage patterns. By viewing the finished drawing, the restrictions imposed by slopes and the development patterns that are in tune with the site generally become more apparent.

From a hillside the long views are generally considered the most valuable. A site analysis should include the identification of the long views and any obstructions or limitations to them. The development of the site should proceed with the maintenance and optimization of the long views. Undesirable views should also be identified and addressed in the analysis.

The approach to the site, as well as the actual means of access onto the site, are key elements. The best paths of circulation, the minimization of impact on the site to develop these networks, and the extent of required cuts and fills all must be considered. Existing design requirements in ordinances may require revision to make the hillside project work. What works on a flat site may not work on a hillside without extensive earthwork and disturbances. Sight distance for egress to public roads should also be considered.

Other elements of the site analysis should include the identification of canyons, wetlands, rock outcroppings, existing structures, unique habitats, or natural features, as well as neighboring land uses and utility locations. The location of rights-of-way, easements, and other encroachments is also important. Based on the site analysis, it may be found that further research or study is required to determine the stability of slopes, hydrologic conditions, or the extent of wetlands. The site analysis is the foundation of the plan. It will provide the framework from which the planning and design are developed. Flat and low areas may present their own concerns. Boggy or wet areas may be wetlands and restrict development. Sites that are low or flat may be difficult to drain and present design challenges of their own.

The aspect of the site may also be an important factor. Orientation toward the sun may influence how well selected vegetation will perform and will impact the performance of buildings as well. A northern-oriented slope will be cooler than a southern-facing slope, and a southwestern exposure may be quite hot in the summer. The implications of aspect can be translated into energy consumption and other factors of the development. Building orientation may become a more important factor in the future if anticipated global climate changes and energy efficiency concerns become paramount.

The USGS, however, is a source of significantly more than topographic maps. The USGS is able to provide aerial photographs, digital orthophoto quadrangles, and other high-quality sources of site data. Through the Center for Integration of Natural Disaster Information (CINDI), the USGS is able to provide a great deal of information about regional and local site hazards such as earthquakes, landslide risks, groundwater conditions, and flood risk. The USGS also is a source of information about site geology. A series of geologic maps and information of geologic hazards (sinkholes, slides, earthquakes, faults, etc.) based on the topographic quadrangle maps is also available. These maps include known paleontological information as well. The USGS completed a survey of the biological status of the United States in 2000. This survey includes information on endangered as well as exotic invasive species.

USDA plant hardiness zones

The U.S. Department of Agriculture (USDA) updated the plant hardiness zone map, so familiar to growers and planners, in 1990 and reformatted the map in 1998. The new version incorporates new temperature information by using coldest weather data from the years 1970 to 1986. The new map introduces a new zone, Zone 11, which is essentially a frost-free zone. This discussion of the plant hardiness zone map is included in this section on zone analysis to encourage landscape architects and site planners to consider potential impacts of global climate change in their consideration of a site. The new interactive map is accessible via the Web site http://www.ars-grin.gov/ars/Beltsville/na/hardzone/ushzmap.html.

It is estimated that warming trends may have significant impacts over the next 100 years, with notable changes occurring by 2025. These changes may have significant impacts on the performance of designs under consideration today. Although there is no broad consensus as to how to address these concerns in design "on the boards," designers should begin to consider incorporating the most likely scenarios and trends into their work. Trends in climate indicate different concerns for different parts of North America.

FEMA maps

The Federal Emergency Management Agency (FEMA) is best known for the flood maps it has published over the years. Just as the USGS is more than topographic maps, however, the FEMA provides much more to the site analysis process than flood information. The FEMA maintains a Web site that allows the designer to create a fairly site specific map of hazards related to earthquakes, tornadoes, wind, and hail as well as floods. The FEMA Web site (www.fema.gov) includes a number of valuable links including one specifically for design professionals' questions. Unfortunately, the FEMA does not yet provide flood insurance maps online, but these maps may be purchased in either paper or digital form through the Web site. The FEMA does provide information on changes to the existing maps on its Web site.

Vegetation

An assessment of existing vegetation may tell a designer a great deal about a site. Evidence of second-growth vegetation is an indication of past activities that should be reconciled by the analyst with other sources of information. If the site indicates significant disturbance from past activities, there should be a record somewhere of what those activities were. The quality of vegetation is also an important consideration. The presence of good-quality specimens of trees or a valuable population of another type of plant might be important to protect or incorporate into a future design. The presence of water-tolerant plant species may indicate a high water table or frequent flooding whereas poor-quality or stressed vegetation may indicate problematic soil or subsurface conditions.

Prior to making a site visit, the analyst should consult local or state sources for information pertaining to protected plant species. In many cases the location of populations of protected species are mapped by such agencies. The discovery of such a plant population or community could have significant impact on the future use and development of the site.

The existence of certain trees or tree masses may contribute value to the finished project. Mature trees are known to increase the market value of property. A qualified arborist should be asked to assess the condition of specimen trees to determine the relative value of the tree. The location of a tree in the terms of the future development must also be considered. Although a variety of methods exist with which to base an evaluation, they generally have certain elements in common. These elements include the type of tree and the characteristics of the species as displayed by the specimen such as form, color, and shape and the tree's condition.

James Urban, ASLA, has developed a practical and usable approach to tree evaluation, and the method is shown in Table 2.1. While Urban's method was specifically developed for city trees, the fundamental approach can serve as a guideline to evaluating the trees on a given site, particularly during the early site analysis stage.

Current aerial photogrammetry

Aerial photogrammetry provides an accurate mapping of topographic and physiographic features using low-level aerial photography. The topography is interpolated from limited topographic data collected on the ground. Properly prepared photogrammetry will meet USGA *National Map Accuracy Standards* as listed in Table 2.2 and may be significantly less expensive than traditional field topographic methods, especially on large projects or projects with significant topographic variation or many features.

The ability to take aerial photographs may be hampered by vegetation that obscures the ground, and therefore these photographs may be collected only during winter months in some areas. In general, the cost of photogrammetry prohibits its use in the preliminary analysis stage. Many municipalities, however, have photogrammetric information available for review.

Historical aerial photography

Unlike photogrammetry, existing aerial photography can be a valuable source of information for the site designer at a relatively low price. In many places

TABLE 2.1 Urban's Tree Condition Methodology

1. *Excellent condition.* No noticeable problems, branching regular and even, normal-sized leaves, normal color.
2. *Good condition.* Full grown with no tip dieback, many minor bark wounds, thinner crowns, slightly smaller leaf size or minor infestations.
3. *Fair condition.* One or more of the following: (a) minor tip or crown dieback (less than 10%); (*b*) small yellowed or disfigured leaves, thinner crown; (*c*) significant limb wounds; (*d*) recent large branch removed that minimally affects shape; (*e*) large insect infestation; (*f*) any problem that should be repaired without long-term effect on the plant's health.
4. *Poor condition.* Any of the following: (*a*) crown dieback from 10% to 25%; (*b*) significantly smaller, yellowed, or disfigured leaves; (*c*) branch removal that affects the crown shape in a significant way; (*d*) wounding to the bark that will affect the tree's health.
5. *Very poor condition.* Any problem that is so significant that it grossly affects the shape or the health of the tree. Trees that have little hope of survival.
6. *Replace.* Some green may be seen, but the tree is not going to survive.
7. *Dead.*

TABLE 2.2 USGA National Mapping Program Standards

USGA National Map Accuracy Standards

With a view to the utmost economy and expedition in producing maps which fulfill not only the broad needs for standard or principal maps but also the reasonable particular needs of individual agencies, standards of accuracy for published maps are defined as follows:

1. *Horizontal accuracy.* For maps on publication scales larger than 1:20,000, not more than 10% of the points tested shall be in error by more than $\frac{1}{30}$ in, measured on the publication scale; for maps on publication scales of 1:20,000 or smaller, $\frac{1}{50}$ in. These limits of accuracy shall apply in all cases to positions of well-defined points only. Well-defined points are those that are easily visible or recoverable on the ground, such as the following: monuments or markers, such as benchmarks, property boundary monuments; intersections of roads, railroads, etc.; corners of large buildings or structures (or center points of small buildings); etc. In general, what is well defined will be determined by what is plottable on the scale of the map within $\frac{1}{100}$ in. Thus while the intersection of two road or property lines meeting at right angles would come within a sensible interpretation, identification of the intersection of such lines meeting at an acute angle would obviously not be practicable within $\frac{1}{100}$ in. Similarly, features not identifiable upon the ground within close limits are not to be considered as test points within the limits quoted, even though their positions may be scaled closely upon the map. In this class would come timberlines, soil boundaries, etc.
2. *Vertical accuracy.* Vertical accuracy, as applied to contour maps on all publication scales, shall be such that not more than 10% of the elevations tested shall be in error by more than one-half the contour interval. In checking elevations taken from the map, the apparent vertical error may be decreased by assuming a horizontal displacement within the permissible horizontal error for a map of that scale.
3. The accuracy of any map may be tested by comparing the positions of points whose locations or elevations are shown upon it with corresponding positions as determined by surveys of a higher accuracy. Tests shall be made by the producing agency, which shall also determine which of its maps are to be tested, and the extent of the testing.
4. Published maps meeting these accuracy requirements shall note this fact on their legends, as follows:

   "This map complies with National Map Accuracy Standards."
5. Published maps whose errors exceed those aforestated shall omit from their legends all mention of standard accuracy.
6. When a published map is a considerable enlargement of a map drawing (manuscript) or of a published map, that fact shall be stated in the legend. For example, "This map is an enlargement of a 1:20,000-scale map drawing," or "This map is an enlargement of a 1:24,000-scale published map."
7. To facilitate ready interchange and use of basic information for map construction among all federal map-making agencies, manuscript maps and published maps, wherever economically feasible and consistent with the uses to which the map is to be put, shall conform to latitude and longitude boundaries, being 15 min of latitude and longitude, or 7.5 min, or $3\frac{3}{4}$ min in size.

SOURCE: From the United States Geological Survey, http://rmmcweb.cr.usgs.gov/public/nmpstds/nmas647.html.

there are a variety of sources for historical aerial photography. Private firms may have generations of aerial photography taken on speculation or on contract. Many communities also have aerial photography collected over years. Some state geological surveys and the USGS also have historic aerial photography available for purchase. The American Society for Testing and Materials (ASTM) has developed the *Standard Guide for Acquisition of File Aerial Photography and Imagery for Establishing Historic Site-Use and Surfacial Conditions* (ASTM D5518-94e1). The guide can assist in the identification of sources of existing aerial photography as well as provide information regarding the specifications of such photography. The sources referenced by the standard guide are limited to public sources.

Public sources of photography are helpful, but many private sources exist as well. Private firms may be willing to work with the designer to enlarge and prepare special prints of existing photographs. Enlarging the photography may provide a valuable planning and analysis tool; however, photography firms may be reluctant to enlarge photography to the scales useful for site planners because of the inherent distortion and inaccuracy that can be anticipated in the resulting print. The most accurate part of a photograph is taken at the center of the lens. The curvature of the lens results in minor distortions toward the edges and corners of the picture. The distortions are minor at the original scale, but they increase as the photograph is enlarged beyond the intended scale. Such enlargements are of limited use, but they may be adequate for preliminary planning purposes and are particularly useful when making presentations to people who cannot read plans.

Enlarged aerial photographs sometimes reveal site features not clearly visible at ground level such as drainage patterns, sinkholes, and the remains of historic structures. The use of old aerial photography may reveal features that have been obscured by later site activities or development. The use of an aerial photograph is also helpful in presenting the site analysis data to clients and others who may not be comfortable reading plans. Examples of aerial photographs used to determine site conditions in the past are shown in Figs. 2.1 through 2.3.

USDA soil surveys

The soil surveys published by the U.S. Department of Agriculture (USDA) are a compendium of valuable information. The site survey contains information about topography, aspect, incidental physiographic information, and water-related issues as well as general information about climate and local history. Soils are classified by series, and these types are further refined into detailed soil map units. The soil descriptions include information on slope, depth to bedrock, soil texture, erodability, and rock and drainage characteristics. Experience has found soil maps to be generally accurate, but occasionally field observations indicate soil conditions at odds with the survey. In such cases local NRCS offices are usually be helpful in resolving the discrepancy.

Although soil borings and test pits may be used eventually, the site analysis may use existing sources of information such as the local soil survey or previous soil analyses. In addition to describing the character of the soil, the soil survey includes information about different management techniques, engineering characteristics, and uses for the land. Among the most important parts of the soil survey for site designers are charts describing the engineering and development capabilities of the land (see, for example, Fig. 2.4). Each local soil survey includes a description of how the survey was made and notes on how to read the survey.

Hazardous soil conditions

Expansive soils occur in every state of the United States. Impacts of expansive soils may be extensive cracking of sidewalks, foundation failures, retaining wall failure, and so on. Expansive soils are defined as described in Table 2.3.

Liquification is associated with earthquakes. It refers to the condition in which solid ground can turn mushy when soils are vibrated. Under certain conditions soils lose all bearing capacity, and buildings and bridges can slip or sink (as in quicksand) or buried structures (such as tanks) can float to the surface. These conditions have been associated with fine- to medium-grained sands and silts found in loosely packed layers. In general, the greater the soil density, the lower the liquification risk. A clay content of 15 percent or more is believed to be adequate protection from liquification (Borcherdt and Kennedy 1979).

Another form of liquification is found in quick clays. These are clays that can become "quick"—that is, they can liquefy. Confined to northern states and Canada (New York and Vermont have had quick-clay failures), these are very fine, flourlike clays formed as sediments in shallow waters and later raised above sea level. Collapse of quick clays has been associated with high water content, as the material weight exceeds its shear strength, resulting in slope failure.

Hydrology

The presence of water on the site and the general pattern of drainage are important concerns of the site analysis. Water is often the key feature of a site.

TABLE 2.3 Recognition of Expansive Soils in the Field

Under Dry Conditions
■ Soil is hard and almost rocklike; difficult to impossible to crush by hand.
■ Glazed, almost shiny surface where previously cut by shovel or scraper.
■ Very difficult to penetrate with pick or shovel.
■ Ground surface displays cracks occurring in a more or less regular pattern. Crack width and spacing are indicative of relative expansion potential in horizontal plane.
■ Surface irregularities such as tire tracks cannot be obliterated by foot pressure.
Under Wet Conditions
■ Soil very sticky. Exposed soil will accumulate on shoe soles to a thickness of 2–4 in when walked upon for a short distance.
■ Soil can be molded into a ball by hand. Hand molding will leave a nearly invisible powdery residue on hands after they dry.
■ A shovel will penetrate soil quite easily, and the cut surface will be smooth and tend to be shiny.
■ Freshly machine scraped or cut areas will tend to be smooth and shiny.
■ Heavy construction equipment such as bulldozers and compacting rollers will develop a thick soil coating, which may impair their function.

SOURCE: Reprinted with permission from Gary Griggs and John A. Gilchrest, *Geologic Hazards, Resources and Environmental Planning,* 2d ed (Belmont, Calif.: Wadsworth Publishing Company).

Figure 2.1 Aerial photograph of site showing conditions in 1963.

Figure 2.2 Aerial photograph of same site showing conditions in 1970.

Figure 2.3 Aerial photograph of same site showing conditions in 1988.

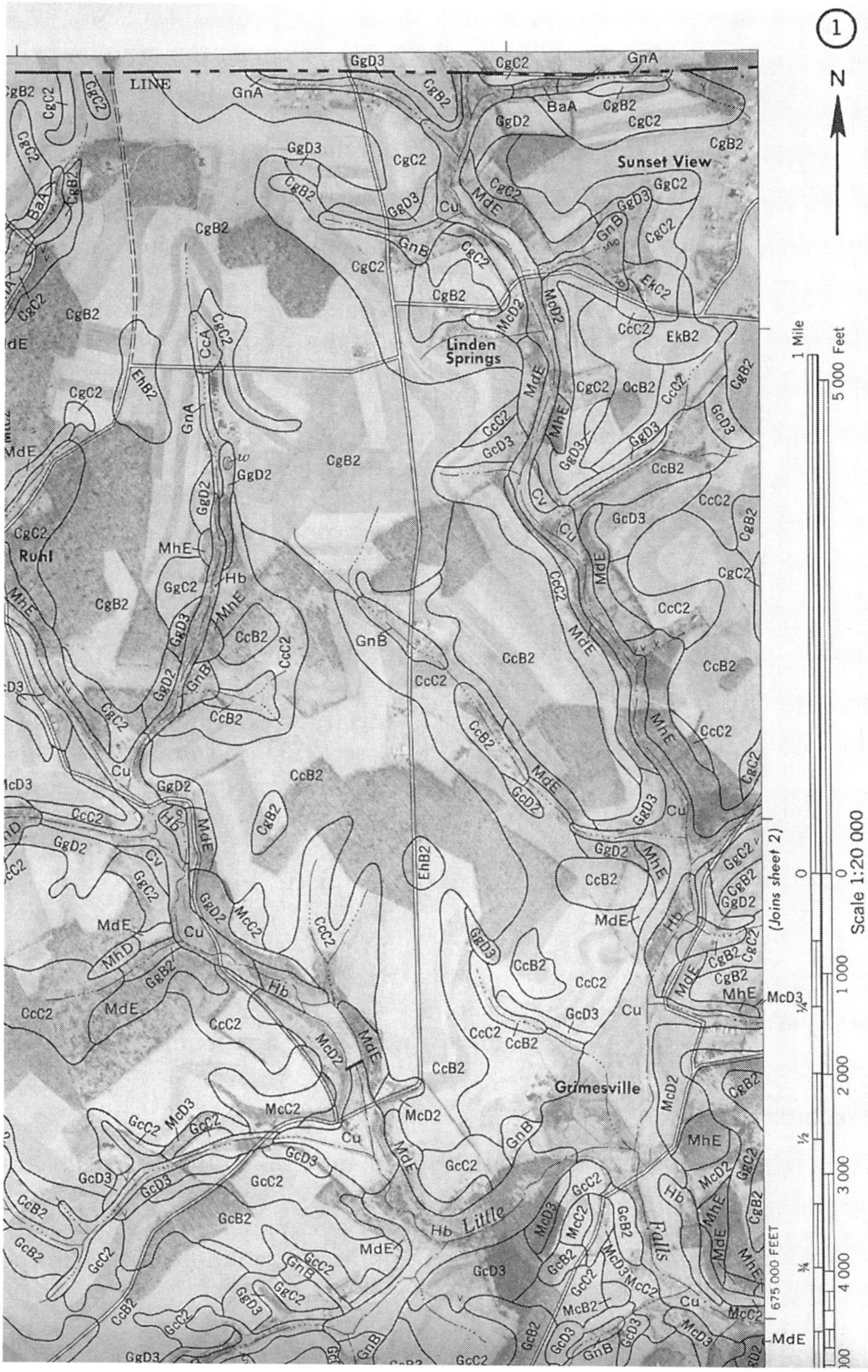

Figure 2.4 Typical USDA soil survey map.

Waterfront on a lake or the ocean or the presence of a stream or pond is considered to bring added value to a development, but it also brings concerns. The presence of a surface water feature may be coincidental with a fairly high water table or shallow geological features. Drainage patterns should be carefully observed in the field as well as from the published sources of information. The presence of associated wetlands and floodplains must also be considered and preliminarily located. The location and extent of riparian zones should be noted. The location of water features and other hydrologically linked features of the site should be carefully observed and evaluated.

Springs and seeps are important to locate and identify in the site analysis process. Very often these features are located on USGS maps or the USDA soil surveys, but their analyst should confirm their presence in the field. It may be appropriate to consider local off-site hydrology as well. The analyst should consider storm water drainage including drainage from other sites onto the subject site. Of particular concern are the volume, concentration, and quality of the runon storm water. Sites located along streams in the lower reaches of a watershed may be concerned with conditions higher in the watershed. The site analysis will also begin to identify storm water management strategies. The drainage pattern of the site and the presence of water features will indicate the likely location of storm water collection facilities.

The site analyst should consider the sensitivity of hydrologic features to development. Erosion and sedimentation during and after construction may represent a serious threat to surface water quality and habitat. If significant measures will be required to protect surface waters, these should be discussed in the site analysis. Many states have programs designating streams and lakes of high quality and providing for special protection measures for these waters. It should be determined if receiving waters are of high quality or restricted and how the status might affect the project.

In addition to sedimentation issues, the advent of the nonpoint source pollution programs of the National Pollution Discharge Elimination System (NPDES) have required municipalities to reevaluate storm water management schemes. The need to establish Total Maximum Daily Loads (TMDLs) for impacted waters may result in more stringent design requirements in the coming years.

Local records and history

Land use planning and development and regulation are generally an issue and a concern for local government. Local governments very often have substantial information about a site. As discussed in preceding sections, aerial photography, mapping, and other physiographic information is often available from local governments.

Zoning. Of all of the local sources of information, zoning regulations are probably the most important. Zoning regulations provide a prescription for how development is to be done in a community. The general conditions of development are described in terms of what development is encouraged and where in the community it will be. Zoning maps provide an overview of the community's vision for itself, showing not only how a site may be developed or used but also how surrounding sites might be.

Zoning regulations may contain design criteria such as parking configurations, lot sizes, setbacks, road widths, road profile restrictions, and sign requirements. Local regulations may also include specific performance requirements such as noise, solar access, or pollution loading restrictions. Zoning ordinances restrict development only in the sense that they provide for the limits and conditions of development, but they facilitate development by providing developers with a guidance document. Having a clear evaluation of the zoning particulars of a site is a critical requirement of a complete site analysis.

Occasionally zoning may include overlay zones that have important implications for land use. Overlay zones such as steep slope restrictions, watershed protection, historic preservation, or aquifer protection zones may severely limit land development activities or require a higher order of performance from the design, construction, and operation of a site.

Land development regulations. The scope of land development regulations varies widely from place to place. Very often these regulations reflect an evolution of practices as much as they are a reflection of a cogent regulatory process. Local ordinances are most valuable because they provide a glimpse into the experience of a municipality by reflecting its concerns and bias. Some ordinances are very prescriptive while others are concerned more with performance. In any case, understanding the local land development ordinances is second only to understanding the zoning regulations.

Land development regulations typically include the requirements for local street design, open space, lighting, subdivision standards (to be considered in conjunction with the zoning requirements), minimum landscaping, and similar site development parameters. The primary differences between zoning and land development regulations lie in the underlying authority. While local officials may have the authority to waive or modify provisions of the land development ordinance on a case-by-case basis, zoning regulations are enforceable and cannot be waived without justification and a formal hearing process. Although procedures exist to provide for variances and exceptions to zoning ordinances, these are formal procedures that offer little latitude to zoning hearing boards.

Zoning requirements of initial concern include the permitted uses, density allowances, minimum lot sizes, setbacks, and open space specifications. Care should be taken to consider the effect of wetlands, floodplains, or other site conditions that might influence the useful area in terms of density on the proposed site. Some zoning ordinances require special setbacks between different types of uses such as a buffer area between residential and commercial land uses. The requirement for buffers, screening, and open space should also be noted.

Utility mapping. Location of utilities is made possible using maps provided by local utility companies. The increase in the use of geographic information systems (GISs) has helped to provide reasonably accurate utility data in most places; however, utility maps are generally not considered accurate, and locations should be confirmed in the field for design purposes.

Historical value. Historical societies and agencies may also have important site information. The identification of historic and archaeological elements of a site is very important. Most states have regulations protecting historic or archaeological materials and sites. Discovering that a site has a historical feature or value is a critical piece of data in the early analysis. Sources of information regarding these features include local and state historical agencies and societies as well as local government records. Other sources include early USGS maps and libraries. Sometimes local names for features such as bridges and roads might be indicators of some historical or cultural element of value. Historical sources often have informative value as well. Place and road names often provide insight into former conditions and uses. "Swamp Road," for example, could suggest seasonal flooding or wetland conditions not in evidence at the time of a site visit.

Local historic and cultural values are sometimes hard to discern. Sources for information may address the physical area of value but not address the community attachment to less tangible values such as views or local character. These values are often unwritten and informal, but they may represent a significant, albeit unofficial, community interest that should be addressed. Though more difficult to identify, analysts should be sensitive to such community values.

Infrastructure

The location of surface and subsurface utilities is also completed in the site analysis. The analyst should identify the locations, capacity, and access to all necessary utilities, as well as the requirements for connections. Of particular importance might be moratoriums on sanitary sewer or water connections or exorbitant connection fees. Equally important is the consideration of interferences between utilities either on the site or in bringing the utilities to the site. Access to public water and sewers should be evaluated. The capacity of existing water and sewers may be of a concern in some communities and should be evaluated at these early stages.

The capacity of road networks to accommodate proposed traffic is also a concern. Are local roads of a type and design sufficient for the proposed project? Are turning radii adequate? Will traffic signals and other improvements be necessary? Requirements to upgrade public highways may be prohibitive for some projects.

Assessing "Fit"

Fit is a difficult criterion to define conclusively. It is, however, like quality—you will recognize it when you see it. In some places gauging fit is as simple as reading the zoning and local development plans; in other communities, fit is a more difficult assessment to make. In general, fit is determined by how the project design and function fit into local zoning goals, land development plans, the physical aspects of the site itself, the neighborhood, occasionally the region, and finally the values and needs of the community itself. In some respects it could be argued that these are listed in order of increasing difficulty to assess and to accommodate.

What are the program requirements?

The process of collecting site information is much the same for every project, but the analysis is always performed in the context of a proposed use or project. It is necessary to have an understanding of the proposed project to conduct the site analysis. In most cases the designer must rely on the client and experience to form a working understanding of the proposed project. Projects with a poorly defined program should be addressed cautiously by the professional. Experience suggests that such projects often have a high risk of failure associated with them; disappointed clients and unpaid invoices seem to accompany poorly defined or considered projects. Occasionally designers are asked to evaluate a site for its possible uses, in which case a series of analyses is done presuming different uses and parameters, but in most instances the analysis is conducted with an end use in mind. The analysis must consider the fundamental elements of a given project such as siting of proposed buildings, access to and from the site, lot layout, parking requirements, vehicular and pedestrian circulation, and a general strategy for storm water management. Physical development constraints such as slopes, wetlands, and floodplains must be accounted for in a preliminary fashion. Site analysts should extend their efforts to consider the off-site issues as well. These concerns may include traffic issues, local flood or storm water concerns, or infrastructure issues.

Permitting and administrative requirements are particularly important in contemporary site development. Knowing which permits are necessary and the expected lead time required to obtain them is often a critical element in a project. The professional should attempt to assess the desirability of the project to local government and people as well.

ADA and pedestrian access

The Americans with Disabilities Act (ADA) became law in the United States in 1990. Under the act a person with a disability is entitled to the same access and accommodations as the public in general. As a result, building and site owners are required to remove any barrier wherever such an accommodation is considered "readily achievable." The readily achievable test can be an ambiguous one for existing buildings, but for new construction it is clear, and all public-accessible designs must incorporate ADA principles and requirements. To enable compliance with the act, in 1998 the *Americans with Disabilities Act Accessibility Guidelines for Buildings and Facilities* (ADAAG) were developed and distributed by the Architectural and Transportation Barriers Compliance Board. More can be read about ADAAG on their Web site http://www.access-board.gov/adaag/html/adaag.htm. Standard design guideline graphics may be found at http://www.access-board.gov/adaag/html/Adfig.html.

While many of the design conventions of ADA have become commonplace, site designers may want to consider forming a preliminary analysis of the accessibility issues that may be encountered on sites. ADA issues in open space, recreational facilities, historic landscapes, or steep sites may present particular design challenges. The site analysis stage is not too early to be thinking about these issues and their impact on the design.

Community standards and expectations

Community standards and expectations are usually unwritten and often ambiguous, but sometimes they are very important considerations in the site analysis. Site designers may intuitively be able to assess the expectations of a community by observing what has been accepted as acceptable in the past: What does the community and neighborhood around the site look like? Standards for plantings, architectural elements, styles, materials, treatment of pedestrians and vehicles in existing design are all standards and expectations that often exceed written ordinances. A community interest that might be impacted by the project such as a loss of locally used open space or a loss of access to other land might engender resistance to a proposal. Anticipating and addressing these expectations in the early phases of design may contribute significantly to the project's acceptance by the community.

Environmental Conditions

Site analysis has necessarily expanded to include at least a cursory assessment of the environmental conditions evidenced on a site. "Environmental" in this sense refers to the narrow considerations of impacts caused by past industrial or commercial activities. An analyst should be aware of conditions that may indicate environmental contamination.

Another environmental aspect of growing concern to site designers is the impact of environmental trends such as global climate changes and its anticipated impacts and the growing demand to incorporate sustainability into site development. In particular, site designers working in coastal areas, areas subject to tidal influence, areas with important hydrologic characteristics such as wetlands or cold-water fisheries may wish to consider the anticipated impacts. Designers may need to incorporate the impacts into their selection of plant types, for example.

Environmental site assessment

The negative legacy of our past industrial waste disposal practices and experiences, such as the case in Love Canal, New York, prompted lawmakers to pass environmental laws to protect the public and to compel landowners to pay for the cleanup of their property. Today prudent real estate buyers and nearly all lenders require an environmental site assessment (ESA) of a property before committing to a purchase. As with any aspect of real estate development, planning is the key to managing this process—understanding the tools and hiring the right people. ESAs are risk-assessment processes used in the planning and feasibility stages of real estate development. Assessments are used to evaluate all types of property—virgin land, recycled land, and renovations—for conditions that are indicative of possible environmental contamination. The presence of actual contamination could trigger liability for the costs of site cleanup and restoration for the owners and users of the impacted property. By identifying the conditions prior to purchase, a buyer can avoid or minimize the exposure to the costs of remediation. Lenders want to limit their exposure to lawsuits and liability for cleanup responsibilities and will demand full disclosure of any known contaminants or conditions. The information in the site-assessment report should identify any recognized environmental conditions that exist and list what further steps might be required. Environmental site assessments are also performed in conjunction with applications for liability protections or release under various brownfield statutes and regulations.

The most common and widely accepted site-assessment protocols are those developed by the American Society for Testing and Materials (ASTM). These are consensus standards developed by practitioners and users of ESAs. The standing ASTM committee meets periodically to consider and occasionally revise the standard guidelines to reflect the state of the practice (see App. B). The ASTM has developed a variety of assessment protocols focused on various assessment activities. A partial list of the assessment standards is provided in Table 2.4.

Why perform a site assessment?

Environmental site assessments have become common practice because of the risk purchasers assume when they take ownership of a property. Under the federal Comprehensive Environmental Response Compensation and Liability Act (CERCLA), a landowner is liable for the environmental conditions on a piece of property whether the individual or company had any knowledge or involvement in causing the condition. This liability can include the costs of cleanup as well as damages to third parties.

The law provides buyers with several avenues of defense from this liability. These include "acts of God" and the "innocent-landowner" defense. The innocent-landowner defense is available to parties that can demonstrate that prior to acquiring a property, they had no knowledge of or reason to know of any adverse environmental conditions. They would demonstrate that they undertook an investigation into the historical use and current condition of the property and could find no indication of environmental contamination. This investigation would have to meet a standard of "due diligence" or customary commercial practice. Buyers of commercial property and lenders have learned to minimize their risk by engaging an environmental professional to complete an investigation. The consensus standard has emerged as a means of evaluating this good commercial practice. Site professionals may have an additional interest

TABLE 2.4 ASTM Standards for Site Assessment

E1528-00	Standard Practice for Environmental Site Assessments: Transaction Screen Process
E1527-00	Standard Practice for Environmental Site Assessments: Phase 1 Environmental Site Assessment Process
E1903-97	Standard Guide for Environmental Site Assessments: Phase II Environmental Site Assessment Process
D6235-98a	Standard Practice for Expedited Site Characterization of Vadose Zone and Ground Water Contamination at Hazardous Waste Contaminated Sites
E1984-98	Standard Guide for Process of Sustainable Brownfields Redevelopment
E1861-97	Standard Guide for Use of Coal Combustion By-Products in Structural Fills
D5746-98	Standard Classification of Environmental Condition of Property Area Types for Defense Base Closure and Realignment Facilities
E2091-00	Standard Guide for Use of Activity and Use Limitations, Including Institutional and Engineering Controls
D5730-98	Standard Guide for Site Characteristics for Environmental Purposes With Emphasis on Soil, Rock, the Vadose Zone and Ground Water
D5745-95-99	Standard Guide for Developing and Implementing Short-Term Measures or Early Actions for Site Remediation
E1923-97	Standard Guide for Sampling Terrestrial and Wetlands Vegetation
E1912-98	Standard Guide for Accelerated Site Characterization for Confirmed or Suspected Petroleum Releases
E1689-95	Standard Guide for Developing Conceptual Site Models for Contaminated Sites
E1624-94	Standard Guide for Chemical Fate in Site-Specific Sediment/Water Microcosms
D6429-99	Standard Guide for Selecting Surface Geophysical Methods
D6008-96	Standard Practice for Conducting Environmental Baseline Surveys
D5928-96	Standard Test Method for Screening of Waste for Radioactivity
D5745-95-99	Standard Guide for Developing and Implementing Short-Term Measures or Early Actions for Site Remediation
D5717-95e1	Standard Guide for Design of Ground-Water Monitoring Systems in Karst and Fractured-Rock Aquifers

in the ESA because of the potential of a late discovery of an environmental condition to disrupt the design and development process. Further, site design professionals may elect to fold elements of the site assessment, a transaction screening, into their own analysis of the site.

Format of a site assessment

Typically a transaction screen, or Phase I environmental site assessment, should be conducted before a title is transferred. A transaction screen may be performed by a person with knowledge of land and real estate. The Phase I environmental site assessment requires the services of an environmental professional. Although no standard definition or credential exists for an "environmental professional," in general, the standard of practice indicates that this person should have a combination of education and experience that is appropriate for the type of work to be performed. A site design professional, however, has adequate knowledge of land and real estate to conduct a transaction screen analysis. The professional can purchase a preprinted checklist from the ASTM that provides the entire standard guideline E-1528. Using the checklist, the site professional is able to walk through a cursory site-assessment process as part of the site analysis. Information collected in the screening process could contribute to the site analysis by identifying additional concerns that might impact the proposed use. The outcome of the site assessment may be to recommend that the client conduct a Phase I ESA.

The transaction screen may be used to provide guidance as to whether a Phase I is called for, but very often lenders require the Phase I as a minimum acceptable level of investigation. The screening process is a straightforward evaluation of the property and is usually most appropriate for properties where no development has occurred. However, in spite of these limitations, the site professional should consider adding the screening to the typical site analysis process. Some lenders have an in-house screening process, but the *ASTM Transaction Screening Guide,* Table 2.5, is the most commonly used format (*ASTM Standard Guide 1528*).

The Phase I site assessment

Several factors contribute to deciding whether to perform a Phase I ESA. First, if the buyer is a professional developer or a person familiar with real estate, there is some likelihood that he or she would be held to a higher standard of inquiry than an individual home buyer. This is probably true of site design

TABLE 2.5 ASTM Transaction Screening Guide, Level of Inquiry for an Environmental Screening

Has the site been filled in the past?
Is there any knowledge that the fill could contain hazardous materials or petroleum waste products?
Is the property in an area currently or historically used for industrial or commercial activities?
Is the property zoned for industrial or commercial uses?
Are adjacent properties used for industrial or commercial activities?
If there are existing or previous commercial or industrial uses, was there any indication that hazardous materials may have been used, generated, stored, or disposed of?
Does the site drain into a municipal collection system?
Do adjacent properties drain on to the site?
Are there reasons to suspect the quality of runoff from adjacent parcels?
Are there transformers on the property?
Is an on-site well required for water supply?

professionals as well. Second, if a site has ever been used for industrial or commercial activities, it should be assumed there is a greater chance that hazardous materials may have been used or stored on the property. This increased risk would compel a greater level of inquiry. Finally, many lenders will require a Phase I as a minimum level of inquiry.

The Phase I environmental site assessment process is usually completed by a qualified environmental professional. Although some states have defined the minimum qualifications for performing an ESA, most states have not. To determine if a state has minimum qualifications for environmental professionals, the state's environmental agency should be contacted. The ESA process requires interdisciplinary skills, and therefore it is difficult to prescribe a specific set of narrowly defined qualifications. Perhaps the best indicators of an environmental professional's qualification is in the combination of specific experience and education. Experience that is specific to the type of property or issues to be assessed should weigh more heavily than other experience. When evaluating education and training, consider the academic background of individuals but also review their commitment to continuing education and training. The ESA is a relatively new process and one that continues to evolve so that staying current with the latest standards and guidelines is critical for the environmental professional.

The *ASTM Standard Practice for Environmental Site Assessments, Phase I Environmental Site Assessment,* E-1527, provides clear guidance with which to undertake an ESA, but it also allows for the exercise of the judgment and discretion of the environmental professional. The expressed purpose of the ASTM standard practice is to establish a standard that will allow property buyers and developers to meet the requirements established by the laws and courts to minimize the risks of environmental liability associated with buying property. The standard can also be used to evaluate the final work product of the environmental professional. A checklist of the key points of the ASTM Standard may be used to measure the completeness of the report and work effort. It should be noted that this checklist is not a part of the ASTM standard guideline.

The Phase I ESA is designed in principle to be a cost-effective overview of a site that should identify indications of recognized environmental conditions. To keep the cost of the investigation at a reasonable level, the typical Phase I ESA involves no collection or testing of samples, and it is limited to information already available through public sources, interviews, or first-hand observation. This approach allows a buyer to determine if there is an indication of a problem or an increased risk with a particular property. By limiting the scope of the ESA, the cost is minimized, but the conclusions of the environmental professional are therefore drawn from limited information. For this reason the environmental professional may be unable to conclude that contamination is or is not present, and he or she may instead state that he or she can conclude only that there are indications of this condition or circumstances that could indicate contamination.

The ESA report should include copies of the notes collected during interviews, the database review summaries, maps, aerial photos, and any other reasonable documentation referenced in the report. It is important to note that the environmental professional is expected to exercise good judgment in the completion of the ESA. In some cases the environmental professional may elect to modify the ESA guidelines. While these changes are to be expected, deviations from the standard should be noted and explained to the site designer's satisfaction.

TABLE 2.6 Phase I Environmental Site Assessment Quality Assurance Review

This Phase I Environmental Site Assessment (ESA) Guidelines Review checklist is to be completed for the quality-assurance purpose of verifying the substantive compliance of an ESA report with the ASTM Standard Practice for Environmental Site Assessments, Phase I Environmental Site Assessment Process, E-1527. Except where noted otherwise, this review is based entirely on the report and does not include an independent confirmation of information.

Records Review

Does the report reference ASTM E-1527?

Was the ESA conducted by an environmental professional?

Is a résumé or statement of qualification attached?

Were proper minimum search distances (MSDs) used in the records search?

- Federal NPL Site List (1 mi)
- Federal CERCLIS list (0.5 mi)
- Federal RCRA TSD list (1 mi)
- Federal RCRA generators list (property and adjoiners)
- Federal ERNS list (property only)
- Equivalent state lists
- State landfill lists (0.5 mi)
- State leaking underground storage tank (LUST) list (0.5 mi)
- State registered underground storage tank (UST) list (property and adjoiners)

If proper minimum search distances were not used, was justification for each reduction and the new minimum distance provided?

Did the environmental professional provide an opinion as to the significance of any listing as a *recognized environmental condition* within the minimum search distances?

Was a current U.S. Geological Survey (USGS) 7.5 Minute Topographic Map used as the source of the physical setting data?

Identify the sources used to determine the history of the site and surrounding areas:

- Aerial photographs
- Local historic maps
- Historic USGS topographic maps
- Fire insurance maps
- Tax files
- Local records
- Interviews
- 50-year chain of title

TABLE 2.6 Phase I Environmental Site Assessment Quality Assurance Review (Continued)

Site Walkover
Did the environmental professional report any obstructions or obstacles that would prevent a thorough site reconnaisance?
Was the exterior of the property visually and physically observed and the description included in the report?
Was an inspection of the interior of the buildings conducted including accessible common areas and a representative sample of occupant areas?
Was information from a prior ESA used in the report?
Were changes between the earlier ESA and current observations noted?
Were the uses and conditions of the site reported?
Was the owner's representative present during the site visit?
Were interviews conducted?
Did the owner provide any additional documentation regarding the site?
Does the report include references to site conditions not visually and physically observed by the environmental professional?
Does the report include:
A description of the current site use and conditions?
A description of the adjoining property uses and conditions?
A description of the topographic and hydrologic conditions?
A general description of the structures?
Is the source of potable water identified?
The locations of roads and parking areas described?
Past uses of the property discernible?
Does the report include a conclusion or recommendations?
Based on this review, does the ESA meet the standard guidelines?

Brownfields

Brownfields are abandoned or underutilized properties that are environmentally impacted or are perceived as being impacted from past industrial or commercial activities. Such sites may present a designer with a wide range of unfamiliar site restrictions and conditions. Site planning on such sites must address the contamination or the mitigation strategy selected to protect the users and the environment. Normal practices of site development and storm water management may be restricted. In the past, site planning proceeded on the assumption that a site was clean. In the event of an impacted site, the designer was usually not involved in the remedial action design; sites were cleaned up, after which the redevelopment occurred as if on a clean site. To be effective, participants in the brownfield redevelopment project—landscape architects and site engineers—should be conversant with the environmental professional and understand the value and limitations of the site-assessment process. A site analysis checklist is given in Table 2.7. Further, a knowledge of the state of practice in site remediation technologies will increase the opportunity for collaboration and innovative site design and enable site designers to work closely with environmental professionals in the interests of the client and the environment.

TABLE 2.7 Site Analysis Checklist, Administrative Issues

Site Condition
Developed
Existing buildings or structures
Former uses
Known site conditions
Character and/or condition of exiting roads
Points of access and egress (approximate site distances)
Expected road improvements
Visibility into and out of site
Security considerations
Neighboring property uses
Existing rights of way or easements on property
Other encumbrances (condominium or community association?)
Zoning Regulations
Zone identification, permitted use? Special exception?
Minimum lot size
Front setback
Back setback
Side setback, one side, total
Permitted uses by right
Permitted uses by special exception
Maximum coverage
Parking requirements
Overlay zoning
Sign requirements
Right-of-way width
Cartway width
Curb requirements
Sidewalk requirements
Fence regulations
Storage requirements
Landscape ordinance

TABLE 2.7 Site Analysis Checklist, Administrative Issues (Continued)

Land Development Regulations
Street profile requirements
Site distance requirements
Slope restrictions
Storm water requirements
Landscaping requirements
Lighting requirements
Utilities
Access and/or distance to and connections requirements:
Natural gas
Telephone
Electricity
Cable television
Public water
Sanitary sewage
Traffic
Condition of local roads
Access to site
Internal circulation constraints
Impact on neighborhood
Topography
General topographic character of site
Areas of steep slope
Aspect and/or orientation of slopes
Site access
Slope stability
Soils and/or Geology
Soil types
Depth to bedrock
Depth to groundwater
Seasonal high water table
Engineering capabilities class of soils (density, Atterberg limits, compressibility)
Existing indication of slope instability and/or site erosion
Sinkholes
Fault zones

TABLE 2.7 Site Analysis Checklist, Administrative Issues (Continued)

Hydrology
Sketch existing drainage pattern, off site and on site
Presence of surface water features
Quality of surface waters
Floodplains
Wetlands
Riparian zones or floodplains
Springs
Wells
Aquifer
Anticipated drainage pattern
Character and quality of receiving waters
Vegetation and/or Wildlife
General types of existing vegetation
Quality of vegetation
Presence of known protected species
Presence of valuable specimens or communities
Presence of exotic and/or invasive species
Historic or Cultural Features and/or Community Interests
Known historical features
Unique natural features or character
Existing parks or public areas
Existing informal public access and/or use on the site
Community character such as architectural style and/or conventions
Local landscaping
Local materials
Environmental Concerns
Past site uses
Neighboring site uses
Evidence of fill, dumping, or disposal
Evidence of contamination (stained soils, stressed and/or dead vegetation, and so on)
On-site storage
Impact of site development on local water and air quality

Chapter

3

Site Grading

Land development involves disturbing the existing condition of a site in favor of a different condition, and the disturbance is usually directed by a design. From all appearances the range of what is good design appears to vary as much as the character of the sites being developed. Increasing concerns with sustainable site development will compel design professionals to give greater consideration to the predevelopment environmental function of a site and to seek ways to retain that function to the degree it is possible. Concern for the environmental and natural function of a site is not limited to the development of green sites. Undisturbed and pristine sites may have a higher functional quality than severely impacted former industrial sites; however, it is not unusual to find existing important functional elements even on environmentally compromised urban sites.

The layout and grading scheme of a site should consider and address the physical characteristics of the site including the functional aspects of the landscape. Ideally the new features such as roads or buildings will fit onto the site in a manner that will minimize the need for large cuts and fills. This requires the plan to accommodate the site and the arrangement of the features in a manner that maximizes the integrity of each of them. By minimizing the disturbance and the excavated area at the design level, the designer begins to mitigate the impact of the development. The design should retain as much of the original terrain and character of the site as is feasible. Roads should be parallel to contours as much as possible, and buildings should be located on the flatter areas of the site to minimize grading. Disturbed areas should be kept as small as possible, and strips of existing vegetation should be left in place between disturbed areas if possible. By grading smaller areas individually, the amounts of time and area of exposure and disturbance are minimized. The time of disturbance should be managed to minimize the risk of erosion and to maximize conditions to stabilize the site.

Engineering Properties of Soil

After obtaining general information relating to the topography, depth to bedrock, and hydrologic character of the soils, the soil survey is conducted to determine the development capabilities of the soil. Grain size distribution is an important factor in how a soil will behave under different conditions. The variations caused by grain size distribution, clay mineralogy, and organic content in the presence of water are issues for engineers; therefore, there are different classifications used by geologists than are used by engineers. The Unified Soil Classification System was developed by the U.S. Army Corps of Engineers to provide a relatively simple and reasonably accurate description of the physical characteristics of soil that are important to site development (see Tables 3.1 and 3.2). The classification is based on grain size, from coarse to fine, or the amount of organic matter in the soil. There are 12 soil classifications: four coarse-grained soils, four fine-grained soils, and four combinations of fine- and coarse-grained soils. The classification also includes three organic soils. A *coarse-grained soil* is one in which over half of the soil is sand sized or larger. In a *fine-grained soil,* half of the soil is silt or clay. Within these categories there are subcategories according to the distribution of soil particle size.

Soil grain sizes are assessed under the Unified Soil Classification System using a series of sieves (see Table 3.3). Other tests such as the Atterberg limits contribute to understanding and classifying the soil. Classification is done in accordance with the ASTM 2488 Standard Practice for Description and Identification of Soils (Visual-Manual Procedure) and the ASTM 2487 Standard Test Method for Classification of Soils for Engineering Purposes. In general, coarse-grained soils (GW) are preferred for subgrade and base materials,

TABLE 3.1 Unified Soil Classification System Symbols

Soil type	Symbol	Description
Clay soils	C	—
Silts	M	—
Sands	S	—
Gravels	G	—
Organic soils	O	—
High liquid limit	H	Water content > 50%, high plasticity (very cohesive or sticky clay).
Low liquid limit	L	Water content < 50%, low plasticity.
Well-graded soils	W	Particles of all sizes.
Poorly graded	P	Grain distribution is important because it affects consolidation and settlement.

Adapted from Harlan C. Landphair and Fred Klatt, Jr., *Landscape Construction,* 2nd ed., Elsevier, New York, 1988.

TABLE 3.2 Unified Soil Classification System

Unified soil classes	Shear strength	Compressibility	Workability	Permeability when compacted
GW	Excellent	Negligible	Excellent	Pervious
GP	Good	Negligible	Good	Very pervious
GM	Good to fair	Negligible	Good	Semipervious to impervious
GC	Good	Very low	Good	Impervious
SW	Excellent	Negligible	Excellent	Pervious
SP	Good	Very low	Fair	Pervious
SM	Good to fair	Low	Fair to impervious	Semipervious
SC	Good to fair	Low	Good	Impervious
ML	Fair	Medium to high	Fair	Semipervious to impervious
CL	Fair	Medium	Good to fair	Impervious
OL	Poor	Medium	Fair	Semipervious to impervious
MH	Fair to poor	High	Poor	Semipervious to impervious
CH	Poor	High to very high	Poor	Impervious
OH	Poor	High	Poor	Impervious
Pt	Highly organic soils, not suitable for construction			

although poorly graded gravels (GP) and silty gravels (GMd) may be used under some circumstances. Soils designated SM or SC are good for athletic surfaces and playing fields.

Porosity is the amount of pore space in a soil, which is related to grain size distribution and consolidation. *Permeability* refers to the rate at which water will freely drain through a soil. Clay soils usually have high porosity but low permeability and may settle considerably when loaded with a foundation, but they have lower compressibility and higher strength.

TABLE 3.3 Soil Fraction Distribution

Soil	Particle size
Fine (silt, clay)	Less than no. 200 sieve
Fine sand	No. 40–no. 200 sieve
Medium sand	No. 10–no. 40 sieve
Coarse sand	No. 4–no. 10 sieve
Sand	No. 4–no. 200 sieve
Fine gravel	3/4 in–no. 4 sieve
Gravel	3 in–no. 4 sieve
Cobbles	3 in–12 in

Soil strength refers to a soil's ability to resist deformation, which is a function of the friction and cohesion in the grain-to-grain contact in a soil. Sand dunes are able to stand at the angle of repose because of the grain-to-grain friction. *Cohesion* is the measure of the capacity of soil particles to stick together, and high cohesion is most often associated with clays. *Shear strength* is the measure of the frictional resistance and cohesion of a soil. To test shear strength, a four-bladed vane is driven into the soil and then turned using a wrench that measures the force (*torque*) necessary to turn the vane. The shear strength of the soil is the force applied at the time of failure. In situ field tests are preferred because soil is in its natural condition.

Bulk density refers to the weight per volume of any unit of soil. As a rule of thumb, the higher the bulk density of a soil, the greater the support it can provide for a foundation. Materials with low bulk densities do not provide a solid foundation for construction.

The Atterberg Limits and Soil Classification method quantifies the variations in soils caused by grain size distribution, clay mineralogy, and organic content. The Atterberg limits are actually two measures: the liquid limit and the plastic limit. These procedures measure the water in a soil at the point at which the soil begins to act as a liquid or begins to flow as a plastic. Water is measured as a percentage of the weight of the soil when it is dry.

The *liquid limit* (LL) is the moisture content at which a soil tends to flow and will not retain its shape. It is determined in a liquid-limit cup in which a molded wet soil patty is placed. A V groove is cut through the patty with a tool designed for that purpose. Using a hand crank, the cup is repeatedly lifted and dropped until the soil flows to close the groove. When the moisture content is sufficient to close the groove at up to 25 drops such that the soil "flows," it is said the liquid limit has been reached.

The *plastic limit* (PL) is the moisture content at which a soil deforms plastically. The soil is rolled into long threads until the threads just begin to crumble at a diameter of about 3 mm. If a soil can be rolled into finer threads without cracking, it contains more moisture than its plastic limits; if it cracks before 3 mm is reached, it has less.

The numerical difference between the LL and the PL is called the *plasticity index* (PI). The PI gives the range of moisture in which a soil behaves as a plastic material. Some clays can absorb water several times their own weight and would be said to have a large range of moisture content in which they behave plastically and before they start to flow. A PI over 15 is a good indicator of an expansive soil.

The Balanced Site

In general, the most economical grading plan is one in which there is a minimum of earthwork and the amounts of cut and fill are in balance. There are several factors that influence the balance. For example, soils with a high plasticity index or with a high organic content may have to be removed and replaced under building pads or under other site structures. Some soils have a tendency to expand when excavated, and some soils "bulk" significantly when disturbed. To design a balanced site, the professional needs geotechnical information regarding the soil's character, the bearing capacity of the soil, and its bulking factor, as well as the depth and character of the bedrock. Volumes have traditionally been calculated using a variety of methods and tools such as the average-end method; however, most designers today use a computer to determine volumes.

Site grading proceeds from a conceptual grading plan that attempts to balance the site and to locate the structures or program elements to maximize the site. From the initial design concept, the grading plan undergoes a series of iterations, each one bringing a greater level of detail to the design until the grading plan is final. The final grades adhere to appropriate grading standards (see Table 3.4). In many places grading standards are included in local ordinances and development regulations. Some government agencies and large development companies may have their own standards with which to guide the design. The final grades incorporate concerns for safety, comfort, and access as well as drainage and local concerns such as ice.

Hillside Developments

Each hillside is unique. The combination of slope, soil, hydrology, geology, vegetation, aspect, and proposed use determines the physical constraints and opportunities for development. In general, it is more expensive to develop a

TABLE 3.4 Typical Grading Standards

	Grading standards		
Element	Minimum, %	Preferred, %	Maximum, %
Lawns	1.0	2–8	10
Athletic fields	1.0	1	2
Mowed slopes	5.0	10	25 (mower safety)
Unmowed slopes	—	25	Angle of repose
Planted slopes	1.0	5	10
Berms	5.0	10	25
Crown of	—	—	—
Unpaved street	1.0	2	3
Paved street	2.0	2.5	3
Road shoulders	1.0	2–3	10
Longitudinal slope of	—	—	—
Local streets	0.5	1–10	20
Driveways	0.5	1–10	20
Parking lots	0.5	2–3	20

hilly or steep site because of the additional costs of grading. Another factor driving up the cost is the lower density of hillside development compared with similar flat sites. In spite of the higher costs of hillside development, however, buyers are attracted to such sites because of the long views and interesting terrain.

There are some fundamental elements that most successful hillside developments have in common (Fig. 3.1) For example, it is often necessary to have differing street widths to minimize site development costs and to maintain the character of a site. Finished grading tends to mimic the natural condition as much as possible, and building sites are selected on the basis of physical conditions. The methods of optimizing the site begin with a careful analysis of the site as discussed in Chap. 2. Hillsides are unique, and their analysis must address and identify those aspects of a specific site that are conducive to successful development. Views, slopes, soil conditions, access, utilities, and individual home sites must be evaluated in terms of development costs and market values.

The finished grading of the site should mimic the original terrain. This is especially true if the original character of the site was considered an important element of the project. If the views and terrain are features that are to attract prospective buyers, then it is important to maintain the sense that the sites are undisturbed and are as "natural" as possible. The most important aspect of this is the quality of the grading. The project shown in Figure 3.1 is a successful

Figure 3.1 Photograph of a hillside development.

development of a fairly high density residential neighborhood on a steep site, and it has retained the site's character. The extra effort is compensated by the greater market value. New slopes should be graded to appear natural; thus, they should have uneven, irregular, rounded, or undulating surfaces. The regular crisp, straight slope and grading of the typical site is inappropriate for this type of project. Detailed grading work is often overlooked when in fact it is the foundation for the appearance and character of the entire site. It is this particular aspect of site development that underscores the importance of using talented, able professional contractors. Slopes with irregular inclinations, rather than a single grade across the entire face, will appear more natural. To increase the natural appearance of a slope, the distance between the top of slope and the toe should vary according to the different slope lengths.

Minimizing the Impact of Site Grading

The most important element in minimizing the disturbed area is the design itself. Site layout and design should be accomplished such that they effectively synthesize the development program or objective and minimize the amount of disturbance and of impervious area. As the site is regraded to provide the necessary shape and surfaces on which to construct the proposed site elements, the impacts of the earthwork increase. One of the most common and significant impacts of the grading work itself is erosion, which then often generates sediment pollution in streams and lakes. Another common impact is blowing dusts, which can accumulate and pollute nearby water systems. The changes in site grades resulting from the earthwork can cause water to drain in new and different patterns. The temporary construction drainage pattern is often neglected in the project planning, and this oversight can become a serious problem if it is not managed properly. The impact could cause off-site damage to wildlife habitat and surface water quality. Other negative impacts may include punitive fines, restoration costs, increased project costs, and both immediate and future public relations problems.

The simplest construction project can foster a wide range of emotional reactions within a community. The potential loss of wildlife habitat or tree masses often upsets people living in an area to be developed, and they will resist the new development strenuously. This reaction can arise regardless of the real habitat value of the development area. Sometimes the impetus for opposition comes from the construction activity by itself. To prevent construction-phase damage to open space and green areas, early identification of these habitat areas and drainage patterns must be completed in the planning stages and accommodated throughout the construction phases.

Critical habitat areas or areas that are to serve as buffers to such areas should be clearly marked in the field, and equipment operators must be instructed as to the purpose of the marks. Tree masses that are to be saved should be identified and protected by fences or barriers to isolate them from the busy construction activities. The most common environmental impact on disturbed sites, besides the initial loss of ground cover, is the temporary influence of storm water runoff in the forms of erosion, sedimentation, loss of soil, and the degradation of downstream water. The clearing of the vegetation disturbs the relationship between the vegetative cover and the soil. In the absence of the vegetation, the soil is more prone to erosion. Water is unable to soak the ground as well as it did before, and it can be very difficult to reestablish vegetation. The loss of vegetative cover means the loss of plant surfaces that intercept and then deflect the energy of the falling rain before it contacts the soil. Without the plant root network to keep the soil structure and the rainwater in place, the soil loses its intact resistance to the erosive forces of wind and rain.

The design and management of sites usually address the long-term protection of sites from erosion and storm water damage, but they often forget the temporary construction condition. Often it is the site contractor that is left to deal with the dynamic, often complex, storm water runoff conditions that exist as a result of interim conditions during construction. This can be an expensive experience, requiring time and money to repair and maintain temporary features. To prevent such problems, the designer of the site-grading scheme should also consider the various interim conditions that will exist during construction, and he or she should formulate at least general strategies for how these conditions will be managed.

Another aspect of grading to be considered is related to the form rather than the function of the new grades. The grading of the site is often designed and completed without consideration of the long-term visual impact and appearance of the new shape of the land. Equipment operators rather than designers often have the final say in how a site will look and how people will appreciate the design. In fact, the grading is the foundation for the appearance of a site, and in that way it is the basis for how the site is seen and appreciated by the ultimate users. A poorly conceived grading plan of a site's final form will have a great impact on the success of the site, physically and emotionally. Final elements that are out of scale or uninteresting may be rejected by people in favor of spaces that are inviting, comfortable, and interesting. The appearance of the final form of grading is as important as the function. Most people find outdoor spaces that are natural in appearance to be the most visually interesting and appealing.

Slopes that are to be mowed should not exceed a 3:1 slope although 4:1 is preferred. New cut or fill slopes should not exceed 2:1. On steeper slopes that exceed 15 ft in height, it may be necessary to include a reverse bench or runoff diversion to convey runoff away. The reverse bench should be designed and built to collect runoff and covey it to a stabilized outlet. Benches are designed with a reverse slope of 5:1, and they must be wide enough for construction and maintenance equipment. Figures 3.2 and 3.3 provide greater detail of reverse bench construction.

New slopes that are to be reseeded should be graded in a manner that is conducive to the establishment of new plants. This requires the surface to be

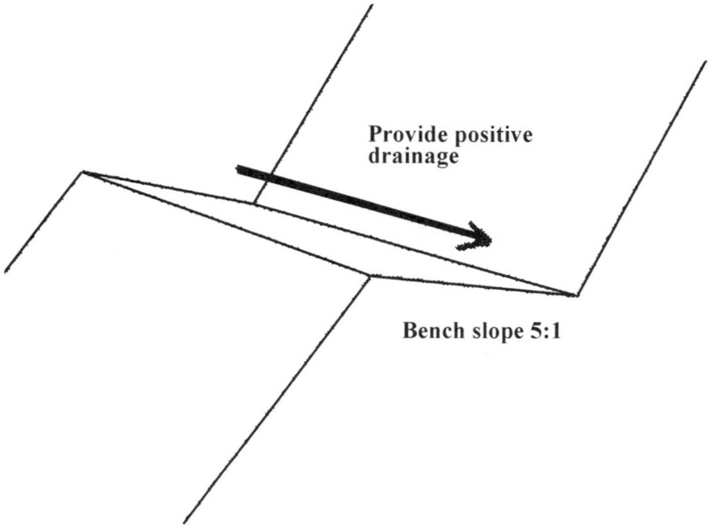

Figure 3.2 Reverse bench detail.

Figure 3.3 Photograph of a reverse bench.

roughened to create the microsites necessary for seeds to take root and establish themselves. Slopes are sometimes roughened or tracked using construction equipment. Figure 3.4 shows a slope partially roughened using tracking equipment. Chapter 7 addresses revegetation in greater detail.

Figure 3.4 Photograph of roughened surface.

Minimizing the amount of disturbance on a site

The layout of a site must consider and address the physical characteristics of a site. The new features such as roads or buildings must fit onto the site and minimize the need for large cuts and fills. This requires that the plan accommodate the site and the arrangement of the features in a manner that maximizes the integrity of each of them. By minimizing the disturbance and the excavated area at the design level, the designer begins to mitigate the impact of development. In addition, the design should retain as much of the original terrain and character of the site as is feasible. To achieve this, roads should be parallel to contours as much as possible and buildings should be located to minimize grading. If flattened places on the site are in short supply, perhaps the buildings can be designed so as to take advantage of the site relief. The disturbance and earthwork should be limited to necessary areas only. Disturbed areas should be kept small, and strips of existing vegetation should be left in place between disturbed areas. Grading should be scheduled so as to minimize the time of exposure and the risk of erosion, thus maximizing the growth conditions in which the vegetation may restore itself.

By minimizing the amount of area that is to be disturbed, the amount of runoff increase can be reduced, and the facilities necessary to handle the runoff can be reduced also. The reduced runoff translates immediately into a decreased risk of erosion and a smaller requirement for storm water facilities. The areas of preserved vegetation may act as adequate buffers between disturbed areas to reduce the amount of active erosion and sediment protection required. Likewise, the less clearing and grubbing that is done, the greater the preservation of infiltration capacity on the site. Although some inconveniences may occur during construction, there are substantial cost savings involved in the reduction of disturbed area.

There are any number of successful projects throughout the country that have adopted this approach to development. Restricting the amount of land to be disturbed—that is, limiting the disturbance to barely more than the footprint of the building—can contribute a great deal to the character and value of a site. Such projects generally include a requirement to use native material and natural exterior finishes to enhance the sense of minimal site disturbances and natural appearance. While adhering to these principles would enhance nearly any project's development, doing so is imperative for developments to be marketed as "natural."

Using grade changes effectively

Variations in grade can serve many purposes in site design. Beyond providing well-thought-out transitions from one elevation to another, grading may be used to reduce noise and to provide a visual separation between features or adjacent properties. The separation provided by a change in grade implies a greater distance between objects than may actually exist. Designers can incorporate grade differences and use this perception of distance to increase visual interest and to create a feeling of expanse that may not otherwise exist.

A low-planted berm between buildings, for example, tends to give a feeling of greater distance when viewed from inside either of the buildings. The berm shown in Fig. 3.5 has effectively screened the residential area from an adjacent highway. Even in areas with little natural relief, subtle combinations of

Figure 3.5 Photograph of a berm.

graded berms and vegetation can effectively separate incompatible uses or undesirable views.

Taller berms or changes of grade are useful in sound control (Fig. 3.6). When considering earth berms or grade changes for purposes of sound control, it is important to remember that in most instances, the closer to the source of the noise, the more effective the berm. Berms should be designed so that the source of the noise is visually isolated from the receiver, and the berm should be continuous. Although a series of hummocks might be more interesting to see, they will not be as effective as a sound barrier for noise reduction. The length of the barrier should be at least as long, but preferably twice as long, as the distance from the source to the barrier. Planted berms should use plant materials of varying heights to create a dense buffer. The use of simple screening plantings may visually screen the source, but such plantings are not as effective as a dense mixed planting. The effectiveness of the mixed planting is a function of its depth and the various textures and surfaces that act to deflect and absorb sound. Vegetated screens are discussed more fully in Chap. 9.

Designers should also consider the location of proposed buildings and other site features to effectively screen sound or to create distance from sources of sound. For example, buildings should be located so that they back up to the sources of sound and act as a sound barrier, and parking areas should be located so that they are a buffer from the sound sources. Sources of noise associated with site development should be considered as either temporary construction noise or postconstruction noise. Some communities have noise ordinances that

Figure 3.6 Photograph of a berm used for sound control.

specifically deal with construction noise, but for the most part these issues are not specifically design issues. The most common postconstruction noise complaints are associated with highway or traffic noise. Numerous steps can be taken to influence traffic noise, but only a few of them are at the disposal of site designers. The site designer's options to influence highway noise include lower speed limits, roadways laid out so as to reduce starting and stopping, and minimum grades. In practice, vegetation makes a fairly poor noise screen. The best practice is to use grading—that is, to raise or lower the road surface. Sound barriers have had success but may also create other problems with sound "reflecting" off the wall or creating "valleys" of poor air quality. Perhaps the best approach is a vegetated slope that provides numerous absorbing surfaces and the mass to screen noise even though such screens require space (Fig. 3.7). Where there is inadequate space or distance between the source and the impacted site, it may be necessary to use structural sound barriers (Fig. 3.8).

If berms are used to screen a view, careful planning and field measurement must be undertaken to assure that the area is effectively obstructed. The screening of an unwanted view may be easily accomplished by using berms and plant materials, but effective buffering requires some planning and evaluation. Often greater effectiveness can be achieved for a lower cost by staggering the islands and mixing the plant materials by size and species. This approach is generally

Figure 3.7 Photograph of a berm between a highway and a residential development.

Figure 3.8 Photograph of sound walls.

more attractive and appears more natural. A well-planned mixture of plants, staggered islands, and undulating berms is nearly always a site-enhancing feature.

Site stabilization

There are two distinct types of stabilization on disturbed sites: temporary and permanent. *Temporary stabilization* generally is used on a portion of a site that has been disturbed and is to be left in a disturbed state for some time prior to final grading and stabilization. Examples of such areas are soil stockpiles and temporary access points. The means of temporary stabilization include vegetation, geotextile fabrics, and/or stone. Temporary stabilization methods are generally inexpensive to purchase, install, and remove. The rule of thumb used in most areas is that if an area is to remain in a disturbed condition but with no further activity for more than 20 days, temporary stabilization is called for. The guideline must be tempered by local conditions, time of the year, and other relative information.

Permanent stabilization is the finished surface of the developed site. This will include vegetation, paving, geotextiles, and stone, as well as combinations of these. In most cases the permanent stabilization of a site will be accomplished either by vegetation or paving. Vegetation is the least expensive cover material to use in most applications; however, in areas of high traffic (pedestrian or vehicular), paving is the obvious choice. Where occasional traffic might occur, such as in maintenance roads or emergency access ways, a combination of vegetation and paving might be desirable. A number of products are available to use in turf as vehicle support systems. The advantages of minimizing paving are reduced runoff and a smaller supporting network of pipes and detention basins.

Paving is required for general- or heavy-use parking lots and cartways. The traditional impervious paving of concrete or asphalt concrete is giving way to wider applications of pavers, permeable paving systems, and even stabilized soil for minimal-use areas. These alternatives reduce the amount of runoff from a site and allow more runoff to be collected to recharge aquifers.

For areas outside of parking and cartways, vegetative cover is usually used. As we have already discussed, one effect of construction activities is the destruction of soil structure, which decreases the soil's ability to support plant growth. Soil structure is determined by the way in which soil particles are arranged into aggregates in combination with organic matter and microorganisms. The aggregates include pore spaces for the movement of water and air through the soils. The loss of soil structure increases erodability and reduces permeability. Before vegetation can be expected to grow and become established in this difficult environment, the soil and the site must be properly prepared. Although preparation does not immediately restore the soil structure, it does provide the elements necessary for the soil to "heal" itself over time.

Mulches

Mulches are generally recommended for all revegetation efforts. The choice of materials is so broad and the variety of characteristics so great that careful consideration needs to be given to the selection of a mulch (see Table 3.5). The complexity of the choice aside, the role of mulch in the vegetation plan should not be overlooked. To different degrees each mulch material has the following attributes: It insulates soil to affect temperatures, it provides runoff protection, it reduces evaporation, it encourages infiltration, and it holds seed in its place. Different materials perform these tasks with different degrees of success. In

TABLE 3.5 Comparison of Mulch Materials

Material	Advantages	Disadvantages
Straw	Low cost, available, absorbent, light color, short application distance, 3000–8000 lb/acre, biodegradable	Must be anchored in place, cost of nets or tackifier, allows weed growth, can be a fire hazard
Wood fiber mulch	Holds seeds and plants in place, can be hydroseeded, inexpensive, stays on slopes, available, 1000 lb/acre	Does not resist erosion or protect from rainfall
Netting/fiber	Resists erosion, protects from rainfall, absorbs water, holds moisture, provides good slope protection	Expensive, installation must be in contact with soil

addition, the means of application and the availability of particular types of mulch may be important considerations. Cost is also a consideration: Some material can be purchased and installed for as little as $1800/acre (wood fiber) while other material is much more expensive, as much as $18,000/acre (jute matting).

Slope Stability

Constructing new slopes presents a series of issues during and after construction. The stability of the slopes is the paramount concern. Slope design begins with understanding the character of the soil and the subsurface conditions. The shear strength of the slope materials will be a determining factor in how steep a designed slope may be without additional structural support. Shear strength is a combination of the grain-to-grain friction between soil particles and the cohesive forces that act to hold soil particles together. As the slope is made steeper, shear stress increases and the ability of the soil to resist gravity decreases. In general, graded slopes that do not exceed the angle of repose of a dry frictional soil should be stable.

Causes of slope failure

The grading operation usually involves removing the vegetative cover, the roots of which may serve to mechanically stabilize the slope. Any change in a slope that increases the slope angle will act to destabilize the slope as it increases the slope loading without increasing the strength of the slope (see Table 3.6). The weight of the soil and the added weight of water acts to increase the stress by increasing the load on soil particles farther down the slope and, perhaps, compressing the lower soils until failure occurs. On projects requiring the creation of steep slopes, a stability analysis should be performed by a soil scientist or soil engineer. Slope failures can occur for a variety of reasons both natural and human. Natural causes of failure include slippage along existing soil transitions or soil structural weaknesses.

Instability in slopes can be addressed by either increasing the resistance of the slope to failure or by minimizing the causes of failure. The causes of failure can be addressed either by avoiding the unstable area or by modifying the

TABLE 3.6 Common Causes of Slope Failure

1.	Overloading slope (weight of buildings or roads)
2.	Increasing fill on slope without adequate drainage
3.	Removing vegetation
4.	Increasing the slope grade
5.	Increasing slope length by cutting at bottom of slope
6.	Changing surface drainage route
7.	Changing in subsurface drainage route

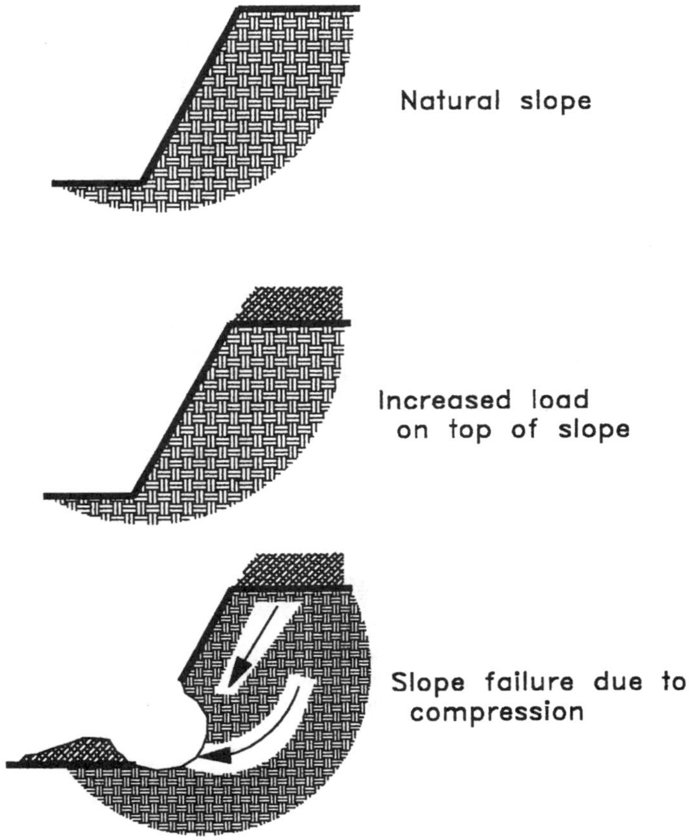

Figure 3.9 Slope failure caused by an increased load on top of the slope, which compresses the underlying material.

design in ways such as those illustrated in Figs. 3.9 through 3.12. Changes in surface conditions will alter the drainage conditions on the surface and subsurface. These changes may in turn impact the stability of the slope by increasing the amount of water in the slope material or by causing the erosion of the surface material. Providing adequate surface and subsurface drainage may be required. In many cases slope failure caused by changes in subsurface drainage is difficult to predict without fairly intensive study, and so these problems may emerge and have to be solved after the site has been altered. The location of facilities or appurtenances on fill or in the zone of influence for a slope should also be carefully evaluated.

Retaining walls

It is often not practical to consider reducing the weight or location of features, and so it is necessary to increase the slope's resistance to failure. Methods of

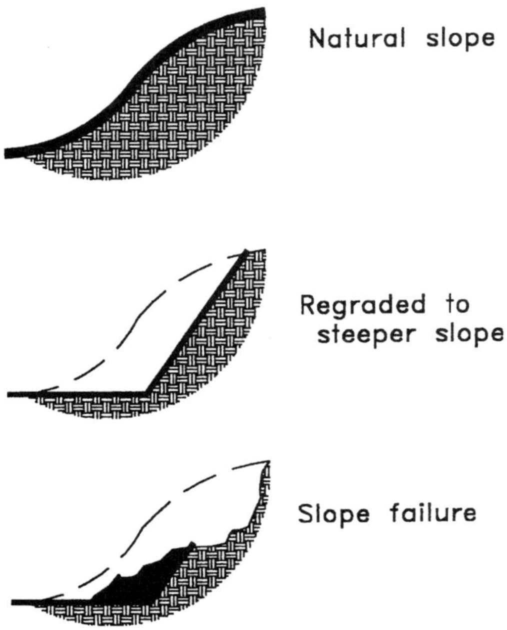

Figure 3.10 Slope failure caused by regarding a slope so that it is steeper.

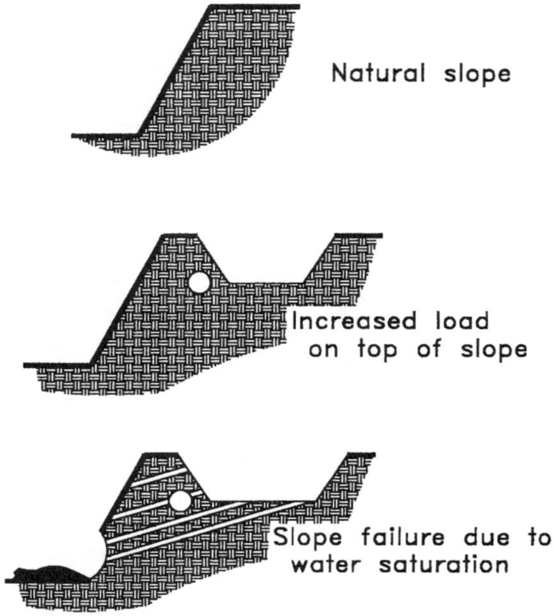

Figure 3.11 Slope failure caused by increased load on top of the slope, which leads to water saturation of the material below the slope.

increasing slope resistance vary from building retaining walls to stabilizing the soil using thermal treatment (heating the soil to the melting point). Although new methods of chemical and thermal treatment have emerged, these methods are generally considered to be experimental and have not been widely used. The most widely used methods are basically variations on the retaining walls or pilings such as the method shown in Fig. 3.13 or the cantilevered reinforced retaining wall shown in Fig. 3.14. New methods include slope stabilization using anchors or interlocking concrete block walls, and using three-dimensional geosynthetic materials as shown in Figs. 3.15 and 3.16. Buttresses are sometimes used as alternatives to the other methods.

To keep changes in grade small, timber and dry laid stone retaining walls have been used successfully (Figs. 3.17 through 3.20). Proper installation is the key to all of these methods, but small retaining walls are often built incorrectly without regard to proper stabilization, footing, soil bearing, batter, and so on. Timber and dry laid stone walls depend on the depth below grade to resist overtopping by the retained earth. Retaining walls may be considered as either flexible or rigid construction. For purposes of this discussion retaining walls are no more than 8 to 10 ft in height, and the maximum surcharge is 2 ft. Taller walls begin to have greater and more complex influences than those discussed here and should be designed by a structural engineer.

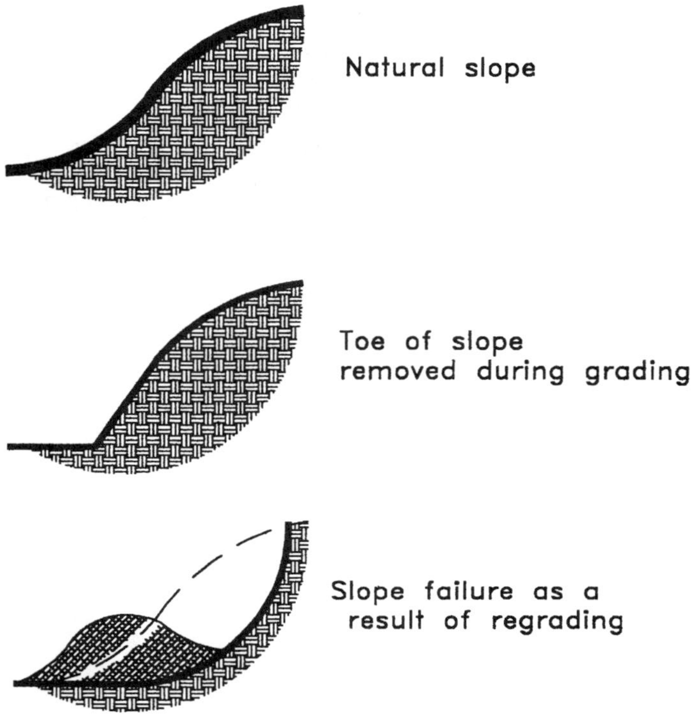

Figure 3.12 Slope failure caused by the removal of the toe of the slope during the initial grading of the slope.

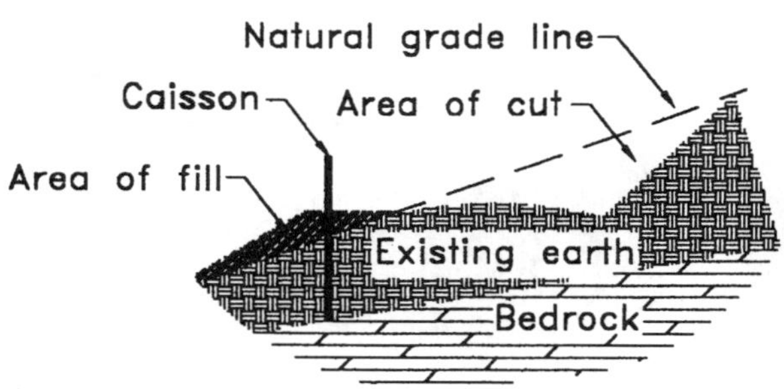

Figure 3.13 Caisson and soil buttress for destabilized slope detail.

Retaining walls are often designed with a batter; that is, they recede away from vertical by a specified amount. Batter is useful to offset the feeling of over-topping from tall, vertical retaining walls, and it helps to hide small imperfections and variations in the wall. In smaller flexible walls, the batter helps to hide and absorb seasonal bulges and movement that might occur, and it contributes to the wall's stability. Although it is determined on a case-by-case basis, a batter of 6:1 is commonly used for flexible walls, and it is somewhat less for rigid walls.

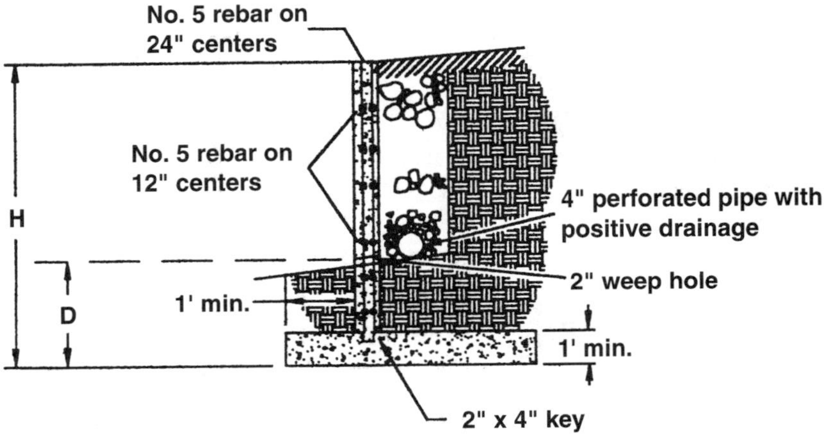

Figure 3.14 Cantilevered reinforced-concrete retaining wall detail.

Figure 3.15 Photograph of interlocking blocks.

All retaining walls require a suitable foundation. Flexible retaining walls are usually fairly small in elevation, and, given a suitable compacted base, they may not require footers to extend below the frost line. As a rule, retaining walls should extend a minimum of 2 ft below grade, or half the above grade height or to the frost line, whichever is greatest. A certain amount of settle-

Figure 3.16 Photograph of interlocking walls.

ment and perhaps seasonal movement can be tolerated in flexible walls. Rigid walls of concrete or masonry construction are used where greater changes in elevation are necessary, where flexibility cannot be tolerated, or where the mass of the wall is used to retain the earth.

The use of gabions to stabilize slopes has become a more common and cost-effective solution (Figs. 3.21 through 3.25). *Gabions* are manufactured wire-mesh baskets that are assembled on the construction site and filled with stone. The gabion is a flexible and permeable structure that can be used to construct retaining walls, toe of slope buttresses, and stream bank protection revetments, and it can be used as a weir in storm water and erosion control. Gabions are installed on a surface that has been leveled and compacted.

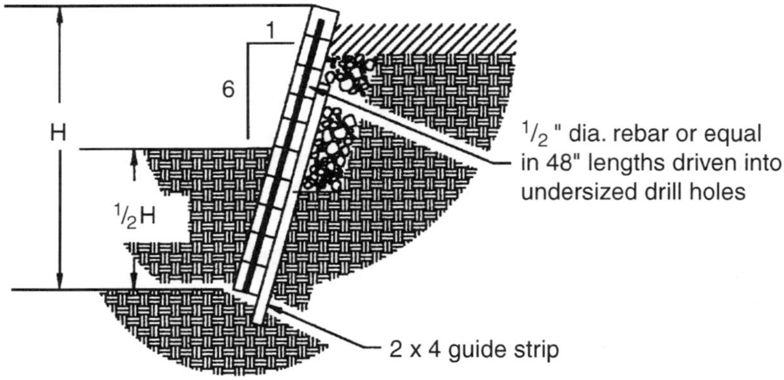

Figure 3.17 Horizontal timber wall detail.

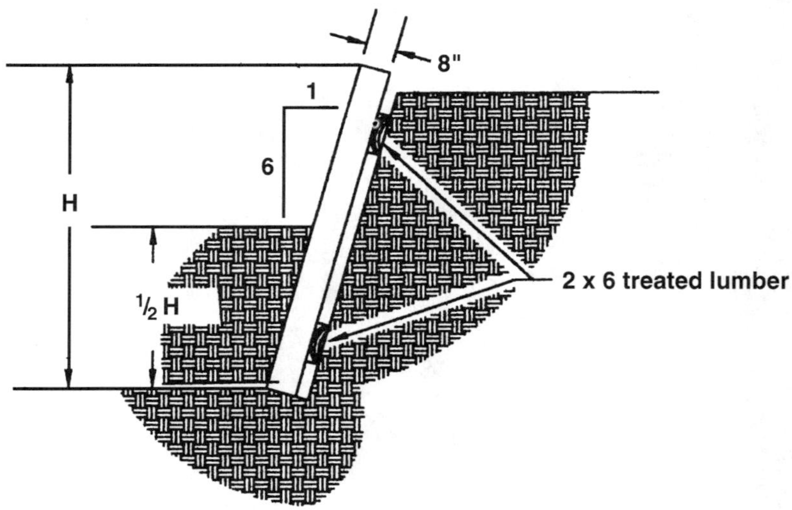

Figure 3.18 Vertical timber wall detail.

Except for revetments, in most application the gabions are constructed and filled with stone in place. Individual gabion baskets are connected to each other using lacing wire or ring fasteners. The minimum standards for lacing wire and ring fasteners are detailed in ASTM A975. Gabion wall design is often conducted with the assistance of the manufacturer of the baskets.

All retaining wall designs must address drainage. Crib walls, gabions, and dry laid stone walls are by definition porous, but masonry and concrete walls must be designed with the means to drain water away from the wall to avoid damage or failure. Water should be directed away from both the top and the bottom of the structure through the use of positive drainage. Weep holes of sufficient diameter should be installed to relieve hydrostatic pressure from behind the wall.

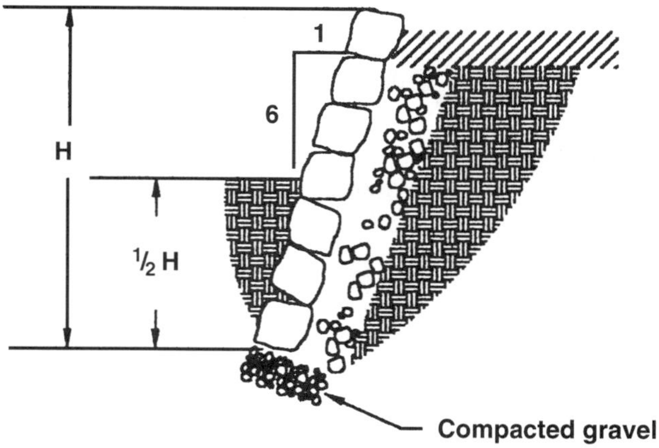

Figure 3.19 Dry laid stone wall detail.

Figure 3.20 Photograph of dry laid stone wall.

Erosion and Sediment Control

Erosion is the uncontrolled transportation of soil either by wind or water. In most site construction cases the primary short-term concern is erosion due to an unstabilized soil surface and the impact of precipitation and runoff. How erosion works is generally understood, and it is the mitigation of these mechanisms that is the focus of erosion control. In general, erosion begins with the

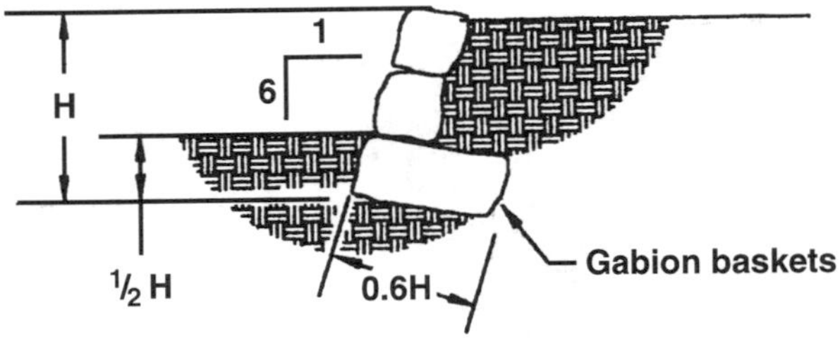

Figure 3.21 Gabion retaining wall detail.

Figure 3.22 Photograph of gabion wall.

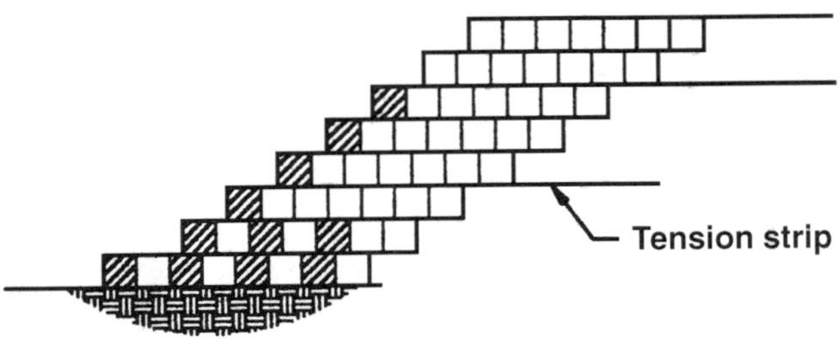

Figure 3.23 Slope stabilization using three-dimensional geosynthetics detail.

Figure 3.24 Photograph of stabilized slope.

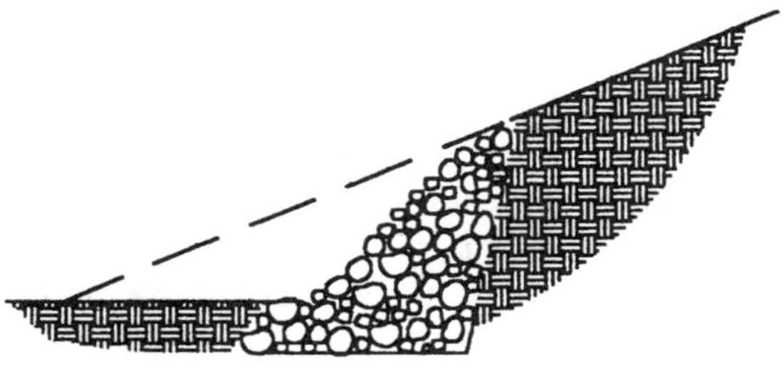

Figure 3.25 Slope stabilization using rock buttress detail.

loosening of soil particles through freeze-thaw or wet-dry cycles, or the impact of falling rain. Erosion is separated into different types by the manner in which it is moved rather than by the cause.

Splash erosion is simply the result of raindrop impact on unprotected soils. Through the repetitious hammering of raindrops, soil particles are gradually moved down hill. This is a process of concern to builders with sites that have unprotected soils that are exposed to the weather. The larger the raindrops and the greater the slope, the farther down hill the soil particle will move and the greater the risk of erosion. As this process develops, soil is broken up and the process of erosion is accelerated. Even on flat slopes the destruction of the soil structure is detrimental, resulting in a hard soil crust when it dries. The crust limits infiltration and increases runoff and further erosion. In soils without structure, it is difficult to establish vegetation, exacerbating the erosion cycle.

Sheet erosion occurs where there is a uniform slope and surface and runoff flows in a sheet. Erosion in these instances usually is limited to the loose soil particles. Sheet erosion rarely occurs in any other than a limited form in the field. Sheet flow tends to concentrate into more defined flows as it is channeled by the irregularities of a site.

The channelized flow result in the types of erosion most think of when the subject comes up: rill and gully erosion. *Rill erosion* is characterized by small, even tiny channels that often abrade and intertwine, while *gully erosion* is identified by the large channels, which are obviously damaging. Where a rill is at worst only a few inches deep, a gully can be as deep as 10 ft or more.

The impacts of erosion and sediment extend from the esthetic impacts to the easily quantified cost of dredging reservoirs to recover lost capacity. The U.S. Army Corps of Engineers spend an estimated $350 million annually to dredge rivers and harbors in the United States. Sediment-filled rivers, reservoirs, and harbors cannot be used for shipping or recreation. The loss of soil as an agricultural resource can have a direct impact on the productivity and feasibility of that operation. To replace topsoil in the United States with commercially available topsoil would cost at least $20/yd^3 ($26/m^3) or about $4.6 billion each year. Taking these replacement costs and the dredging costs together make a compelling economic argument for erosion and sediment control. The federal government through the National Pollution Discharge Elimination System regulates discharges from most construction sites. Most states have their own version of these regulations and require builders to meet a minimum set of performance standards (Table 3.7).

The essence of the principles lies in the fundamental difference between the prevention of erosion and the control of sediment. Erosion prevention and sediment control are proactive. While it is not possible to have site development without some earth disturbance, often the amount of disturbance is well

TABLE 3.7 Principles of Erosion and Sediment Control

1. Design development to fit the site and the terrain.
2. Protect and retain existing vegetation to the extent possible.
3. Protect and/or revegetate, and mulch exposed areas.
4. Minimize steepness of slopes to manage both velocity and flow of runoff.
5. Schedule earthwork and construction to minimize soil exposure and enhance stabilization.
6. Protect new swales and drainage paths. Improve stabilization of existing channels for increased flows and velocities.
7. Trap the sediment on the site.
8. Maintain site controls.
9. Develop contingency plans before they are needed.

beyond the area required. Since sediment control features such as filter fences, sediment traps, stone filters, check dams, and sediment basins are designed according to the size of the disturbed area, the smaller the area disturbed, the lower the cost of site control.

Sediment control is, in effect, planned damage control. These efforts are geared entirely toward collecting, directing, capturing, filtering, and releasing sediment-laden runoff, after erosion has occurred. Practically speaking, in most cases erosion control is not among the first things a site designer or a builder is concerned with. Their attention is drawn to many issues in the course of the project, and erosion control is usually dealt with as part of these issues. Erosion and sediment control consists of both temporary and permanent measures. Permanent measures are provided to prevent erosion from occurring after construction is completed. These permanent measures include stabilized and established vegetation and paving.

Initial erosion and sediment control operations consist of the construction of ingress/egress controls, which include tire scrubbers or a stabilized construction entrance that remains in place and in working order until earth-moving activities are completed and a driveway or entrance is stabilized. Erosion and sediment controls should be constructed and stabilized and functional before general site disturbance begins. Only limited disturbance is permitted, so allow for the proper function of sediment basins, sediment traps, diversion terraces, interceptor channels, and/or channels of conveyance.

After completion of the site work and the grading of all disturbed banks, open areas are seeded, fertilized, and prepared in accordance with specifications and cultural requirements of the site. Temporary erosion and sediment pollution controls should be maintained throughout the duration of the work and until the site is stabilized. After a rain, the devices should be checked and inspected for condition and integrity. Devices that require maintenance, repair, clean-out, or replacement shall be addressed.

Silt fences must be installed parallel to existing contours or constructed level alignments. Ends of fences must be extended 10 ft, traveling up slope at 45° to align with the main fencing section. Sediment must be removed where accumulations reach halfway above the ground height of silt fencing. Any silt fence that has been undermined or topped should be replaced with rock filter outlets immediately. In long sections of fence, stone filter outlets might be used where water collects or flows concentrate behind the filter fence. Storm water inlets must be protected until the tributary areas are stabilized. Sediment must be removed from inlet protection after each storm event.

Sediment must be removed from traps when storage capacities are reduced to 1334 ft^3/acre. Most regulations require that sediment be removed from the basins when storage capacities are reduced to 5000^{ft3}/tributary acre. Stakes located in the trap and marked with the clean-out elevation are required in some jurisdictions. The stakes should be placed at about halfway between points of concentrated inflows to the basin risers or outlet. When sediment has accumulated to the clean-out elevations on half the stakes, it must be removed to restore basin capacity.

Any disturbed area on which activity has ceased for more than 20 days must be seeded and mulched immediately. During nongerminating periods, mulch should be applied at the recommended rates. Disturbed areas that are not at finished grade and that will be disturbed again within 1 year may be seeded and mulched with a quick-growing temporary seed mixture. Disturbed areas that are either at finished grade or will be disturbed again but beyond 1 year must be seeded and mulched with a permanent seed mixture and mulched. Diversions, interceptors, swales, channels, sediment basins, and sediment traps are seeded immediately upon the completion of construction. Seeding specifications are best tailored to regional and site requirements.

When applying straw as mulch, all of the straw should be dry and free from undesirable seeds and coarse material, and it should be applied at a rate of 115 to 150 lb/1000 ft^2 or 2.5 to 3.0 tons per acre. Mulched areas should be checked periodically and checked immediately after storms and wind. Damaged or missing mulch should be replaced. A tackifier should be applied after the straw is applied. The tackifier may be asphalt or polymer spray. It should be applied at the rate recommended by the manufacturer and with suitable equipment. In lieu of manufacturer's recommendations, it should be applied at a rate of 0.04 to 0.06 gal/yd^2. Erosion control blankets or netting should be selected to fit the application and site conditions, and they should be installed and used in accordance with the manufacturer's specification.

Sediment basins and sediment traps are used to capture sediment on the disturbed site. In general, a *sediment basin,* or a *silt basin* as it is sometimes called, is a large control device used for drainage areas in excess of 5 acres, and a *sediment trap* is a small control device used for drainage in areas smaller than 5 acres. The size of the contributing area and specific design parameters are sometimes dictated by local or state regulations. Sediment basins are often large enough that they require a substantial space on the construction site. They are usually constructed very early in the construction process, and they remain until nearly the end of the project. The basin should be located so that it will not capture clean runoff along with the runoff from the disturbed area. If at all possible, clean runoff should not be mixed with the sediment-bearing runoff. Typical erosion and sediment control design details are shown in Figs. 3.26 through 3.31. Local jurisdictions may have requirements that differ slightly from the preceding suggestions.

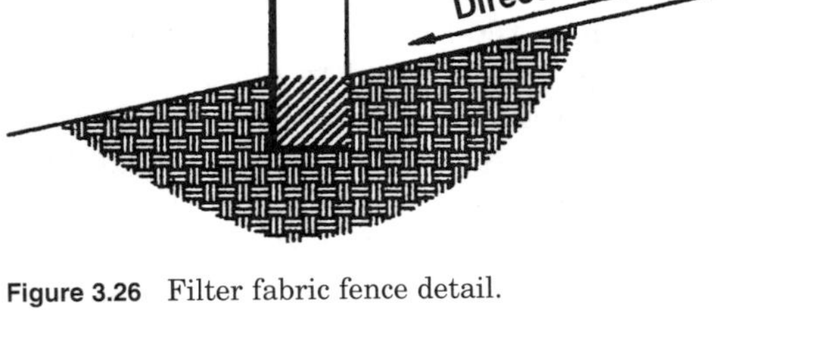

Figure 3.26 Filter fabric fence detail.

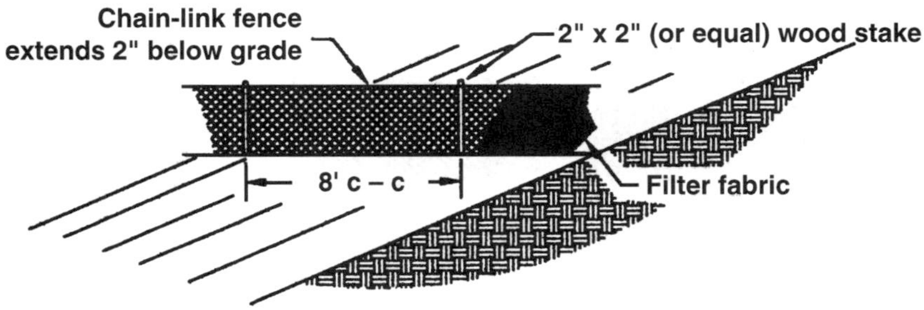

Figure 3.27 Reinforced filter fabric fence detail.

Figure 3.28 Photograph of reinforced filter fabric fence.

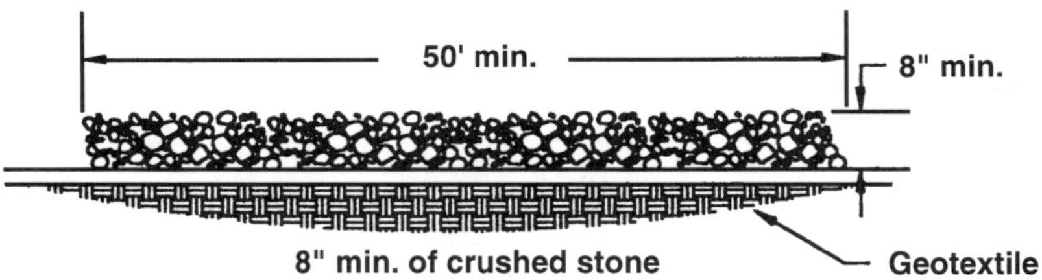

Figure 3.29 Site entrance control detail.

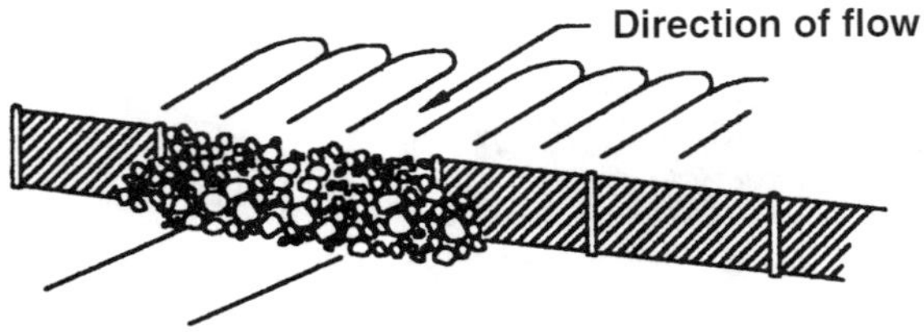

Figure 3.30 Stone filter detail.

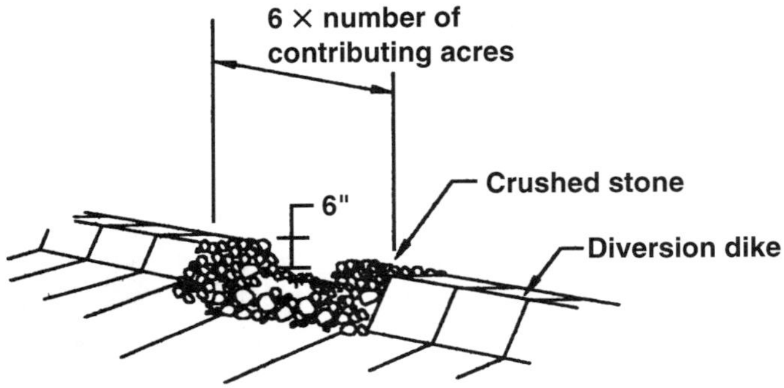

Figure 3.31 Diversion dike detail.

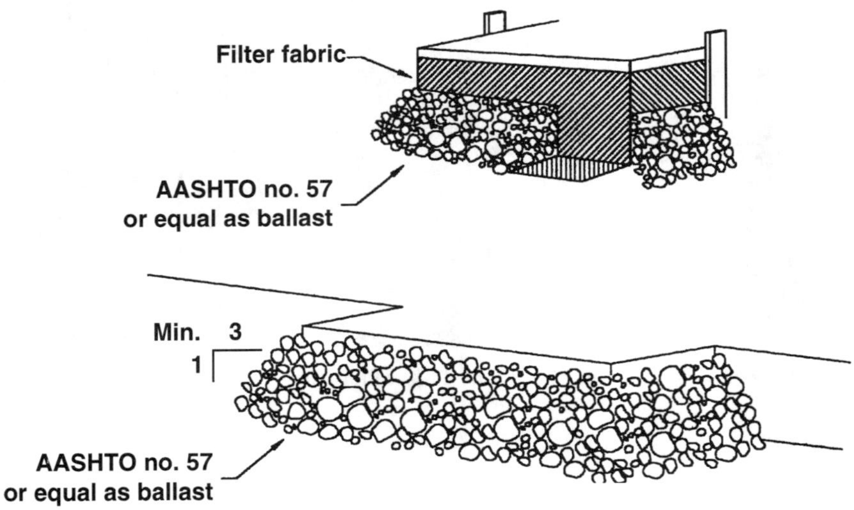

Figure 3.32 Catch basin control detail.

Basins are typically designed to contain a 10-year storm, though local and state regulations may differ (Fig. 3.32). Sediment basin outlets are designed to allow the basin to dewater at a rate slow enough to provide for settlement and fast enough to remain in service and reduce the risk of insect infestation. The

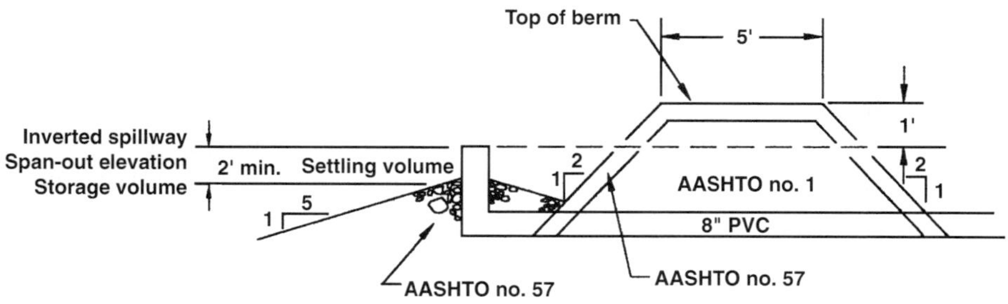

Figure 3.33 Sediment trap detail.

typical principal spillway is designed with a minimum flow of 0.2 ft^3/s, which is equivalent to runoff of 5 in/24 h. Antiseep collars are used in larger basins where berm height exceeds 10 ft or the local soil has a very low clay-to-silt content (unified soil class SM or GM). Settling and sediment storage requirements for basins differ from state to state. Sediment traps are smaller versions of the sediment basin and are used for drainage areas of less than 5 acres (Fig. 3.33).

Dewatering outlets are designed using the formula for flow through an orifice:

$$Q = CA\sqrt{2gh}$$

where Q = flow, $ft^{3/s}$
C = coefficient of contraction for an orifice, usually 0.6 (sharp-edged orifice)
A = area of the orifice, ft^2
g = acceleration of gravity, $32.2/s^2$
h = head above orifice, ft

The equation can be used to determine the length of time necessary to dewater a basin:

$$T = \frac{A\sqrt{2h}}{3600TC\sqrt{g}}$$

where T = time, h
A = surface area of the basin, ft^2

The formula can also be used to determine the orifice size required to dewater a basin within a required time:

$$A_x = \frac{A\sqrt{2h}}{3600TC\sqrt{g}}$$

Site management

Construction site managers must ensure that the site complies with local regulations for erosion and sediment control. In addition, they must ensure that the control measures are actually effective in meeting local objectives and be prepared to supplement the measures specified by the local ordinances to achieve the ultimate control objectives. With increased local training and enforcement comes a greater emphasis on the performance of the grading and erosion and sediment control plans. Unfortunately, erosion and sediment control plans are often the last thing designed and the first thing installed. Installation is haphazard or incomplete, and maintenance is limited to responding to complaints and inspections.

Although the landowner ultimately has responsibility for the proper management and control of the construction site, it is the site manager that has day-to-day control. The design professional is often called upon to modify the erosion and sediment control plan to meet site conditions or to address failures. The development of the erosion and sediment control plan should include the management of the facilities for the duration of the entire project, not simply the start and end of the project. As with any element of a project, it should be planned, responsibility and resources assigned, and performance expectations communicated, and then performance should be monitored and confirmed from time to time. Of the eight causes of failure listed in Table 3.8, the site manager actually has control over only three; compensating for seasonal differences, installation, and maintenance of facilities (Figs. 3.34 and 3.35). Most of the causes of failure are related to design. Even the best management plan cannot overcome a design problem or extreme weather conditions.

Erosion and sediment controls are often designed without regard for the dynamics of a construction site. Designs tend to address specific moments in the course of the site work and not the constantly changing site conditions. The contractor should review the erosion and sediment control plan to be sure that there is adequate room to store topsoil or excess material. If storage is required, is there a practical pattern for the use of heavy equipment?

Temporary drainage conditions may also present a problem if not planned for. The installation of sediment traps and basins may have to consider an interim step or two if significant changes in grade are proposed. Are these interim steps provided for in the plan? Most important to the contractor, does the plan make sense? A dialogue between the designer and the contractor to exchange ideas and solutions can be an important step in the successful erosion and sediment control plan. The designer is in the best position to initiate this meeting. If a

TABLE 3.8 Common Causes of Erosion and Sediment Control Failure

Poor site analysis
Design incompatible with site
Inadequately sized facilities
Wrong materials specified or used
Poor installation
Poor maintenance
Failure to compensate for seasonal differences or extreme weather conditions

Figure 3.34 Photograph of failed filter fabric fence because of poor maintenance.

contractor has not been chosen at that point of the project, a meeting with a qualified contractor may be just as valuable. The closer this working relationship, it seems from experience, the better the site controls work. The site manager has an interest in these early stages because eventually he or she is responsible for its implementation. The entire thrust of the management plan is aimed at controlling the causes of failure and maintaining the integrity of the site controls.

The installation of control features requires adequate information and detailing in the plan. The plan should include construction details for the various facilities that are to be installed. This would include the routine details but also more specific information such as staple patterns on erosion control fabrics or inverted elevations on sediment trap dewatering outlets. The adequate installation of controls begins with understanding the construction details.

The typical erosion and sediment control plan includes a construction sequence and, when appropriate, phase lines. The designer often is required to make assumptions about the project that may not be true later on. The construction sequence should be reviewed and understood. Items that cause conflicts or are no longer accurate should be addressed to the designer so that a revision can be made to the report. Too often these details are overlooked or discounted as unimportant until there is a problem later in the project and the contractor is found to be "out of sequence." This small detail is suddenly disproportionately important.

The site manager must understand the plan before he or she begins. Implementing the plan without comment or revision may be seen as tacit acceptance and approval.

Figure 3.35 Photograph of failed filter fabric fence because of abuse.

A project directory is a simple tool. It is merely a directory of the phone numbers, addresses, and fax numbers of the various people involved in the project. The list normally includes the owner's name, the project engineer, the project surveyor, the municipal engineer, the site manager's name (and an alternate or two), as well as any subcontractors or others that might be important. The list should also include the names and information of the regulatory and enforcement personnel.

The directory should include an identification of what each person listed is responsible for or why the name is listed. Such a directory will help in responding to emergencies and problems much faster and more smoothly. This preparation is a key means of avoiding fines and enforcement actions.

The startup meeting

The project startup meeting is a fundamental element of the management plan. It is the site manager's responsibility to organize the startup meeting. The meeting should be attended by the site designer, erosion and sediment control plan designer, and local enforcement personnel, as well as supervision and staff from the project. It may be appropriate to have others attend such as municipal representatives or environmental regulators.

The agenda for the startup meeting should include introductions of the attendees (a signin sheet is recommended to collect phone numbers), review of the scope of the project, and consideration of what is to be done in the course of developing the site. A site plan should be used to act as a discussion guide. If phases are involved, the delineation and field recognition of phase lines should be discussed. The construction sequence should also be reviewed. The review of the grading operations should include specifically the identifications of areas of significant cuts and fills and sensitive areas such as wetlands or floodplains. The erosion and sediment controls that will be used throughout the project should be reviewed, and maintenance schedules and repair plans should be specified. Contact people for emergency response should be identified. A site walkover should be conducted to familiarize everyone with the startup condition of the site and areas of concern. This is particularly important if there is existing erosion or sedimentation occurring. Minutes should be taken during the meeting and distributed afterward to all the attendees. A copy of these minutes should be kept in the project log.

Once the earthwork has begun and the project is up and running, the site manager will be diverted from the erosion and sediment control plan. A schedule of routine maintenance, developed prior to the startup, is a helpful prompt for the manager to keep the commitment to the plan. By assigning a staff person to follow up on the schedule, the manager can be sure the routine inspections and maintenance items are being addressed.

Routine inspections are scheduled at frequencies that reflect the site characteristics, the time of the year, and the condition of the site. A hilly site that is fully disturbed during the rainy part of the year will justify more frequent inspection than the same site partially stabilized during a dry season. Inspections themselves are relatively inexpensive, requiring only a visual check in most cases to ascertain the condition and any corrective action that might be required. The use of a small tape recorder makes note taking almost effortless.

It is unreasonable to assume that the schedule set out in the beginning of a project will be met perfectly throughout the project. Some flexibility is appropriate in the system. In most cases, slipping the schedule 2 or 3 days is not a problem. Since inspections should be made after every significant rain or melt event without exception, the routine inspection schedule can be adjusted to reflect these events.

From the startup meeting and throughout the project until final stabilization is confirmed, a logbook should be maintained by the person assigned the responsibility to oversee the erosion and sediment control plan. The purpose of the logbook is to record the routine inspections and maintenance as well as the general progress and activity on the site. A well-maintained logbook is a record of performance and compliance with the plan. Records of routine inspections including corrective actions taken and photographs are of particular importance in the logbook. Copies of inspections by regulatory or enforcement personnel, and notes and photographs taken during the inspection should also be included in the log. Information should be included regarding precipitation and other weather conditions that are pertinent to actions and decisions taken and the required inspections after storm events.

It is not unusual during the course of a construction project to have changes made in the erosion control plan. The changes may occur because of a change in the project or a change in site conditions encountered during the construction process. It is common to have changes in the erosion and sediment control plan as well. These changes are often a response to an unforeseen condition such as a concentrated flow of runoff where one was not anticipated. The site manager must have the flexibility to respond to the problem quickly. In fact, anything but a quick response would be inconsistent with the objectives of the plan. Once the response is made, however, a note should be made in the logbook and the owner, site engineer, and regulator should be notified. A copy of the notice should be kept in the logbook.

Chapter

4

Designing for People

Design and planning encompass the process of bringing a vision from an idea to the point of implementation of that idea. Site planning and design require the professional to consider a broad range of concerns in the synthesis of a design concept. There are the physical aspects of the site itself, the vision or program of the client, the designer's own creative inclination, the concerns of the community, and the interests of the end user. The public's interests are represented by a variety of public authorities that exist for the purpose of regulating and overseeing the development process. The land development ordinances are generally intended to act as a set of minimum standards or guidelines. Few ordinances can be applied to a specific site or project without some adjustment or accommodation. The more prescriptive the ordinance, the greater the frequency and scope of the necessary accommodations; the less prescriptive, the lower the minimum standard. Such general standards represent a local view of the minimum requirements for land development and should not be confused with a measurement of design quality. Quality design and development typically need to go beyond the minimum threshold of the local ordinances.

For every developer interested in only the lowest-quality of work and effort, there are others with a broader view of what they do and the quality of their legacy. Because most development companies are businesses and every project is a business activity, site design professionals must balance their clients' objectives and the community's standards and expectations with ensuring a profit for the developing company. The community's expectations and standards are composed of stated, tangible parameters as well as unstated expectations and intangible elements. It is the design professional that ultimately must find a balance between the goals of the client, the standards of the community, and the interests of the end user of the project. The end user is usually not part of the discussions that drive the design process. To arrive at a design solution that will satisfy all of these diverse interests, the designer must be a student of design outcomes and performance as well as design synthesis.

General Site Design Guidelines for Pedestrians

There is no shortage of sources for site furnishings today. The industry provides a range of well-designed and durable materials in many styles from which the designer may choose. Virtually all of these furnishings comply with the accepted standards of human dimensions; however, it remains the responsibility of the design professional to select and specify the materials appropriate to the site. A working knowledge of human dimensions and behavior is necessary. Figures 4.1 through 4.3 provide an outline of human dimensions and design conventions.

Walkways

A fundamental element of design for the pedestrian is the pathway or sidewalk. The peak time for walking is midday (countercyclical to vehicle traffic), and sidewalks should be designed to account for this peak time. Many localities have predetermined minimum standards for sidewalk development in residential areas, but they do not provide guidance for commercial sites or other circumstances in which minimums are not adequate. The sidewalk width must be designed to provide the level of service suited to the user. The parameters of sidewalk width are determined according to the anticipated volume of foot traffic, the speed at which the pedestrians will be walking, and the desired density of traffic (Fig. 4.4). The width can then be determined as:

$$W = \frac{V(M)}{S}$$

where W = the width of the pathway or sidewalk, ft
V = the traffic volume, persons per minute
M = the space module allowed per person, ft^2
S = the walking speed, ft/min

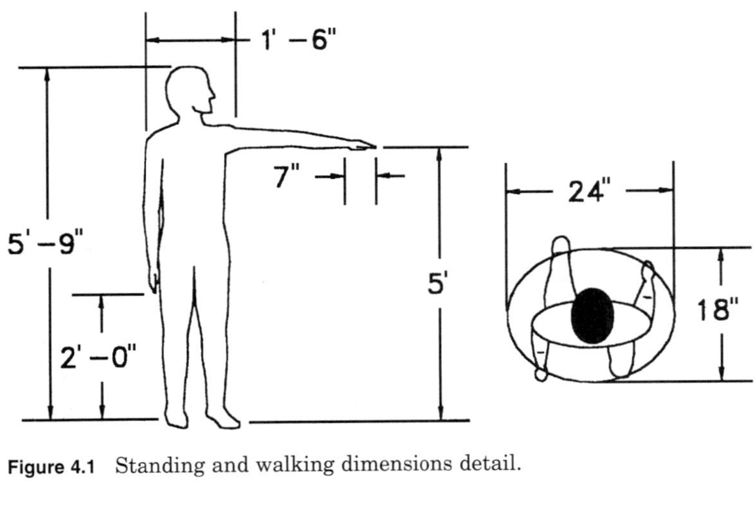

Figure 4.1 Standing and walking dimensions detail.

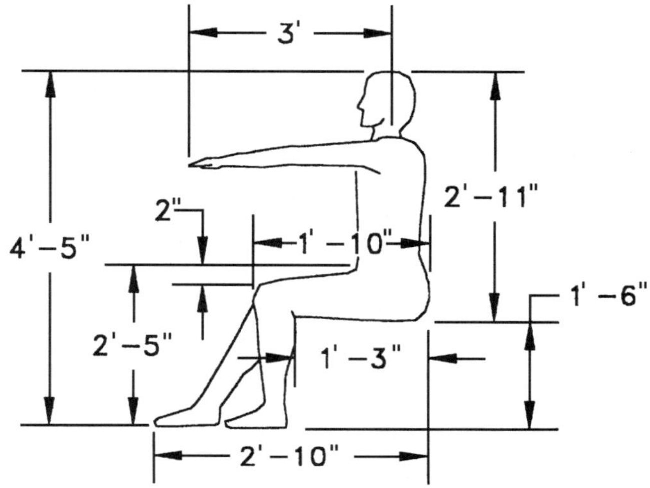

Figure 4.2 Sitting dimensions detail.

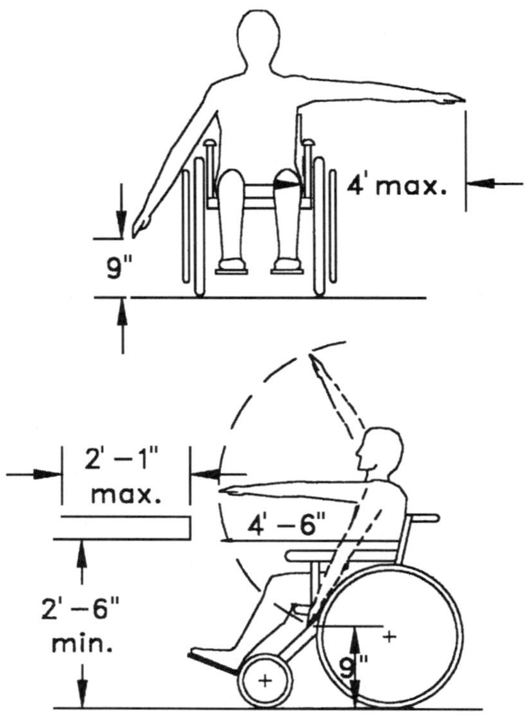

Figure 4.3 Wheelchair use dimensions detail.

Walking speeds vary greatly among people, but a normal average walking speed of 4.0 ft/s is usually assumed. A number of factors influence this speed. For example, older people walk slower than younger people, and most people tend to walk faster in the middle of a block and slow down at intersections. The activity that walkers are engaged in affects their speed as well—for example, shoppers walk slower than commuters. Men tend to walk faster than women. Groups of people will walk slower than individuals. Curbs, islands, circuitous pathways, changes in grade, and even ramps can present barriers of one sort or another to various users. Changes of grade of more than a few percent should be signaled visually and texturally. To determine the appropriate level of service, designers should weigh the anticipated use of the site, the characteristics of the users, and the character of the final design (Table 4.1).

Grades also affect walking speed, level of service, and safety. Sidewalks should be designed with a minimum cross slope of 1 percent to allow for drainage, but the cross slope should not exceed 3 percent. A longitudinal slope of up to 3 percent is desirable, but slopes greater than 5 percent should be avoided in areas where freezing may be an issue. As a rule of thumb in areas where climate is a consideration, any sidewalk with a slope in excess of 5 percent should be considered and treated as a ramp with associated handrails.

When incorporating stairs into an outdoor design, there are often local standards to consider; however, when such regulations are not in place, a rule of thumb to determine tread width is the following:

$$2R + T = 26 \text{ to } 27 \text{ in}$$

where R = riser height, in
T = tread width, in

Table 4.2 lists some general guidelines for designing outdoor stairways. Figure 4.5 gives dimensions for the amount of stair tread that is actually usable and for the nosing (that is, the rounded edge) of the stair tread. Figure 4.6 shows stair treads with painted nosing on each tread.

All site features should comply with the specifications provided in the *Americans with Disabilities Act Accessibility Guidelines for Buildings and Facilities* (see Table 4.3). Ramps should be designed to meet the ADA requirements (Figs. 4.7 through 4.18). Ramps with a slope of between 1:12 and 1:16 should be designed to not exceed a rise of 30 in (760 mm) or a run of 30 ft (9 m). Flatter ramps of 1:16 to 1:20 slope may be designed to a run of 40 ft, but the maximum rise should not exceed 30 in. The minimum clear width of a ramp should be 36 in (915 mm). Ramps shall have level landings at the bottom and top of each ramp and each ramp run. The cross slope of ramp surfaces should be no greater than 1:50. Outdoor ramps and their approaches should be designed so that water will not accumulate on walking surfaces. Landings should be at least as wide as the ramp run leading to it and be a minimum of 60 in (1525 mm) clear. If the ramp changes direction at landings, the minimum landing size should be 60

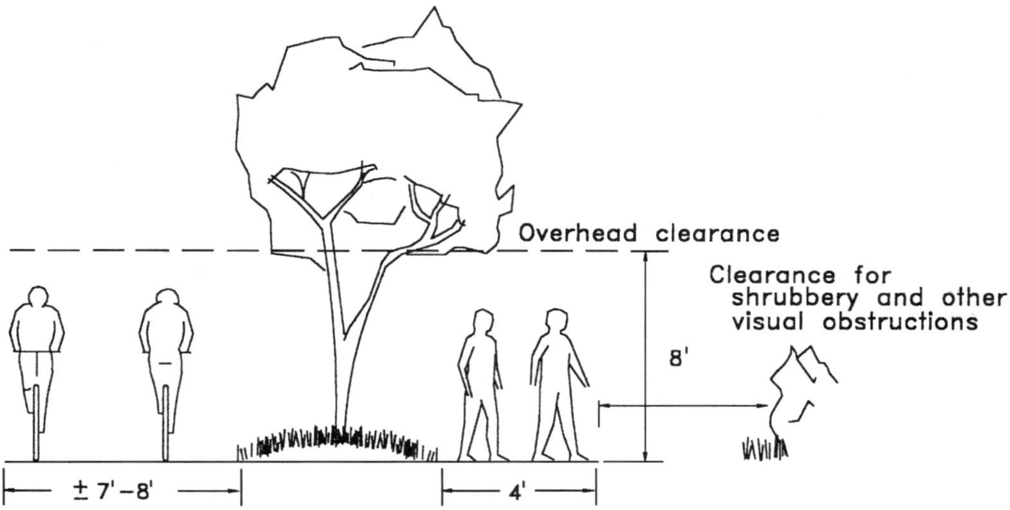

Figure 4.4 Pathway design parameters detail.

TABLE 4.1 Levels of Service for Sidewalks

Level of service	Area per person	Interference
A	130 ft^2	A person can walk at his or her desired speed without interference.
B	40–130 ft^2	Pedestrians are aware of other pedestrians.
C	24–40 ft^2	A pedestrian needs to make minor adjustments to avoid conflicts with other pedestrians.
D	15–24 ft^2	A pedestrian has a limited ability to choose own speed, and he or she needs to make frequent adjustments.
E	6–15 ft^2	The sidewalk is very crowded, speed is reduced, and the shuffling pace occasionally makes changes in direction very difficult.
F	6 ft^2	There can be stationary or shuffling movement only, and there is unavoidable contact with others.

SOURCE: American Association of State Highway and Transportation Officials (AASHTO), *A Policy on General Design of Highways and Streets,* AASHTO, Washington, D.C., 1990.

by 60 in (1525 by 1525 mm). If a ramp run has a rise greater than 6 in (150 mm) or a horizontal projection greater than 72 in (1830 mm), then it should have handrails on both sides. Note, however, that handrails are not required on curb ramps.

TABLE 4.2 Design Considerations for Outdoor Stairways

1.	Outdoor stairs should be made easier to use than indoor stairways because people tend to be moving faster when outdoors.
2.	The use of a single stair should be avoided. A minimum of three steps should be used to clearly signal the change in grade.
3.	A minimum tread height of 4.5 in should be maintained. A maximum tread height of 7 in should also be observed.
4.	Stair treads should be designed with a minimum of 2% positive pitch to provide drainage.
5.	Vertical distance between landings should be 5 ft or less.
6.	Stair design should incorporate visual signals to indicate stair treads and edges.

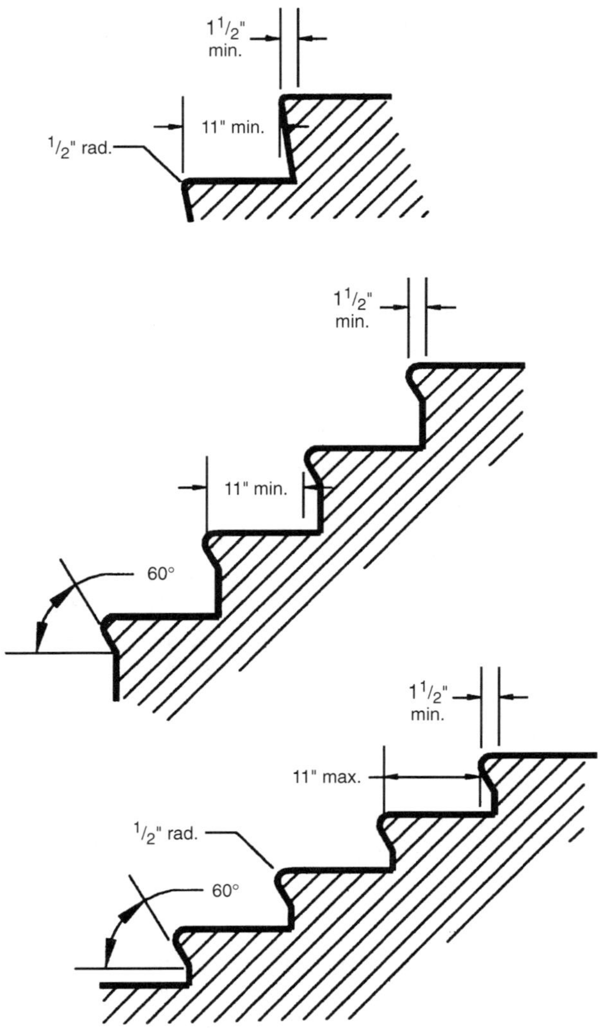

Figure 4.5 Usable tread width and acceptable nosing detail.

Figure 4.6 Photograph of stair treads with painted nosing.

TABLE 4.3 Site Survey for Compliance with the Americans with Disabilities Act

1. Is there an adequate number of accessible parking spaces designated?
2. Do the designated accessible parking spaces meet design minimums?
3. Is there 1 van-accessible space for every 8 accessible spaces, or are all spaces consistent with the universal parking space design?
4. Are accessible spaces marked using the international symbol of accessibility?
5. Are van spaces marked "Van Accessible"?
6. Is there at least one accessible route allowing access to all public facilities?
7. Are depressed curbs provided on the accessible route?
8. Do ramps meet ADA minimums (1:12 slope or less, 36 in wide, 30-in maximum rise, level landing every 30 ft, adequate landing size)?
9. Are overhead and wall clearances adequate?
10. Are surfaces nonslip?
11. Are minor changes in grade beveled to minimize the risk of trip or obstruction?
12. Are handrails provided where appropriate?
13. Are textural or audible signals provided where necessary?

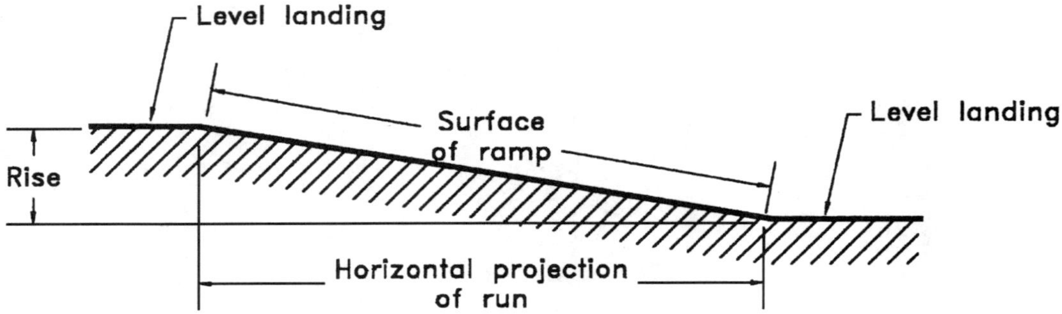

	Maximum rise		Maximum horizontal projection	
Slope	in	ft	ft	m
1:12 to 1:16	30	760	30	9
1:16 to 1:20	30	760	40	12

Figure 4.7 Ramp detail.

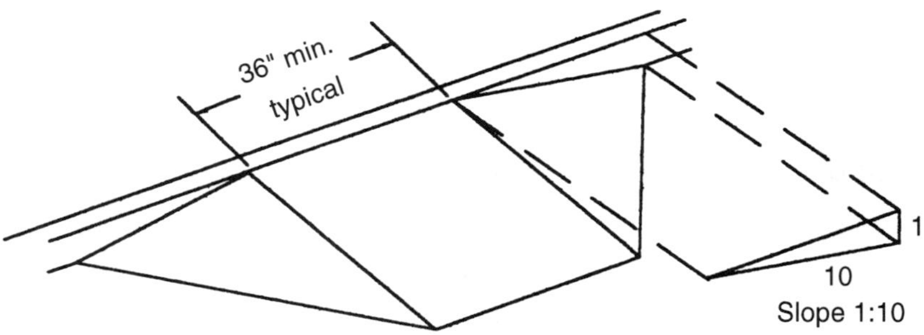

Figure 4.8 Built-up curb ramp detail.

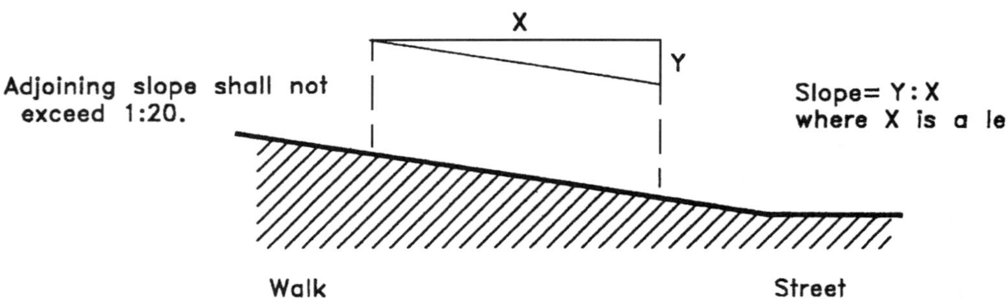

Figure 4.9 Measurements for curb ramp slopes.

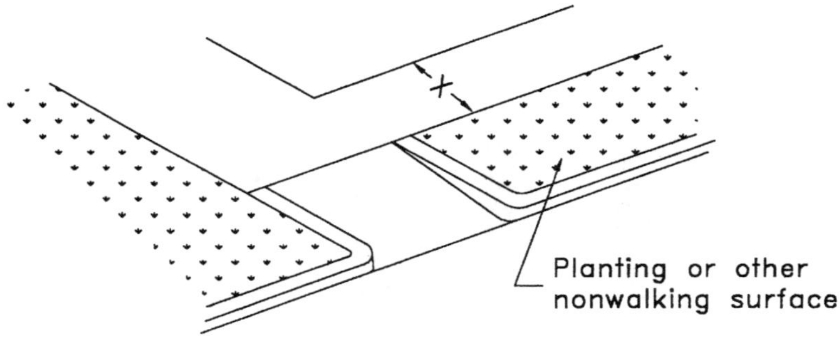

Figure 4.10 Sides of curb ramps, returned curb detail.

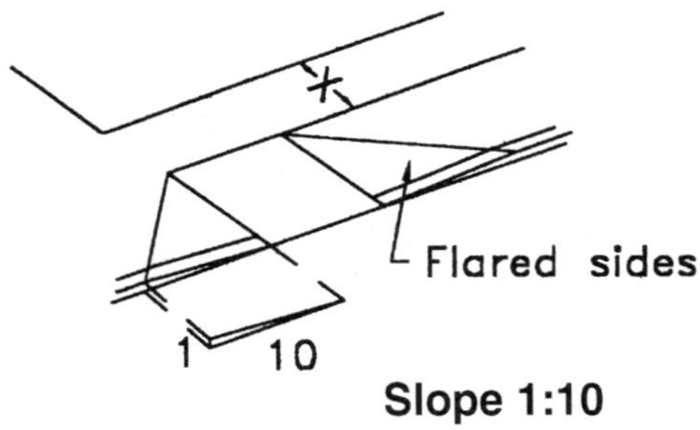

Figure 4.11 Sides of curb ramps, flared sides detail.

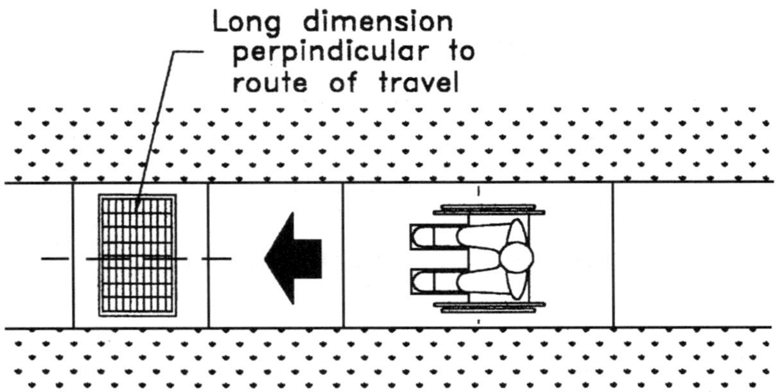

Figure 4.12 Grating orientation detail.

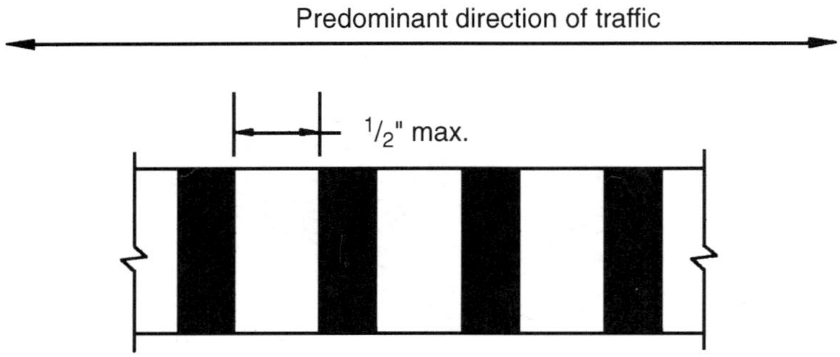

Figure 4.13 Grating opening detail.

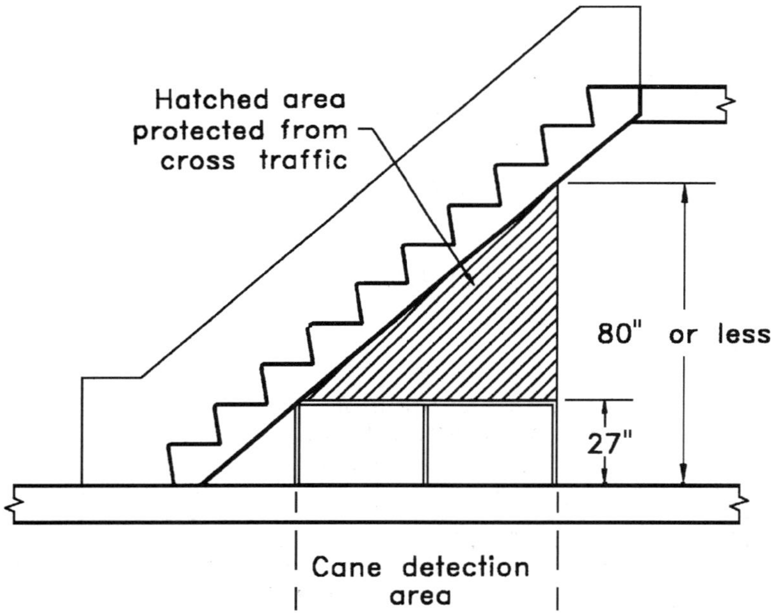

Figure 4.14 Protruding objects and overhead hazards detail.

Handrails should be continuous along both sides of ramp segments. The inside handrail on switchback or dogleg ramps should always be continuous. If handrails are not continuous, they should extend at least 12 in (305 mm) beyond the top and bottom of the ramp segment and should be parallel with the floor or ground surface. The clear space between the handrail and the wall should be 1½ in (38 mm). The top of the handrail gripping surfaces should be mounted between 34 and 38 in (865 and 965 mm) above the ramp surfaces. Handrails should not rotate within their fittings.

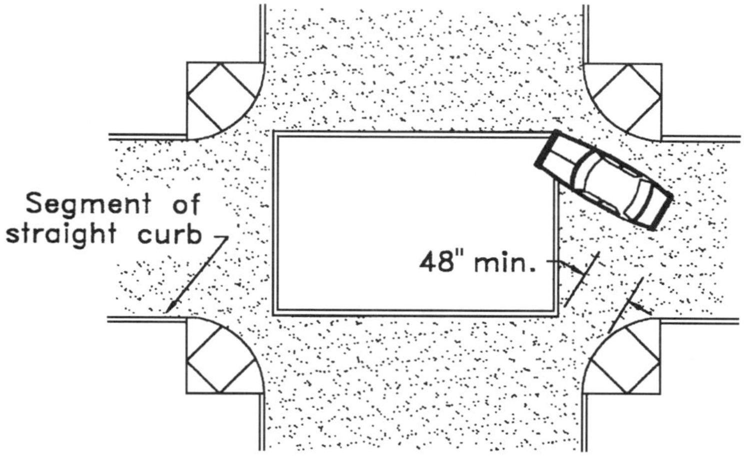

Figure 4.15 Curb ramp at marked crossings detail.

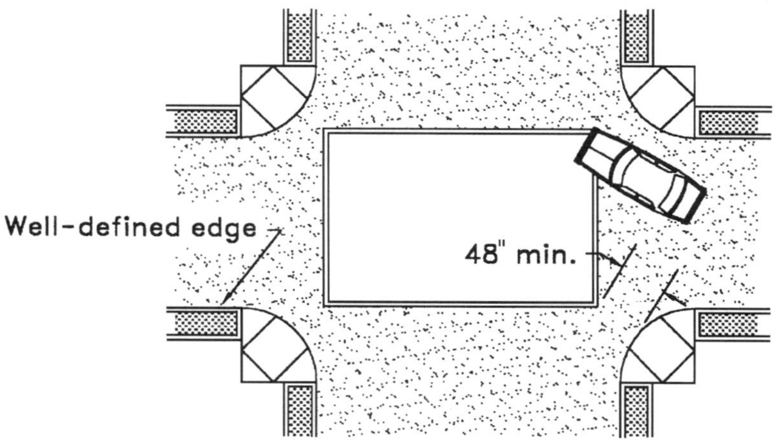

Figure 4.16 Curb ramp at marked crossings detail.

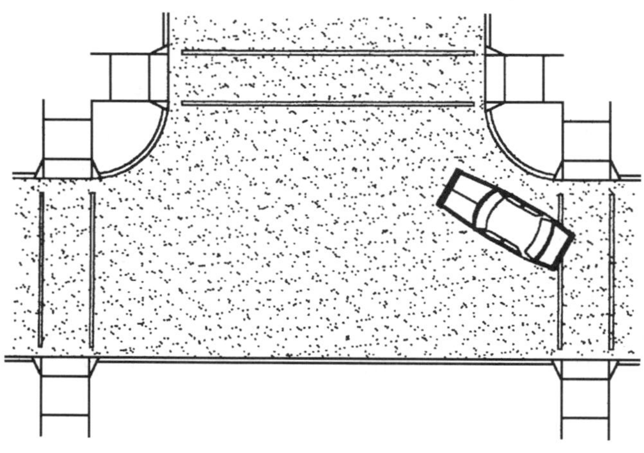

Figure 4.17 Curb ramp at marked crossings detail.

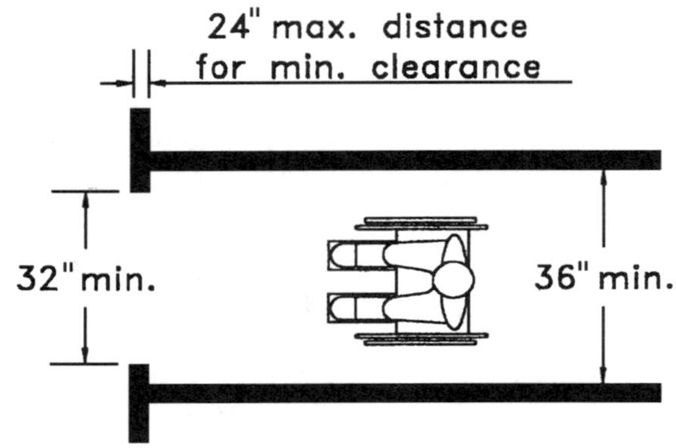

Figure 4.18 Minimum clearance width for wheelchair detail.

Paving materials and design

The choice of paving materials is broad and generally is determined by the nature of the project under design and the preferences of the designer and client (see Table 4.4 and Figs. 4.19 through 4.22). In general the characteristics of concern for paving materials are the installation and life cycle costs, durability, slip resistance, and appearance. Bricks and pavers for pathway and sidewalk paving are described in ASTM C902 by grades and by type. Type I brick is recommended for high-traffic areas such as driveways or entranceways; type II brick is used on walkways and other areas of moderate traffic, and type III brick is used in areas where low levels of traffic are anticipated such as patios.

Open-Space Requirements

It is common practice today for developers to provide open space and recreation facilities as part of residential and commercial projects. As often as not, the developer is required to do so by local ordinance. However, many communities do not have a coordinated or planned approach to incorporating the additions into community life. Instead, the effect of the ordinance may be to create pockets of playground equipment or open space that are unrelated and unconnected to the development of the community at large. Local ordinances are also often unclear as to how to evaluate open space so that passive open space and active open space are not differentiated or there is no qualification or valuation of open space. Without a comprehensive plan, a community may miss opportunities to serve its citizens with the best and most appropriate use and type of open space. Not all open space is of equal value. Sites along busy highways, commercial areas, or industrial zones may not be desirable as open

TABLE 4.4 Materials for Pathway and/or Sidewalk Paving

Material	Type	Characteristics
Stone	Granite	Hard, very dense; difficult to work with; weather resistant; very durable; long-wearing in high-volume areas; should have low ferrous or pyrite content to avoid rapid weathering
	Limestone	Wide variation in color and durability; susceptible to chemical weathering; easier to work with than granite
	Sandstone	Durable; wide range of colors; mostly earth tones; similar to limestone in workability
	Flagstone	Durable; moderate to expensive; may be slippery when wet
	Slate	Durable; expensive; may be slippery when wet
Brick	Sx grade*	Resistant to frost/freeze and thaw; can be used as paving material; high installation cost
	Mx grade*	Not recommended for use where brick will be saturated with water; can be used as paving material only in dry or well-drained situations
	Nx grade*	In general not suitable for paving purposes
Asphalt		Installed in light-duty (usually 2 layers) to heavy-duty (as many as 5 layers) applications; inexpensive; durability often a function of native soil/subsurface conditions and weather; susceptible to damage at edges; susceptible to freeze damage if base becomes saturated; absorbs heat; susceptible to damage from petroleum products
Concrete		Versatile; commonly used as paving material; durable; relatively easy to install; good life cycle costs; multiple surface treatments for enhanced texture and color; usually reinforced with wire mesh or reinforcing bar; thickness determined by function and soil conditions

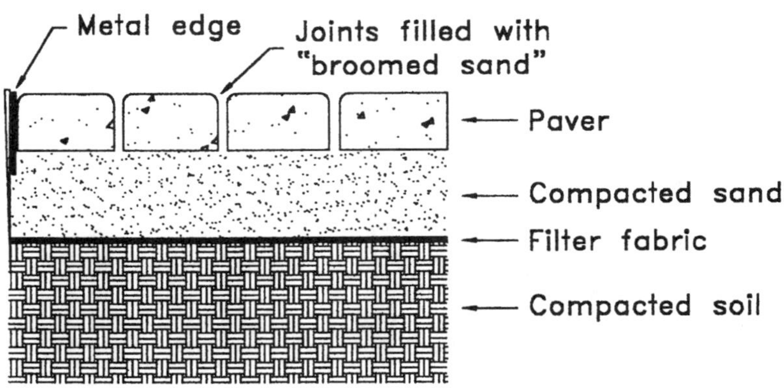

Figure 4.19 Paver installation detail.

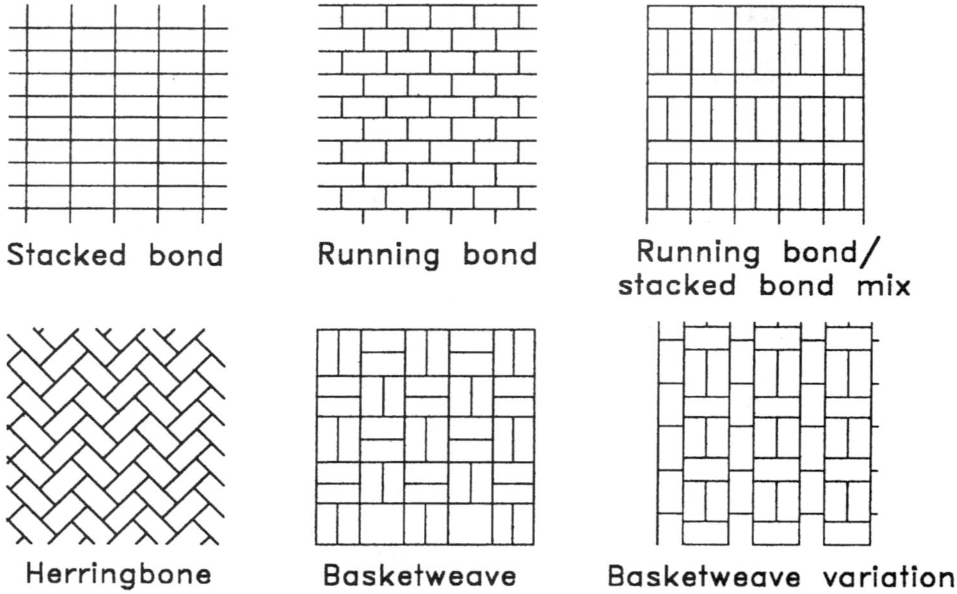

Figure 4.20 Brick bonds and patterns.

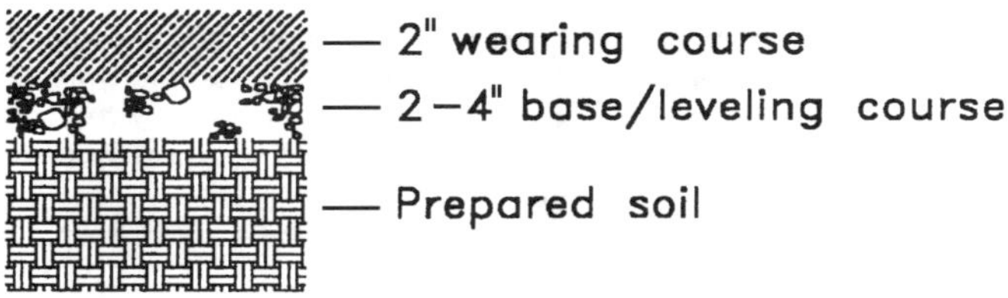

Figure 4.21 Detail of typical asphalt pavement for pathway, light duty.

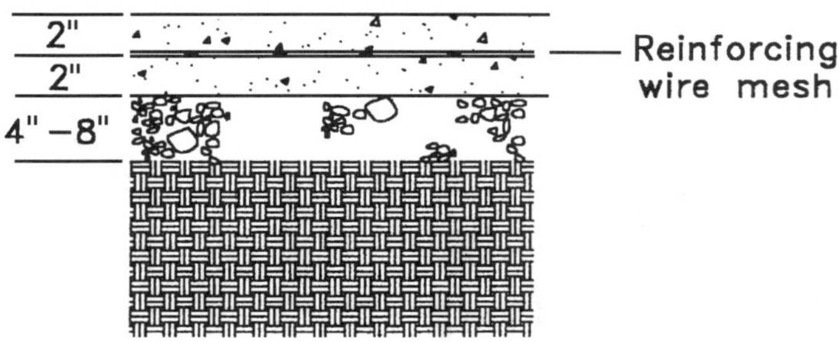

Figure 4.22 Detail of typical concrete pavement for sidewalk, light duty.

space. Areas of wetland, riparian zones, or floodplains may be desirable for some purposes but not for others. It is just as true, however, that certain active-open-space features may be desirable to only a very few residents.

Recreation and open space must be planned with regard to the projected users, the physical space, and the capability of a management entity to maintain the facility. These considerations must be measured in the short and long terms. New facilities should be planned to work in conjunction with existing facilities. The development of complementing facilities maximizes the recreation–open-space dollar, provides a broader choice of activities to the user, and precludes the development of competing and redundant facilities. In developing active or passive open space, developers and communities alike must be concerned about the actual demand, current and future, for those facilities. The demand for a particular type of recreation opportunity or facility should be tied to the target population. If it is to be used exclusively by the inhabitants of the new development, then the demographics of the new population should lead the design. If the facility is to have a broad base of community use, then another set of considerations should lead the design. The unwanted facility is not an amenity; it does not attract users (or buyers, in the case of a new facility). In fact, such underutilized space may be an attractive nuisance that costs more in maintenance and liability insurance than it returns in value to the community.

Active open space must be compatible with the site as well as with the user. An analysis of the site must include existing features such as watercourses, tree masses, topography, adjacent land uses, and areas of historic significance. These concerns that might otherwise restrict development may be effectively coordinated with the open-space and recreation elements of the design. By first assessing the existing qualities and characteristics of the site, the compatibility of the site with a proposed open-space design can be evaluated.

Many studies have been conducted to determine the leisure activities of various age groups within communities. Caution, however, should be used in applying such studies, because the information in such studies has a shelf life. The study measures the preferences of a community at a given time, but the mix of preferences within a population changes with time. Analyzing the needs of a particular community should include (a) the age group (or expected age group, in the case of projections) or the age distribution within the population, (b) the projected number of users within the population, (c) the sources of funding, maintenance, and management and the capacity of the resource to maintain the facility, and (d) the availability and accessibility of existing facilities.

The sizing of facilities is also very important. The proposed active open space must be large enough to serve the user population but small enough to be maintained by the responsible parties. An evaluation of the appropriate size or number of facilities should include a projection of the future users. As the population grows older or younger within a community, the demand for facilities will change. Planning the active open space should include not only a demand analysis of today's users but a projected demand of those users in 10 and 20 years.

The choice between using community-owned land for active or passive open space must take into consideration many factors. All age groups have a desire and an interest in both types of facility. Passive activities include reading, picnicking, sight-seeing, photography, people watching, and strolling (as opposed to walking for exercise). Space for passive activities includes unimproved open space, and parkland, and wildlife habitat, but it can also include space on the fringes of activity areas that allow, even encourage, people watching and nature observation.

Tables 4.5, 4.6, and 4.7 provide an overview of the levels of participation and use of public facilities associated with various recreational activities. In Table 4.6 "average days per year" refers to the number of days survey participants were engaged in outdoor activities. Table 4.8 lists approximate open-space development standards for various activities. Table 4.9, elaborating on the information provided in Table 4.8, gives space requirements for baseball diamonds suitable for softball and baseball and for men, women, and children's games. Finally,

TABLE 4.5 Percentage of Population Participating in Given Outdoor Activities

Activity	Population participating, %
Sight-seeing and/or driving for pleasure	72.5
Picnicking	70.6
Swimming	66.7
Bicycling	47.5
Hiking and taking nature walks	40.9
Baseball	32.4
Fishing	31.9
Boating and canoeing	30.7
Golf	29.0
Camping	26.4
Tennis	24.0
Basketball	22.0
Ice skating	21.2
Football	16.5
Hunting or sport shooting	14.0
Snowmobiling and/or offroad vehicle driving	12.0
Horseback riding	11.1
Snow skiing	5.7
Street hockey	5.0

TABLE 4.6 Outdoor Activity Days per Year by Age Group

Age group	Average days per year
5–9	205
10–19	255
20–29	149
30–44	99
45–64	55
65	15

TABLE 4.7 Activities or Facilities in Order of Demand

1. Bicycle paths
2. Tennis courts
3. Swimming pools
4. Ice skating areas
5. Playgrounds
6. Hiking and walking trails
7. Offroad vehicle trails
8. Ballfields
9. Picnic areas
10. Natural swimming areas

Table 4.10 lists approximate development standards for community facilities for various sizes of the populations to be served within a community.

The age distribution of a user population is important because the pressure on facilities will vary based on the number of users and the frequency of use. Although the greatest proportion of the population tend to use the more passive open space, a greater percentage of the age groups that use active recreation facilities actually participate in sports. (Figures 4.23 through 4.30 illustrate sports and recreation site designers.) Basically in a large population of children, the percentage of individuals who use the available facilities is greater than in an older population group.

Accessibility and open space

The enactment in 1990 of the Americans with Disabilities Act (ADA) has served to increase our awareness of barriers to access to open space and the need to remove or bypass them. Today plans must accommodate the entire population to a reasonable extent. The design and construction industries have been building new structures with a greater freedom of access for more

TABLE 4.8 Community Open-Space Development Standards

Activity	Space required	Area required	Facilities per population
Badminton	1,620 ft^2	20 × 44 ft	1/5,000
Basketball			
Youth	2,400–3,036 ft^2	46 × 84 ft	1/5,000
High school	5,040–7,280 ft^2	50 × 84 ft	1/5,000
Collegiate	5,600–7,980 ft^2	50 × 94 ft	1/5,000
Tennis	7,200 ft^2	36 × 778 ft	1/2,000
Handball	800–1,000 ft^2	20 × 50 ft	1/10,000
Ice hockey	22,000 ft^2	85 × 200 ft	1/2,000
Football	1.5 acres	180 × 300 ft	1/20,000
Baseball			
Little League	1.2 acres	60-ft baseline	1/5,000
Official	3–3.85 acres	90-ft baseline	1/30,000
Soccer	1.7–2.1 acres	225 × 330 ft	1/5,000
Softball	1.5–2.0 acres	60-ft baseline	1/5,000
Golf			
Par 3	50–60 acres	—	
9 holes	50 acres/min	—	1/25,000
18 holes	110 acres/min	—	1/50,000
Playground and/or park	1.5 acres	—	1/1,000
Community park	3.5 acres	—	1/1,000 people

than 10 years, but the ADA extends the requirement to provide access to existing parks and recreation facilities. In fact, parks are specifically identified in the act as a public accommodation that must respond to the minimum requirements of the ADA, and even facilities that existed prior to the passage of the act are subject to the reasonable-accommodation test.

More than 43 million Americans are disabled by numerous different physical and mental impairments, and as our population grows older, the physiological changes of aging will bring access issues to light for even more people. New facilities and major remodeling projects are incorporating at least the minimum standards of access, but it may be more difficult to manage the adaptation of existing facilities. Certainly shrinking budgets and the characteristics of older facilities can combine to limit the resources and opportunities to provide access, but access "in the park" may mean more than simply installing a ramp or handrail in the right place. Reducing barriers to parks and open space involves an understanding of the user's needs and capability, as well the intrinsic value of the site and the desired experiences or programs. The kind and nature of the impairments that affect people are numerous, and the differences among sites and programs are so broad that any analysis must be careful in using standardized solutions or programs to address concerns of

TABLE 4.9 Baseball Diamond Dimensions

	Softball				Baseball		
	Men's		Women's				
	Fast pitch, ft	Slow pitch, ft	Fast pitch, ft	Slow pitch, ft	Little League, ft	Pony league, ft	NCAA†, ft†
A. Pitching distance	46	46	46	40	46	54	60.5
B. Home plate to back stop	25–30	25–30	25–30	25–30	25–40	40–60	
C. Baseline	60	60	60	60	60	80	90
D. Radius of skinned area	60	60	60	60	50	80	95
E. Radius of base area	30	30	30	30	18	24	26
F. Foul line	275 min	225 min	225 min	225 min	200	250	350
G. Coach's box	3 × 15	3 × 15	3 × 15	3 × 15	4 × 8	8 × 16	5 × 20
H. Diameter of pitcher's mound	8	8	8	8	10*	15*	10*
I. Home plate to pocket	275	225	250	225	250	300	400

*The pitcher's mound should be raised 10 in for Little League, 15 in for Pony League, and 10 in for NCAA.
†National Collegiate Athletic Association.

TABLE 4.10 Area Requirements for Community Facilities

	1000 people/ 275 families	2000 people/ 550 families	3000 people/ 825 families	4000 people/ 110 families	5000 people/ 1375 families
	One- or Two-Family Development				
Acres in school site	1.2	1.2	1.5	1.8	2.2
Acres in playground	2.75	3.25	4.0	5.0	6.0
Acres in park	1.5	2.0	2.5	3.0	3.5
Acres in shopping center	0.8	1.2	2.2	2.6	3.0
Acres in general community facilities	0.38	0.78	1.2	1.5	1.9
Aggregate area	—	—	—	—	—
Total acres	6.63	8.41	11.40	13.90	16.60
Acres per 1000 people	6.63	4.20	3.80	3.47	3.32
Square feet per family	1080	670	600	550	530
	Multifamily Development				
Acres in school site	1.2	1.2	1.5	1.8	2.2
Acres in playground	2.75	3.25	4.0	5.0	6.0
Acres in park	2.0	3.0	4.0	5.0	6.0
Acres in shopping center	0.8	1.2	2.2	2.6	3.0
Acres in general community facilities	0.38	0.78	1.2	1.5	0.38
Aggregate area	—	—	—	—	—
Total acres	7.13	9.41	12.90	15.90	19.10
Acres per 1000 people	7.13	4.70	4.30	3.97	3.82
Square feet per family	1130	745	680	630	610

SOURCE: Charles W. Harris and Nicholas T. Dines, *Time-Saver Standards for Landscape Architecture* (New York: McGraw-Hill, 1988), pp. 210–224. Used with permission of McGraw-Hill.

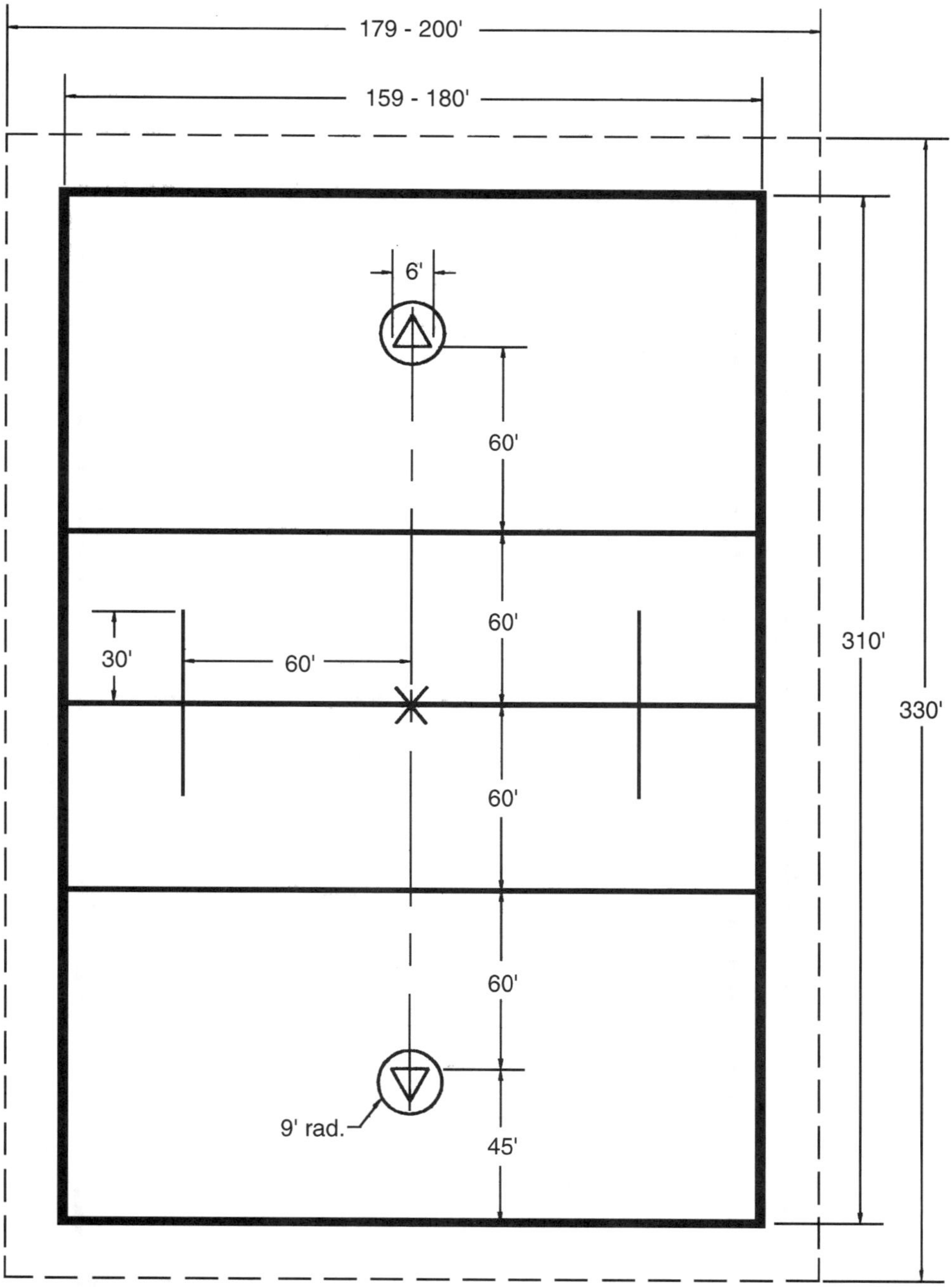

Figure 4.23 Lacrosse field, men's, detail.

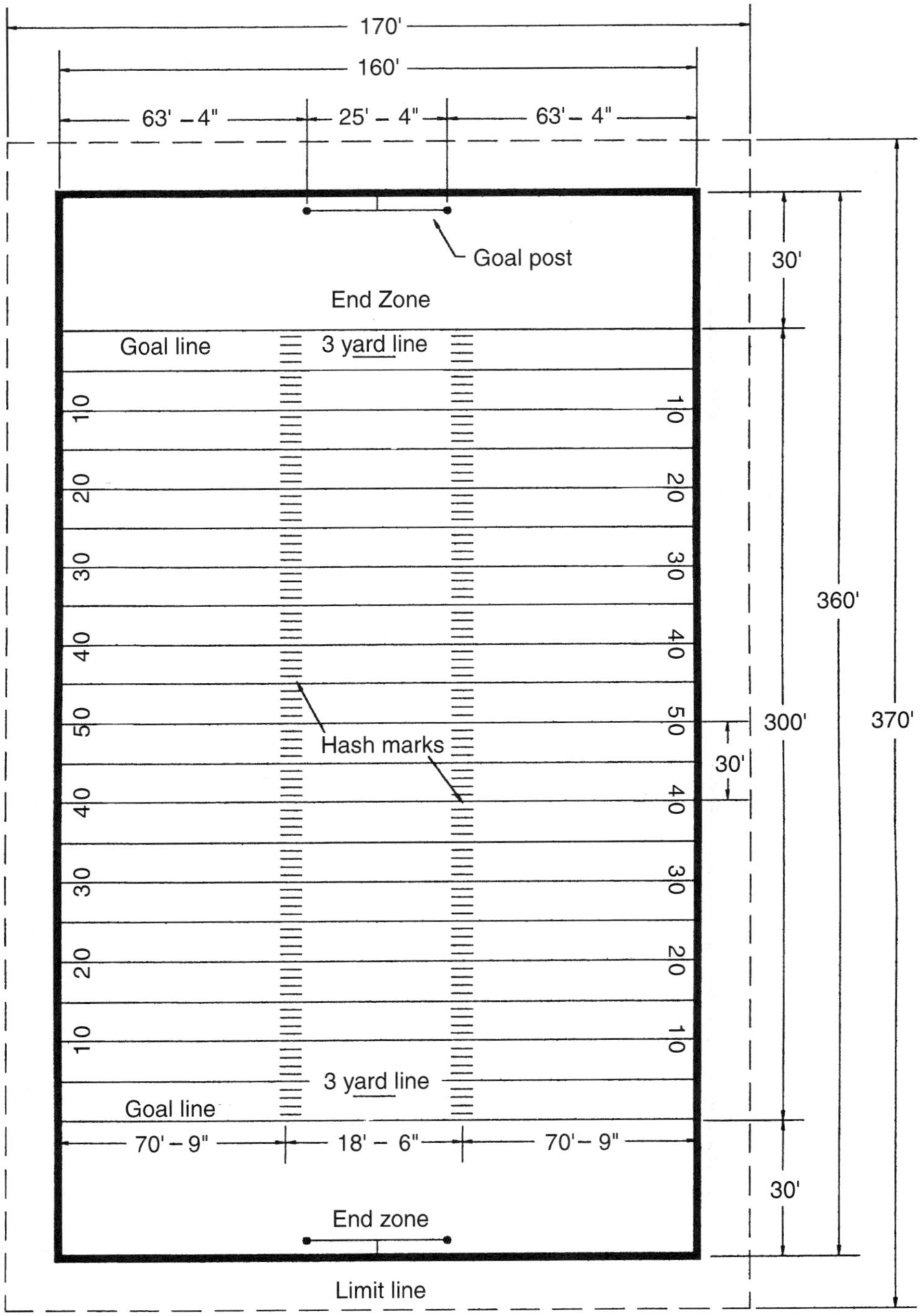

Figure 4.24 Football field detail.

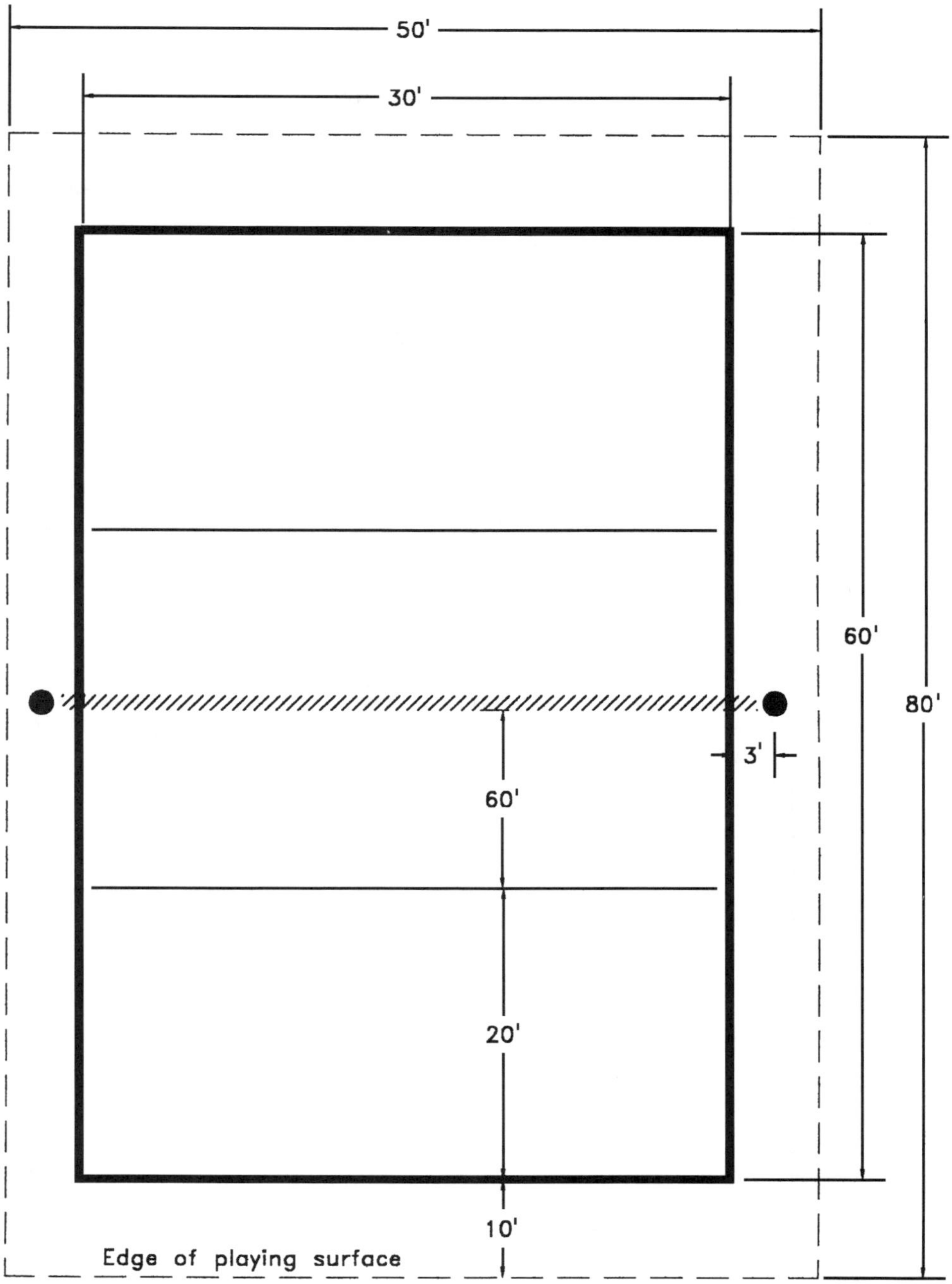

Figure 4.25 Volleyball court detail.

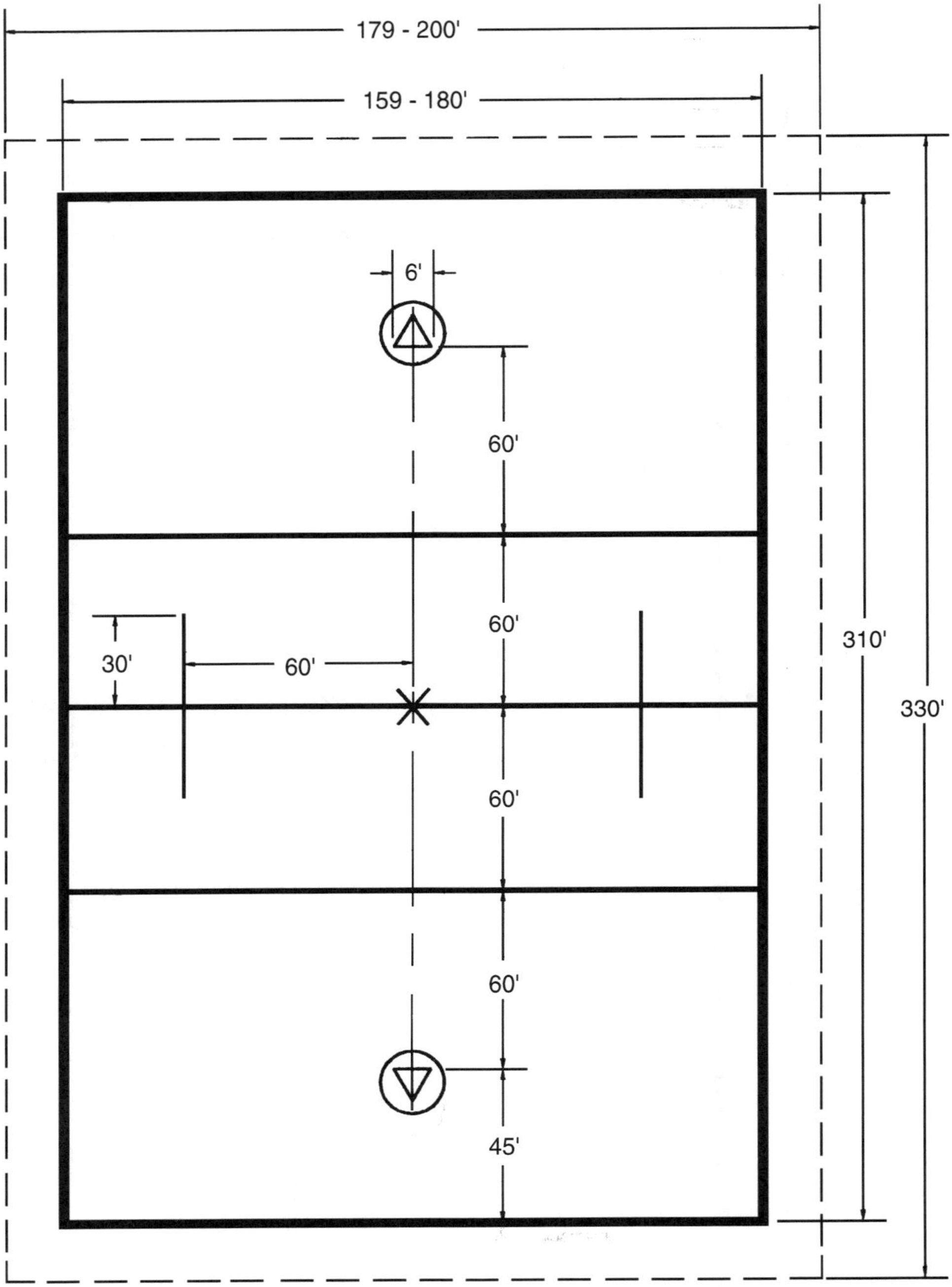

Figure 4.26 Soccer field detail.

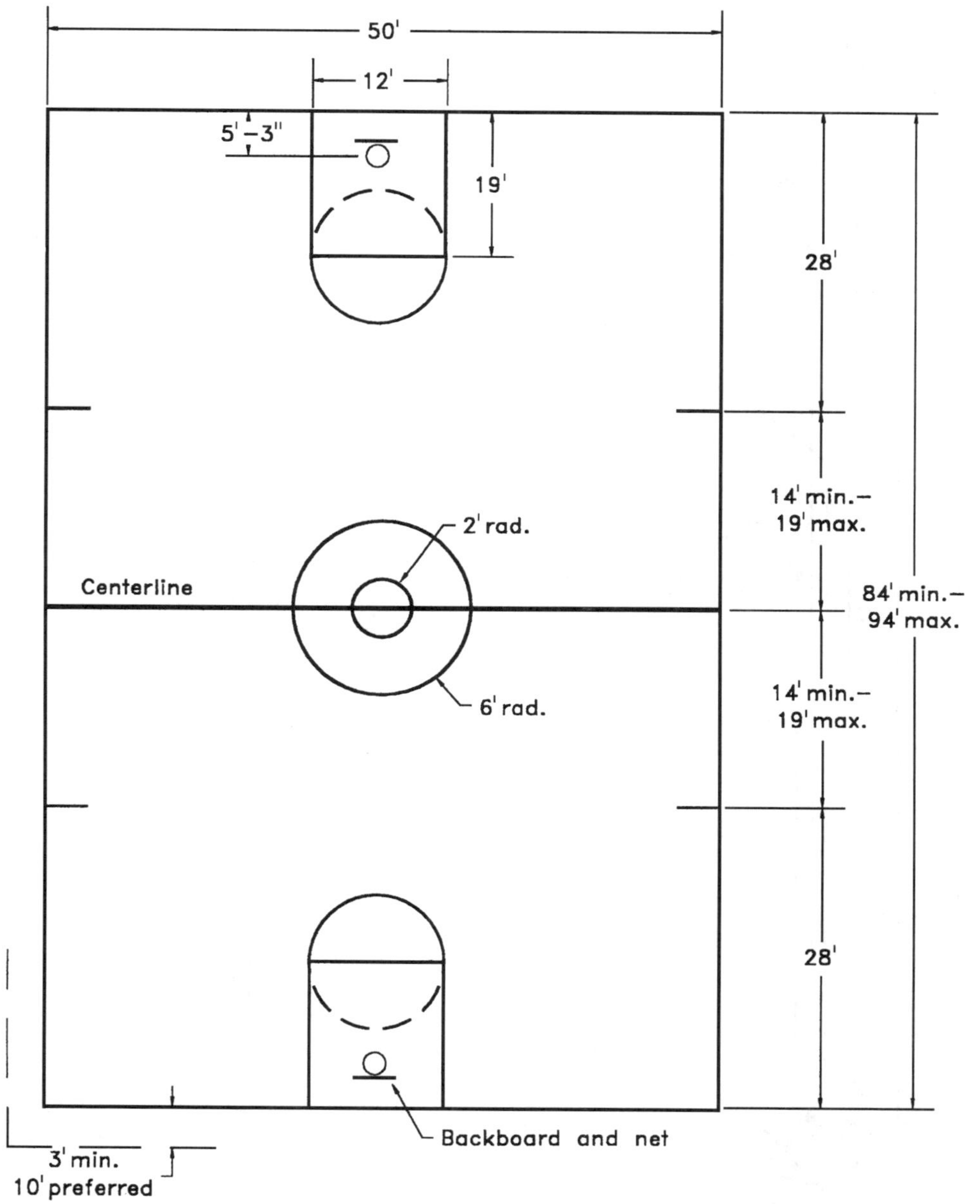

Figure 4.27 Outdoor basketball court detail.

A) Pitching distance
B) Home plate to backstop
C) Baseline
D) Radius of skinned area
E) Radius of skinned area and bases
F) Foul line
G) Coach box
H) Diameter of pitcher's mound
I) Home plate to pocket

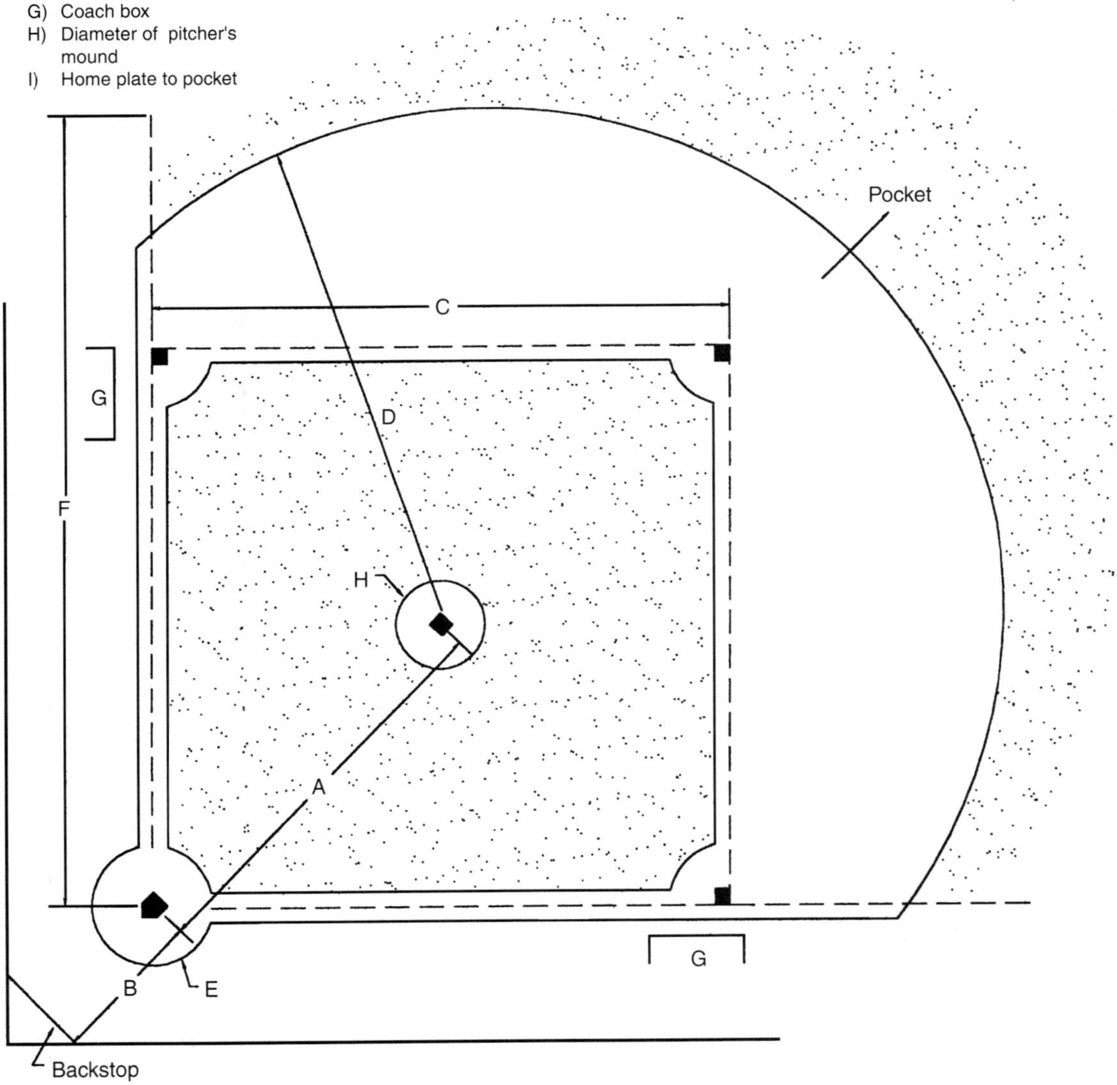

Figure 4.28 Baseball diamond detail.

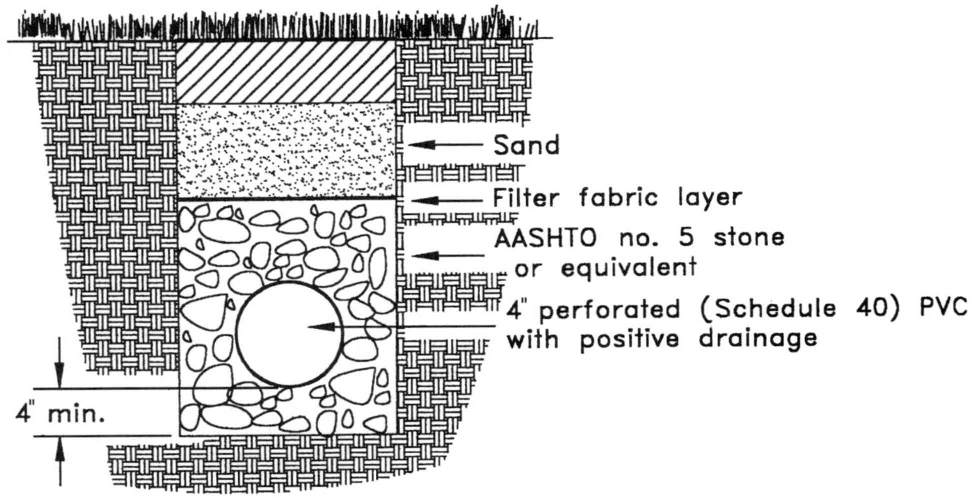

Figure 4.29 Subsurface drainage detail.

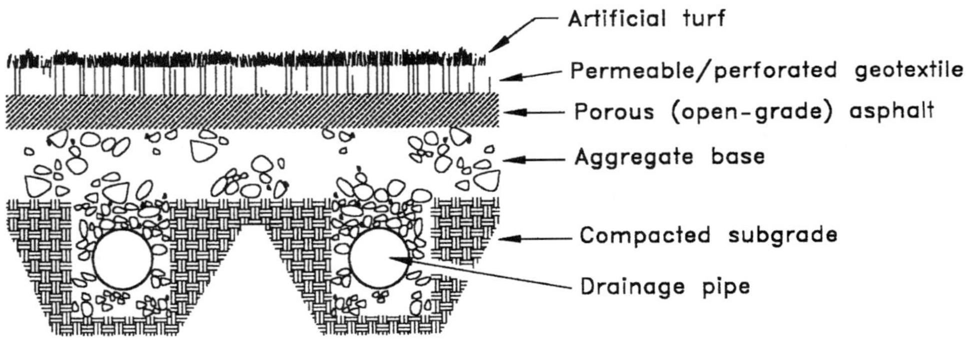

Figure 4.30 Artificial turf detail.

access. In terms of parks and open space, the barriers and limits to access may be different and may affect a broader slice of the population than originally thought. In some cases, the physical limits to parks and open space may be inherent in the character of the sites themselves. It is important to note that many physically challenged people can satisfactorily access and enjoy a site (independently or with minimum assistance) that is designed for the general public. For a valuable perspective, designers should consider working with a physically challenged member of the community when planning an open-space design.

The physical characteristics of a site also influence the degree of accessibility and the methods of providing accessibility. The quality of facilities and the quality of experience should be considered in any design or evaluation. The

National Center for a Barrier Free Environment suggests a systematic, staged approval process to maximize the quality of the facility and the program. A simple system of integrated stages of increasing challenge, as shown in Table 4.11, provides the users with the opportunity to determine their own limits (Fig. 4.31). In this way the facility does not act as the limit to participation.

Parks and open spaces that are able to provide varying degrees of access and challenge serve the general public in that everyone has access but is able to pursue the limits of individual interest and ability. The range of accessibility offers an escalating scale of challenge but provides for a maximum range of access. The details of the mechanics of accessible design have been published and distributed throughout the design and construction industries; the standards for ramp length and height or handrail height are easily determined if they are not already part of everyone's standards. There are, however, other "nonstandard" concerns that should be part of an evaluation. When performing an evaluation, it is necessary to develop a critical eye to assess the facility in terms of users with different capabilities and needs.

TABLE 4.11 Suggested Stages of Accessibility

Stage 1	Provides access to all buildings, secondary facilities, and program.
Stage 2	In addition to all of stage 1, access is provided to "unique" opportunities or features.
Stage 3	In addition to stages 1 and 2, various degrees of access and challenge are provided to secondary opportunities or facilities.

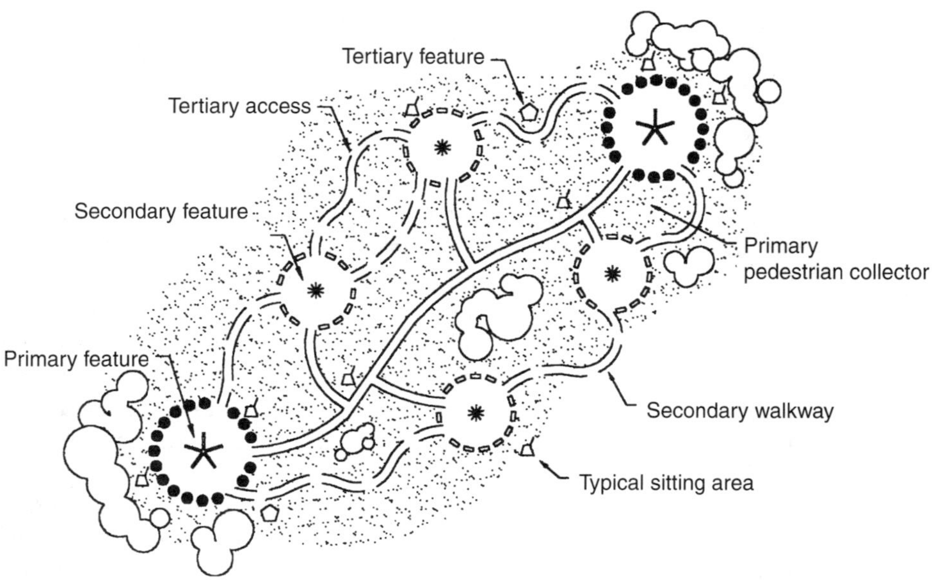

Figure 4.31 Phased integrated access system (see also Table 4.11).

For many users, walks and pathways are more than just a means of going from point *A* to point *B*; the walk is also the experience and it is interesting. Surfaces for stage 1 and 2 walkways should be stable and firm with nonslip textures. Grades for these walks would average about 3 percent, but not exceed 5 percent. Depending on the actual grades and lengths of walks, rest areas with places to sit should be provided at regular intervals. The Minnesota Department of Natural Resources Trail Planning Classification guidelines have been widely distributed and used as a model for designing these elements of walks and pathways.

Open space for older users

The aging of the population of North America presents particular opportunities for site designers, and many firms have already specialized in the design of places especially for older folks. Accounting for the interests and needs of the older person in a design requires some understanding of the effects of aging on the individual (see Table 4.12).

Beyond the obvious design issues, there are steps that can be taken to help make the walkway and park in general more user friendly. Visually impaired users may require tactile signals to receive information on their surroundings. Texture changes at breaks in grade or intersections may also assist elderly users who may have reduced depth perception capability. Installing a handrail at a sudden change in grade or a stair on an outdoor walk sends a clear signal to the user and provides the information in a subtle fashion. Where possible both stairs and ramps should be provided; for many people, walking down a ramp is more difficult than using stairs.

In the stage 1 integrated walk network, pavement, color, and texture as well as signage can be employed to assist the users with way finding and guidance. Construction of barricades to obstruct vehicles must consider the disabled. A cable or chain strung across a pathway can be a significant obstruction, and a system of removable bollards might be preferable (Fig. 4.32). By developing clear simple signs with thematic use of color, letter style, or texture as a means of communication, significant information can be provided with a minimum of detail. The use of color to identify a particular degree of accessibility or stage of a facility is simple, direct, and without stigma. In addition, lettering styles can be made consistent throughout a facility to convey a maximum amount of information in a simple useful form.

Walkways should be visually interesting, but in general, encroachment by trees and shrubbery are to be avoided. As seen in Fig. 4.33, lower limbs should be removed to a minimum of 8 ft of overhead clearance at the walkway and no closer than 1 ft to the edge of the walkway. If it is necessary to have a grate in a walkway, the maximum opening in the direction of travel is $^{3}/_{4}$ in. Larger openings may catch cane tips or bicycle tires.

The design of open areas should give particular attention to way finding. Large undefined areas may be confusing and underused rather than providing opportunities for viewing activities in open-space areas. In evaluating open space, its

TABLE 4.12 Checklist of Physiological Changes with Age and Some Design Implications

Sensory Process and Perception

Age-related sensory losses occur with vision, hearing, taste, touch, and smell. One possible and practical design response to these losses is to load the environment with redundant sensory clues. This includes special attention to:

1. The quality and quantity of light
2. The use of color (Brighter colors and those in the orange-yellow-red spectrum are easier to distinguish.)
3. Contrasts of light and dark shadows and advancing and receding colors as they distort depth perception
4. The intensity and pitch of sounds (Lower-pitched sounds are more easily heard.)
5. Tactual cues that may be more easily "read"

Central Nervous System and Cognitive Functions

Although many cognitive functions do not change with age, concept formation ability and reaction time may be reduced. To facilitate orientation and promote safety, special attention must be given to:

1. Decreased concept formation ability affecting orientation or way finding
2. Slower reaction time
3. Difficulty in distinguishing and interpreting background noises from foreground sounds

Muscular and Skeletal Systems

Muscular strength, agility, and fine-motor control may diminish with age. The reduced resiliency of the skeletal system requires attention to safety, security, and environmental negotiability, as injury may be more devastating for older people. These have special implications for the design of:

1. Ground surfaces and changes in elevation
2. Facilities requiring fine- and/or gross-muscle movement

Temperature Adaptation

The reduced ability to adapt to changes in temperature requires amenities and detailing for temperature moderation and/or control.

Disease

Susceptibility to chronic diseases restrains activity. Special considerations for health-related problems include:

1. Providing easy access to nearby restrooms
2. Providing options for those with various levels of reserve energy
3. Limitations on fine-motor control and gross-muscle movements due to arthritis

SOURCE: From Diane Y. Carstens, *Site Planning and Design for the Elderly* (New York: Wiley, 1985).

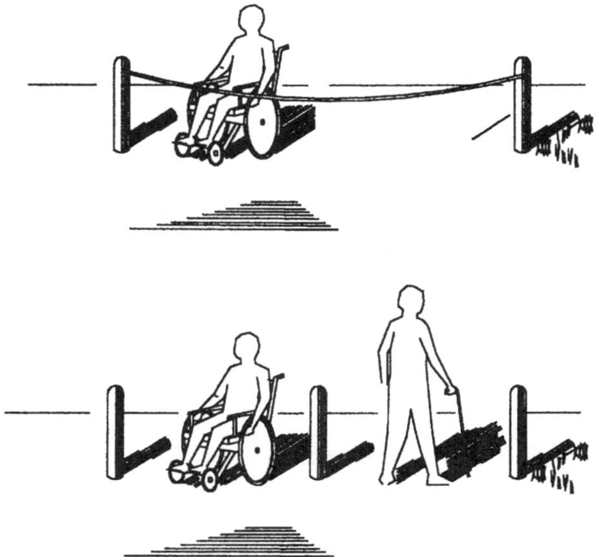

Figure 4.32 Replace cables and chains with bollards.

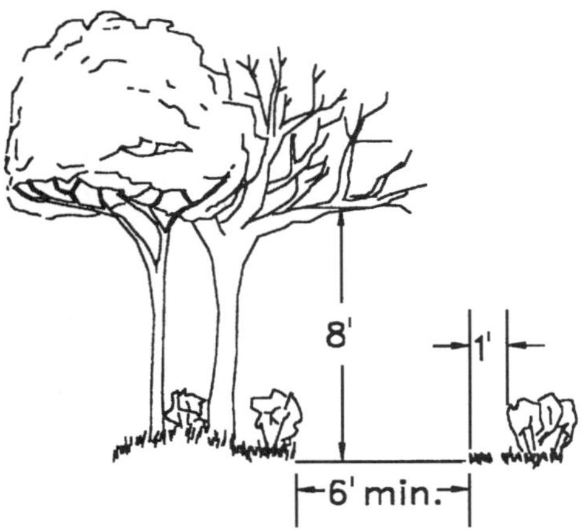

Figure 4.33 Minimum path clearances detail.

purpose should be identified, and it should be evaluated on the basis of that purpose. A hierarchy of space is desirable so that smaller "private" spaces are connected to larger more public spaces. Areas should have edges or boundaries to reduce ambiguity, provide identity, and assist in way finding. The facility should be evaluated to remove or mitigate overhead hazards, as well as trip hazards.

The key to a successful redesign or adaptation is to maximize the access and maintain the quality of the experience. Barriers that restrict users from general access prevent the maximum use of facilities without a corresponding element of enhancement or need for preservation of quality. The thoughtful evaluation will strike a balance between the removal of borders and the maintenance of the character and fabric of a place or experience. Sensitivity to the special needs of older users may not be intuitive.

Playgrounds

The design of play areas and playgrounds should provide a variety of play equipment and special areas for different age groups and activities. The design should provide for shade and sunny areas and places for quiet activity and observation as well as more physically active play. The U.S. Consumer Products Safety Commission estimates that 100,000 children are treated at hospital emergency rooms for injuries suffered at playgrounds (private and public). Most of these children are between the ages of 5 and 10 years old. The majority of these injuries are related to design issues and not supervision issues. The evaluation of existing playground equipment should begin with a routine inspection at the startup and throughout the season; loose parts should be tightened and friction points lubricated (see also Table 4.13).

The ASTM has developed three important specifications for playground designers. The *ASTM F 1487 Standard Consumer Safety Performance Specification for Playground Equipment for Public Use* addresses the safety and performance of equipment; it was revised and updated in 2001. The *ASTM F 1951 Surfacing Standard* and the *ASTM F 1292 Standard Specification for Impact Attenuation of Surface Systems Under and Around Playground Equipment* address surfacing and fall protection. The Americans with Disabilities Act also applies to playgrounds. Selections for playground equipment should be compared to these consensus standards. The U.S. Consumer Product Safety Commission (CPSC) also publishes technical information guides to assist in the evaluation and selection of materials and products.

Ideally the access to the playground should not include direct street access, and it should be located at least several hundred feet from the street. Playgrounds should be sized on the basis of 70 ft^2/child or 21 ft^2/family. A 2000-ft^2 playground

TABLE 4.13 Evaluation of Playground Equipment Potential Hazards

Pinch points or crush points
Sharp edges and catch points
Exposed screws and bolts
Spacing of rings, rungs, rails (choking hazards)
Spacing of equipment
Overlap of fall zones
Hard surfaces
Fall hazards

Compiled from data supplied by U.S. Consumer Product Safety Commission and American Society of Testing and Materials.

for small children will serve about 100 families. Approximately 50 to 60 percent of the area should be turf. Equipment should be spaced to provide safe and comfortable traffic flow around it; generally a minimum spacing is 12 ft between pieces of equipment. Placement and spacing of equipment should avoid overlapping fall zones as well.

Play areas for small children must include benches on which parents may sit and observe their children, and the design should allow for strollers, carriages, and the like. This may require wider sidewalks or paved areas so that standing groups of parents do not encroach onto the traffic pattern. Access to a play area should be limited for security purposes, although care should be given to avoid an institutional feeling that would discourage use. As a rule of thumb, playground equipment that requires participation should be located toward the entrances of a playground because the presence of groups contributes to the security of the facility.

As the target age group of a playground moves from small children to children between the ages of 5 and 12, there are some additional considerations. It is sometimes true that the play area for these older children includes a "tot lot" facility for younger children. The requirements for older children are developed around or in addition to the tot lot. Older children require larger spaces for participatory games and activities, so large surfaced or turf areas need to be provided. The shape and size of these areas deserve particular attention since at this age the games take place over larger areas for which adequate space must be provided. These types of facilities will serve a larger population than the tot lot and are often associated with other facilities such as schools or churches. An area of 5 to 8 acres will serve up to 250 families or about 110 elementary school children. For each 50 families, the size of the area needs to be increased by 0.2 to 0.4 acre. A maximum service population for such a facility would be about 1500 families. Above this service level, additional facilities should be considered to avoid overcrowding and to reduce the distance to the facility for families.

The choice of playground surface material can be a critical factor in determining the injury from the impact of a fall. Materials are selected for their shock-absorbing ability. Head injuries have the greatest life-threatening potential and so are used as the design criteria for surfacing materials. The height of a fall is the next most critical element of playground injury risk. The *critical height* is a term used to describe the approximate maximum height of a fall from which a life-threatening head injury would not be expected (Table 4.14). Critical heights are determined by several different methods including the *ASTM Standard Specification for Impact Attenuation of Surface Systems Under and Around Playground Equipment, F1292.* Surface materials should be selected using the critical height of the specified playground apparatus. The critical height is determined from the highest accessible part of the piece of equipment.

There are many different types of surfacing materials available commercially. Hard surfaces such as asphalt, packed earth, or even turf are not acceptable materials. In general, the available acceptable surfaces are of two types: unitary materials and loose-fill materials. *Unitary materials* are generally

TABLE 4.14 Critical Heights Determination on Selected Playground Equipment

Equipment	Highest accessible part
Swings	Height of swing at 90° from the at-rest position
Slides (including platform)	Top of platform guardrail
Climbers	Maximum height of structure
Horizontal ladders	Maximum height of structure
Merry-go-rounds	Any part at the perimeter on which a child might sit or stand
Seesaws	Maximum attainable height of any part
Spring rockers	Any part on which a child might sit or stand

SOURCE: Adapted from the U.S. Consumer Product Safety Commission.

rubber or foamlike materials installed as either interlocking or joined mats or in some cases poured in place. The performance of these materials varies widely (Fig. 4.34). Specifiers should request current test information for the product to determine its acceptability for a particular application. A disadvantage of using unitary materials is their high initial cost (including the cost of base preparation). Also, some interlocking mats have been observed to curl up at the edges, creating a trip hazard. Unitary materials are also subject to damage by vandals in areas where vandalism is a problem.

The advantages, on the other hand, are significant. These materials have a consistent performance over their life cycle. Unitary materials have a low maintenance cost (vandalism costs excepted). The life cycle costs of unitary materials are often less than loose-fill materials. The material stays in place—it is not moved during play—and no unwanted objects can be hidden. Unitary materials also provide an accessible surface.

Loose-fill materials also include a broad range of products from sand to shredded bark to shredded foam. The advantages of loose-fill materials are primarily related to cost. Loose-fill materials are relatively inexpensive and are readily available, they require limited site preparation, and they are easy to install. The disadvantages include higher maintenance and life cycle costs. They are subject to contamination by precipitation, dirt, and other unwanted materials. Their performance may be affected by displacement by children during play and by weather conditions such as high humidity or freezing (see Figs. 4.35 through 4.37).

Bicycle and Multiple-Use Paths

According to some reports, more than 30 percent of Americans ride bicycles for pleasure. As interest in bicycling has increased over the past 30 years, the interest in bicycle paths and trails has increased as well. Communities across the country have developed or are planning to develop bicycle paths. Bicycle routes are usually one of three types: the dedicated bicycle path system sepa-

Figure 4.34 Photograph of improperly installed playground surface pads.

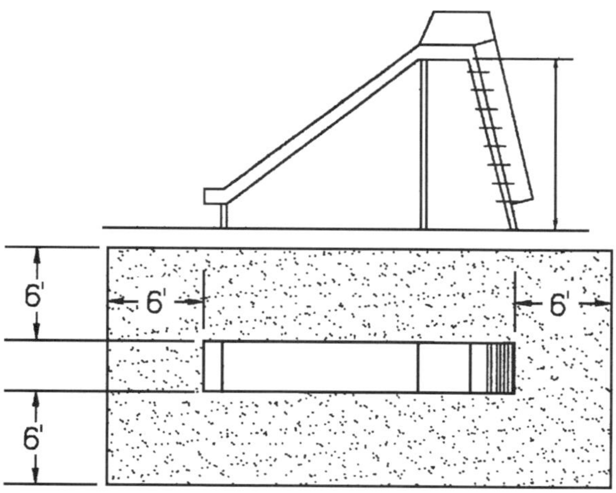

Figure 4.35 Fall zone for slide detail.

rate from streets and automobile traffic, the designated-lane system, and the road-sharing system. The dedicated bicycle path system has expanded significantly since the 1980s with the expanded rail-to-trail networks and the number of large residential developments incorporating bicycle paths. The lane system has also become popular in some suburban areas where wide streets

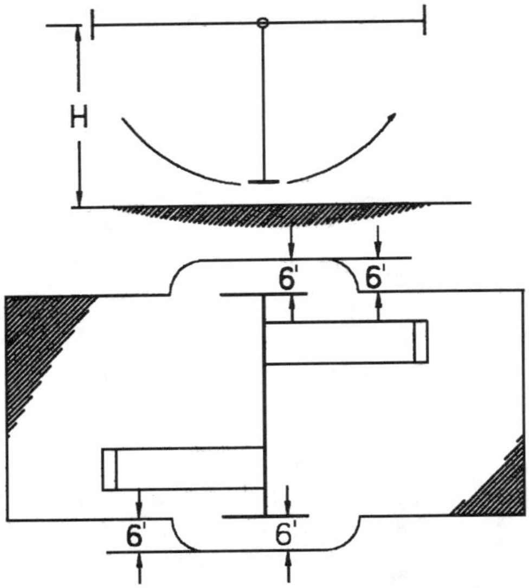

Figure 4.36 Fall zone for single-axis swings detail.

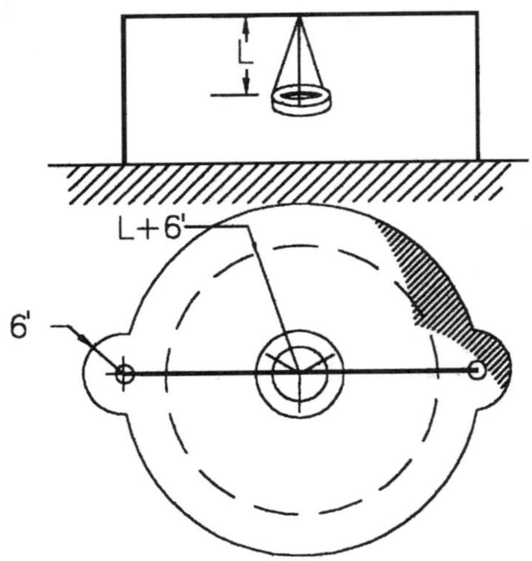

Figure 4.37 Fall zone for multiple-axis swings detail.

provide adequate room. In other communities cyclists take to the streets and share the way with automobiles. The Bicycle Institute of America has developed the Bicycle Friendly Communities program to encourage the creation and maintenance of bicycle routes (see Table 4.15).

Bicycle routes require much the same level of planning and care as street design. Improper planning and installation can result in poor surface conditions and unsafe design. As with streets, an estimate of traffic volume is necessary.

TABLE 4.15 Criteria for Bicycle-Friendly Communities

Primary Criteria
Applicant must meet all of the following:
1. Governing body establishes a written policy designed to develop and maintain "bicycle-safe" streets and pathways.
2. Community budgets and spends $1.00 per capita per year on bicycle facilities and events.
3. Governing body passes an annual proclamation recognizing May as National Bicycle Month and encouraging citizens to observe Bike-to-Work Day.
4. Community establishes Bicycle Advisory Committee and designates bicycle issues contact person on government staff.
Secondary Criteria
Applicant must meet two of the following four:
1. Community police teach bicycle safety in schools, stressing the wearing of helmets.
2. Community sponsors annual cycling event.
3. Community publishes bicycling information, identifying suggested routes and stressing safety.
4. Community provides public bicycle parking facilities and encourages private bicycle parking facilities.

SOURCE: From *Bicycle USA Magazine,* November/December 1994.

Unlike automobiles, however, there is little quantitative information or methods for estimating bicycle volume. Recreational cyclists will often drive to the bicycle route many miles from home. Without such supporting data, designers must rely on the experience of others. Fortunately there is a good deal of experience in the design, construction, and maintenance of bicycle routes in the United States.

In general, bicycle trips are one of three possible types: commuter, recreational, or neighborhood. Commuter trips and neighborhood trips are usually made on public streets either sharing the travel lanes with motor vehicles or riding in a designated bike lane. Separate routes are used primarily by recreational cyclists. The design of such routes must allow for horizontal and vertical alignments, the types of surface materials, signage and markings, bicycle and automobile parking, and associated facilities such as resting places and restrooms. The nature of designated bike routes also varies significantly, from the rail-to-trail routes to more strenuous mountain bike routes. The frequency and location of off-trail rest areas must be determined according to the use and rigor of a given trail. As a rule of thumb, on a rail-to-trail bike route where grades do not usually exceed 3 percent, pull-offs and rest facilities should be provided at least every 2 to 3 mi. For trails that are also used by walkers, a rest area should be installed about every mile or so. Rest areas should be set well off the travel lanes of the path and should be provided with benches, as shown in Fig. 4.38.

As shown in Fig. 4.39, bike trails and pathways are commonly designed to serve pedestrians and others using different means of mobility such as roller

Figure 4.38 Photograph of rest area on a bike trail.

blades. Communicating the rules of the road for these sometimes conflicting uses is best done with clear signage and pavement parking where possible, as shown in Fig. 4.40.

The rail-to trail routes have been so successful and popular because the grades are relatively flat and rarely ever exceed 3 percent, while mountain bike paths may approach 20 percent. In general, bike routes are best if limited to maximum grades of 4 or 5 percent with only short sections at steeper grades. The end of extended steeper sections is an ideal place for a wider path surface and perhaps a bench to allow cyclists to pull off the path and rest. At grades over 5 percent, it is difficult to ride without standing. Extended grades of 8 percent or more require most riders to dismount and walk the bike. Consideration should be given to installing wider riding surfaces on the steeper sections of routes with minimum travel lane widths to allow passing. Separating bicycles and automobiles may be accomplished by providing lanes divided by pavement marking or by constructing lanes separated by barriers, as shown in Figs. 4.41 through 4.45.

Drainage becomes a more important consideration when pathways are paved and drainage is restricted. Provision should be made in the design to assure positive drainage from the path surface. Pooled or standing water, such as shown in Fig. 4.46, represents a danger to cyclists anywhere but especially on curves or turns. Shallow standing water may be a hazard even to pedestrians as well if it freezes. There are many ways to prevent water from pooling on bike paths, and some are shown in Figs. 4.47 through 4.50.

Figure 4.39 Photograph of rail-to-trail project.

Figure 4.40 Photograph of rules of the road on a multiple-use trail system.

Figure 4.41 Bike path detail.

Figure 4.42 Photograph of separate travel lanes.

Seating

There are commercial sources for seating of all kinds from which designers may choose. The actual choice is determined by many factors that are project specific such as the desired style, materials, durability, and availability of the seating. A key concern is, of course, the comfort of the seat. Designers occasionally elect to design seating also. Figure 4.51 shows commonly used seat dimensions, and Fig. 4.52 shows commonly used table dimensions.

Figure 4.43 Photograph of bike route separated from automobile traffic.

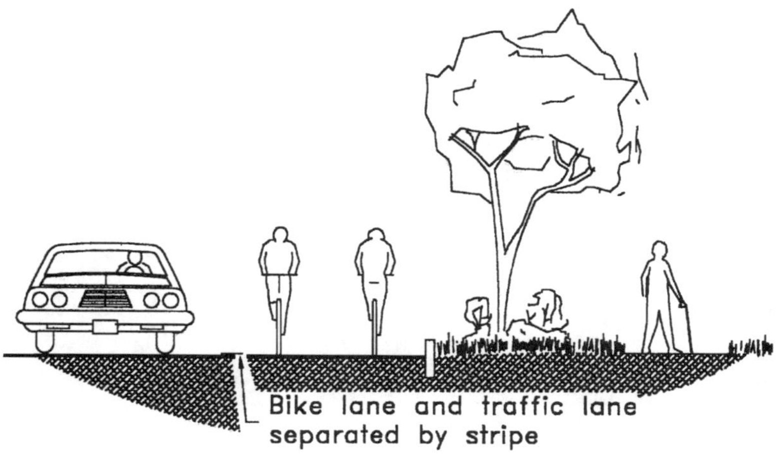

Figure 4.44 Bike lane in street detail.

Site furniture is important for more than the convenience of passers-by. The type of seating helps define the area. Benches facing each other, for example, invite socialization and interaction and attract people to common spaces. Seats near a playground or tot lot encourage adults to bring children and to use the space, which in turn increases surveillance of the play area.

Figure 4.45 Photograph of designated travel lanes in streets.

Figure 4.46 Photograph of path surface with pooled or standing water.

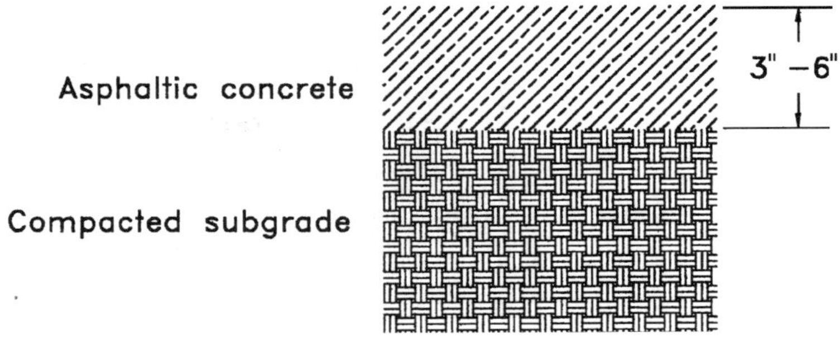

Figure 4.47 Asphalt concrete bike route surface detail (no base).

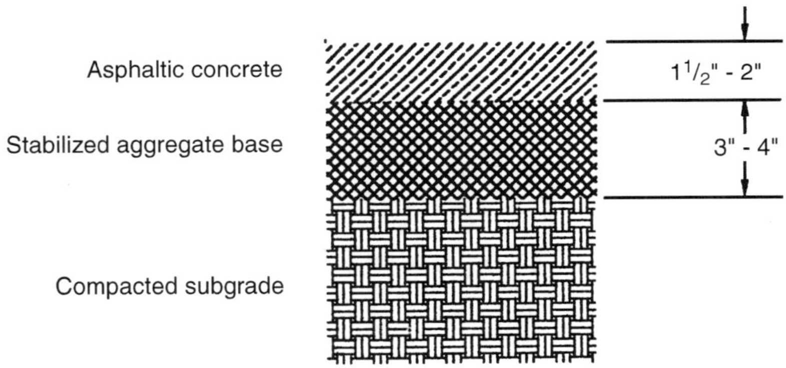

Figure 4.48 Asphalt concrete bike route surface detail (aggregate base).

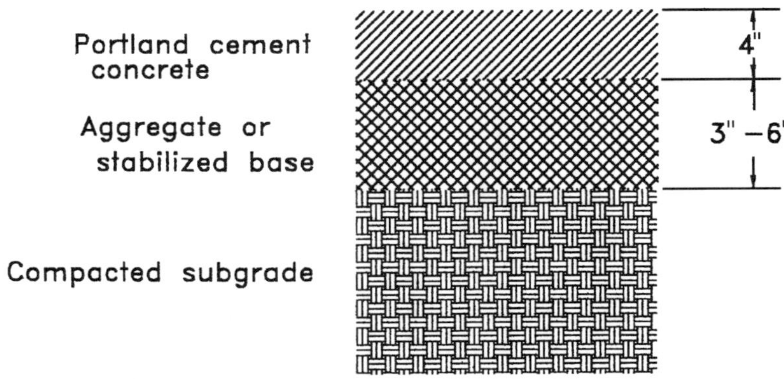

Figure 4.49 Portland cement concrete surface detail.

Walls and Fences

Fences and walls are common site and landscape features, and they are used most often to increase privacy or security as well as to create backgrounds and visual points of interest. For purposes of this discussion, walls and fences will be treated separately. Walls are usually freestanding masonry structures,

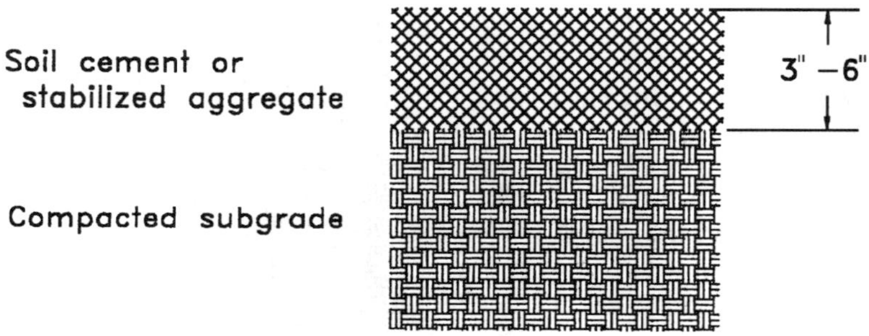

Figure 4.50 Soils cement or stabilized aggregate detail.

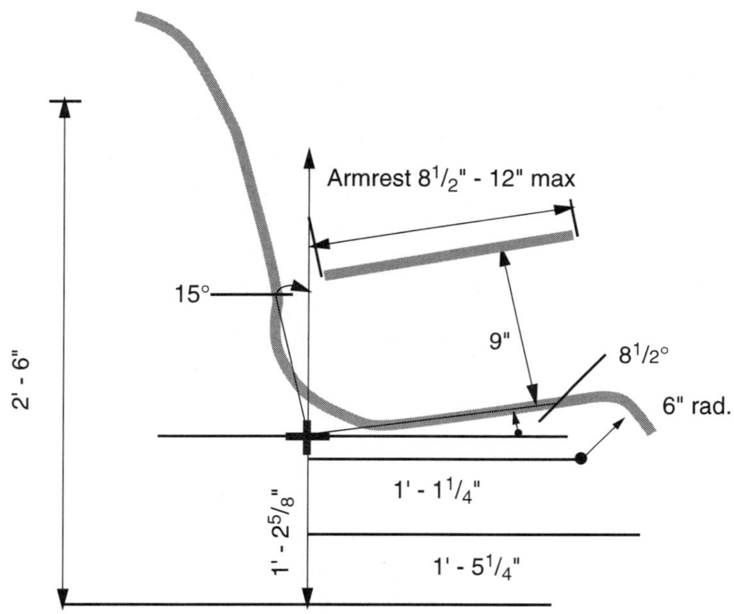

Figure 4.51 Typical seat dimensions.

while fences, more commonly used than walls for residential or esthetic purposes on commercial sites, are made of other materials, usually wooden. The *ASTM F537 Standard Specification for Design, Fabrication and Installation of Fences constructed of Wood and Related Materials* addresses the materials as well as design and construction specifications. Security fences are most often wire or metal and are generally not used for esthetic purposes although many decorative security fences are available.

Fences

The range of designs for fences is very wide. Nearly any type of fence is commercially available today, and few fences are specially designed and constructed

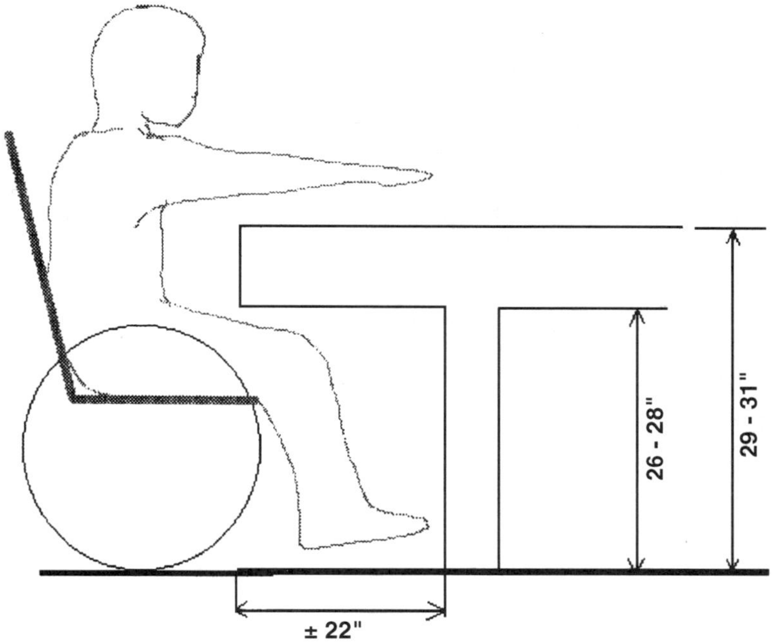

Figure 4.52 Typical table dimensions.

anymore. Nonetheless, specifying fence materials and construction details is important. Fences should be selected based on specific objectives for the project. Fences and walls may be important elements in designing for community safety and security as well. As barriers, they provide guidance to pedestrians, direct traffic, and provide clear demarcation of public and private areas. Even low fences or walls represent a psychological barrier for the casual pedestrian.

Whatever is selected, the fence should be of the proper scale and proportion to meet the design objectives and remain compatible with the project plan as a whole (Fig. 4.53). Fence or wall textures and designs contribute a great deal to their impact. Walls or fences can be made more attractive by introducing elements such as piers or details that break up the monotony of a static unbroken surface. Color also contributes to the fence performance. Lighter-colored surfaces tend to stand out in the landscape whereas darker colors tend to recede and blend in. Fences to be installed on slopes present a somewhat greater challenge. As a rule of thumb, fences should ride parallel with the slope rather than stepping down the slope with each panel horizontal but lower. Solid-panel fences usually cannot be installed parallel to the slope, and it will require additional work to fill or enclose the resulting gaps at the downhill end of each section. Care should also be taken to be sure the selected fence is consistent with local zoning and association requirements.

The key to fence integrity is the installation of fence posts that anchor and support the fence sections. Fence posts may be made of metal pipe, PVC, or

Figure 4.53 Photograph of picket fence corner detail.

wood. Figures 4.54 through 4.58 illustrate wooden fence installations. Wooden posts should be treated and dry. If treated wood is used, the type of wood treatment should be carefully evaluated (see Chap. 1). The dimension of wooden posts should be selected to provide adequate support for the fence; 4 in by 4 in is generally considered to be the minimum acceptable fence post size. Posts should be installed to at least the frost depth in places where frost heave occurs and at a depth adequate to resist the anticipated wind loads of a particular area. All corner posts should be set in concrete to add strength to the installation.

A less common but elegant method of diversion from the English landscape tradition is the *ha-ha,* or *sunken, fence* (see Fig. 4.59). To prevent livestock from wandering off, this technique of making an abrupt change in grade was used to avoid visually cluttering the landscape with walls and fences.

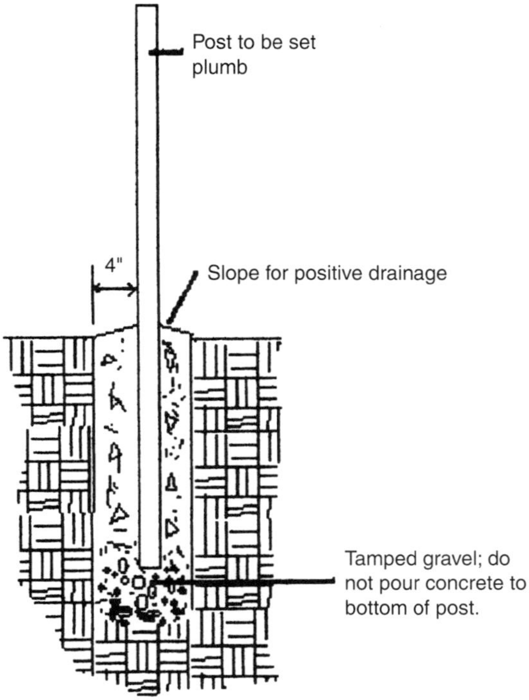

Figure 4.54 Typical wooden post installation.

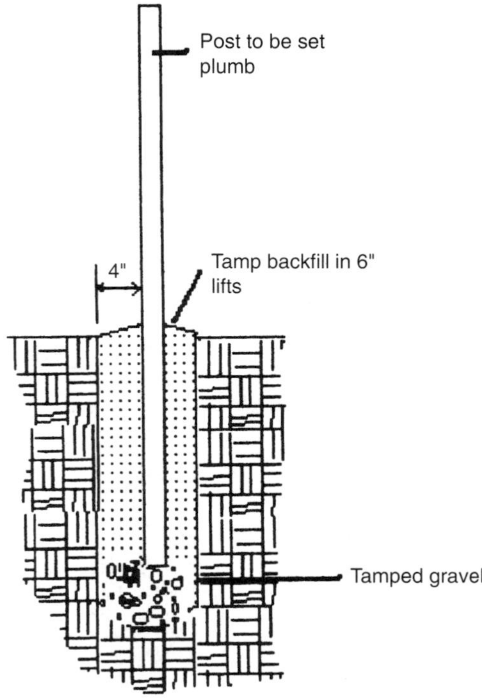

Figure 4.55 Fence post set in tamped backfill.

Walls

For more substantial applications, masonry or stone walls may be desirable. The heavy materials require an adequate supporting foundation. All free-standing walls must be designed to resist overtopping due to wind loads and subsurface soil failures. When wind pushes on the solid surface of the wall, it causes the wall to act as a lever turning on a pivot at ground level. The wall is able to resist overturning by virtue of its weight and the extension of the length of the lever by a footer. Wind loads vary across the nation and are provided or dictated in many local building codes. The weight of masonry materials varies from about 120 lb/ft^3 for brick or cement masonry units (CMUs) to 145 lb for stone. Concrete mortar typically has a weight of 150 lb/ft^3.

To check a wall for its resistance to overturning, it is necessary to determine the wind load for the area in which the wall will be constructed. Typically loads are determined for a 1-ft section of the proposed wall. To determine the wind load pressure P, multiply the height of the wall by the wind load. For example, consider a wall in Buffalo, New York, as shown in Fig. 4.60. The recommended wind load is 30 lb/ft^2. The pressure of the wind load P is determined at the center of moments of overtopping and righting. The overtopping moment M is calculated at half of the wall height above grade plus the depth below grade. For a wall that is 4 ft above grade and 1 ft below grade and 0.67 ft thick, P is 3 ft^2 by 30 lb/ft^2, or 90 lb. The weight of a 1-ft section of the wall

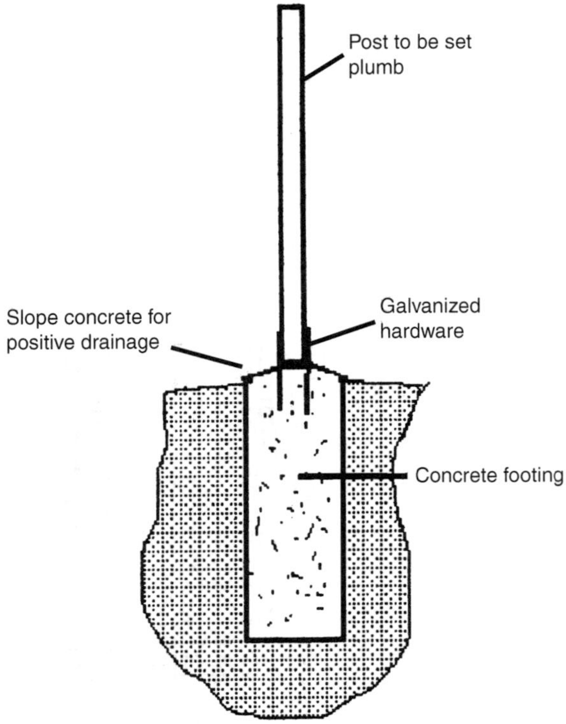

Figure 4.56 Wooden fence post anchored in concrete.

if made of brick would be 412.05 lb. Based on this calculation, a wall 0.67 ft thick would be 4.33 ft (height from the top of the wall to the top of the footer) times 120 lb/ft^3 times 0.67, or 348 lb plus the weight of the mortar at 0.9 ft times 0.67 ft times 150 lb/ft^3, or 90 lb. Finally, 348 lb plus 90 lb yields a weight of 438 lb.

To determine the wall's resistance to wind load, the overturning moment (M_O) and the righting moment (M_R) must be compared. The overturning moment is measured at half the height of the wall plus the depth below grade, 3 ft in the example: $M_O = 120 \times 3 = 360$ lb. The righting moment is measured: $M_R = 438 \text{ lb} \times 0.67 = 293.46$ lb. In this case the righting moment is less than the wind load, so additional stabilization is required.

The calculation above is based on a single section of freestanding wall 1 ft long, and it does not consider other aspects of the wall such as corners, piers, or other support. To prevent overtopping, the wall may also be designed with piers or with sections at right angles to the wall. The lateral support of solid walls is designed using a ratio of the length of the wall between lateral supports (L) to the thickness of the wall (T): L/T. Table 4.16 summarizes the L/T for freestanding walls. The L/T for wind loads for the example is 14. The maximum length of wall between supporting members therefore is: $14 = L/0.67$ ft, or $L = 9.38$ ft.

Figure 4.57 Photograph of arbor.

Eccentric loading on footings may result in footing failure. In most cases it is recommended to keep the weight of the wall in the center third of the footing (see Fig. 4.62). Shifting the wall toward either side of the footing increases the load on that portion of the footing and increases the instability of the wall. In such a condition, there is concern with exceeding the strength of the soil either because of the weight or because of the increased pressure as a result of the wind load.

Serpentine brick walls have been used in gardens since at least the 1700s and are found in many historic gardens. These walls are illustrated in Figs. 4.63 through 4.66. Besides being decorative, the serpentine wall has additional lateral strength because of the configuration. To keep that strength, it is critical, however, that the wall be carefully designed and constructed. The radius of any

Figure 4.58 Photograph of arbor detail.

Figure 4.59 The ha-ha, or sunken, fence.

curved section should not be greater than twice the above-grade height of the wall. The depth of the curve should be at least one-half of the above-grade height. Many historic serpentine walls are built using a simple brick foundation as shown in Fig. 4.63. If the wall is in a location where frost heave is a concern, or where the wall will be bearing weight other than its own, or where it might be bumped by vehicles, a more substantial footing might be in order. In the event that a more substantial wall is required, an 8-in wall may be used with a reinforced concrete footing.

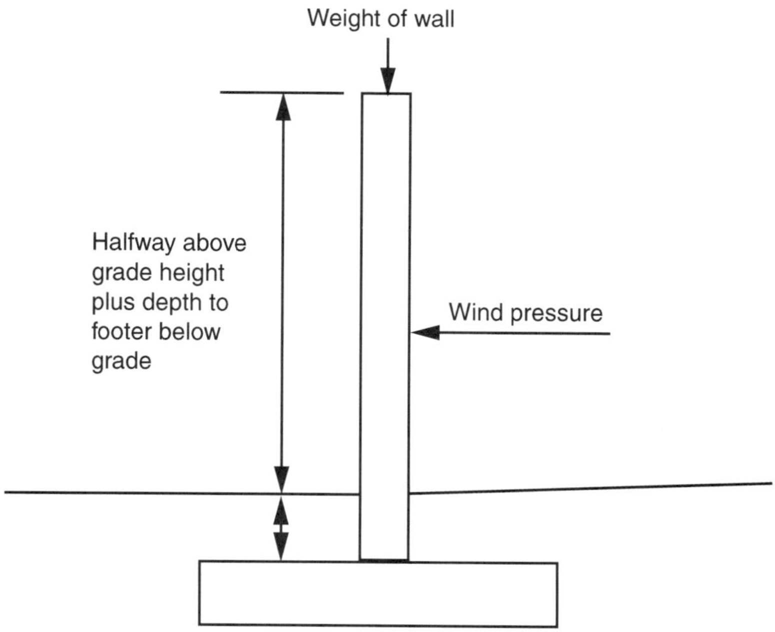

Figure 4.60 Evaluating free-standing walls for overturning.

Figure 4.61 Photograph of brick entrance.

TABLE 4.16 Ratio of the Length of the Wall between Lateral Supports (*L*) to the Thickness of the Wall (*T*) for Freestanding Walls at Given Wind Pressures

Design wind pressure, lb/in^2	Maximum L/T
5	35
10	25
15	20
20	18
25	16
30	14
35	13
40	12

SOURCE: From Harlow C. Landpahir and Fred Klatt, Jr., *Landscape Architecture Construction,* 2d ed. (New York: Elsevier Science Publishing, 1988), p. 208.

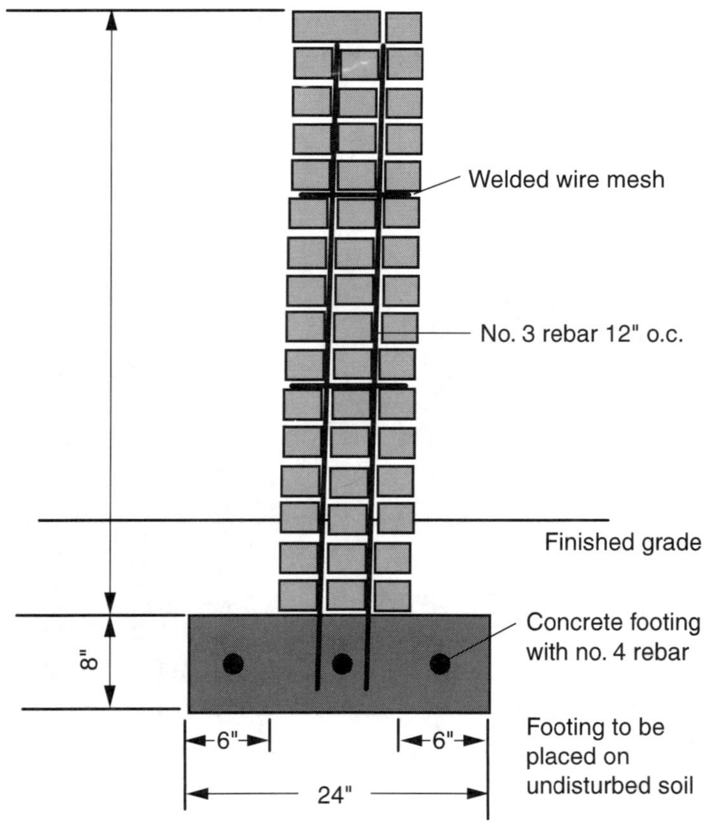

Figure 4.62 Brick wall detail.

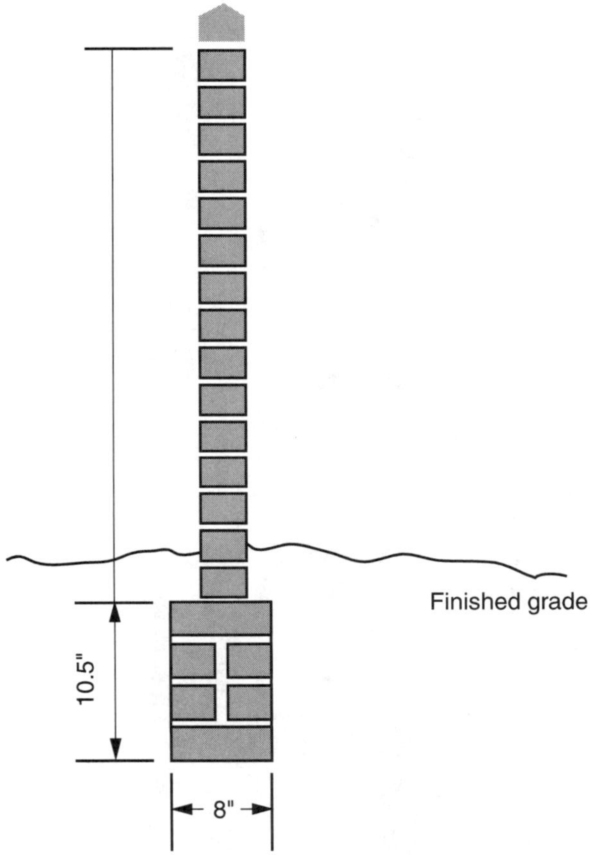

Figure 4.63 Serpentine brick wall cross-section detail.

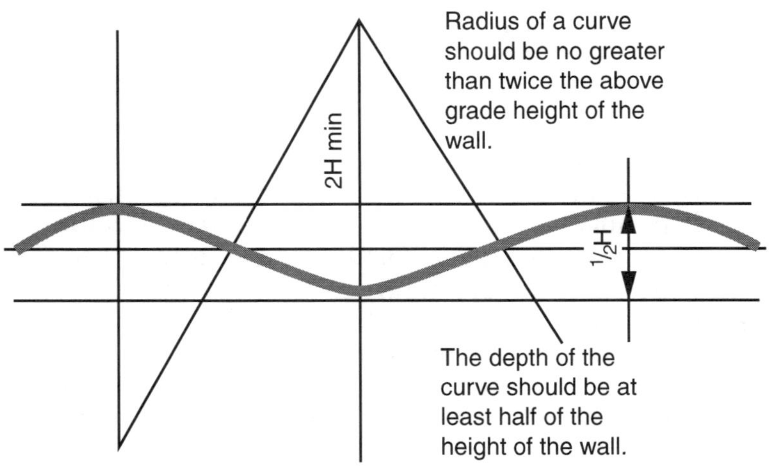

Figure 4.64 Serpentine brick wall layout detail.

Figure 4.65 Photograph of serpentine brick wall.

Signage

The design of signs is a specialty within itself, though many types of off-the-shelf signs are available commercially. For common signs such as those identifying designated handicapped parking or restroom facilities, it may be best to rely on types of signs that are familiar and in common use. The key element of signs is readability at an effective reading distance (see Table 4.17). To determine readability, it is necessary to understand the purpose of the sign. Signs that provide direction or that are meant to draw a person's attention from a distance require larger lettering than signs describing a display or vista that is immediately before the viewer. In many communities, sign and lettering sizes are regulated in the zoning ordinances. In designing and locating signs, it is important to remember that the farther away the desired effective reading dis-

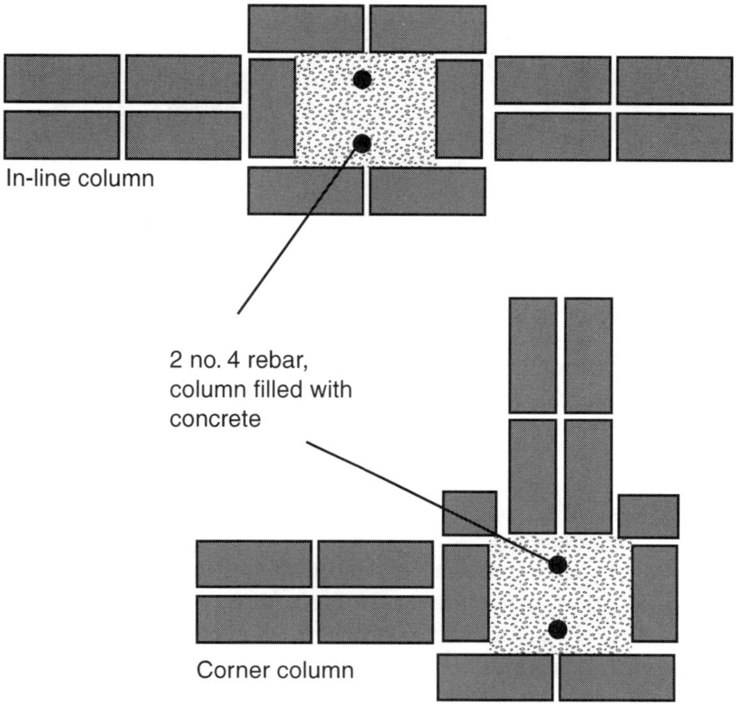

Figure 4.66 Pier detail.

tance, the larger the letters should be and the higher the sign. However, in general, a person is less likely to look up more then 10° to view a sign; therefore, signs that are placed above the viewing distance tend not to be seen. Also, note that it is easier to read light images on dark colors than the other way around.

Signs that use symbols to convey information such as warnings or directions are preferred over those that have information in only one language. Likewise, consideration should be given to ADA concerns for designing signage. It should be determined if the information to be conveyed on the sign is necessary for access to or from an area or facility. In some cases textural signals should be installed with the signs. The familiar universal symbols have made sign selection for many purposes much easier. Sign shape and color are also important considerations. Many signs now use standardized shapes and colors, so care should be taken to not use these combinations unintentionally.

Signs directed toward drivers must be visible and readable from quite a distance away. Common street and traffic signs have been developed with fairly explicit standards of design and installation; however, site-specific signs should allow for the fact that drivers have a very short time in which to read and comprehend the information on a sign. In most instances, several signs in a sequence may be more effective than too much information on a single sign. Information should be organized and presented in a hierarchy of importance, from general to more specific, rather than given as a string of unweighted data.

TABLE 4.17 Effective Reading Distance and Letter Size

Distance, ft	Capital letter size, in	Symbol size, in
30	—	3.0
40	—	4.5
50	1.0	5.0
75	1.0–2.0	6.5
100	1.5–2.5	8
150	2.5–3.0	12
200	3.0–4.0	15

SOURCE: Charles W. Harris and Nicholas T. Dines, *Time-Saver Standards for Landscape Architecture* (New York: McGraw-Hill, 1988), p. 200. Used with permission.

Water Features

Ponds and pools have become very popular landscape features in recent years. Successful ponds and pools are always a marriage of good design and sound construction. Even small variances in construction or incomplete details in a design may result in an unsatisfactory pond. A water feature can bring a great deal to a landscape of any size. People are drawn to water features perhaps more than to any other single landscape feature. One reason for this preference is that there are clearly important psychological and emotional values in well-designed water features. Whether it is the sound of falling water, the turbulence of fountains or falls, or the cooling effect of spray and evaporative cooling, water features are highly valued elements of both the designed and natural landscape.

Pools and ponds

Water features are used in many forms in the landscape, ranging from very natural appearing small ponds to very formal precision water veils. The possibilities are limited only by the imagination and the physical characteristics of water. For purposes of this chapter, water features will be discussed in terms of small pools and ponds. Ponds are illustrated in Figs. 4.67 through 4.70. Ponds include biotic features such as plants and perhaps fish whereas pools have no biotic elements. The key to the water feature is the pool or pond: The presence of fountains or falls or other features are framed within the pool or pond. In the past, concrete has been the most common and popular material for pond or pool construction, but with the development of more sophisticated geotextiles and plastics, fiberglass prefabricated pools and fabric pond liners have become the most common choice. Ponds are constructed using rigid preformed basins or liners, most commonly made of fiberglass, EDP, PVC, or similar materials. The pond liner should be selected for durability and ease of

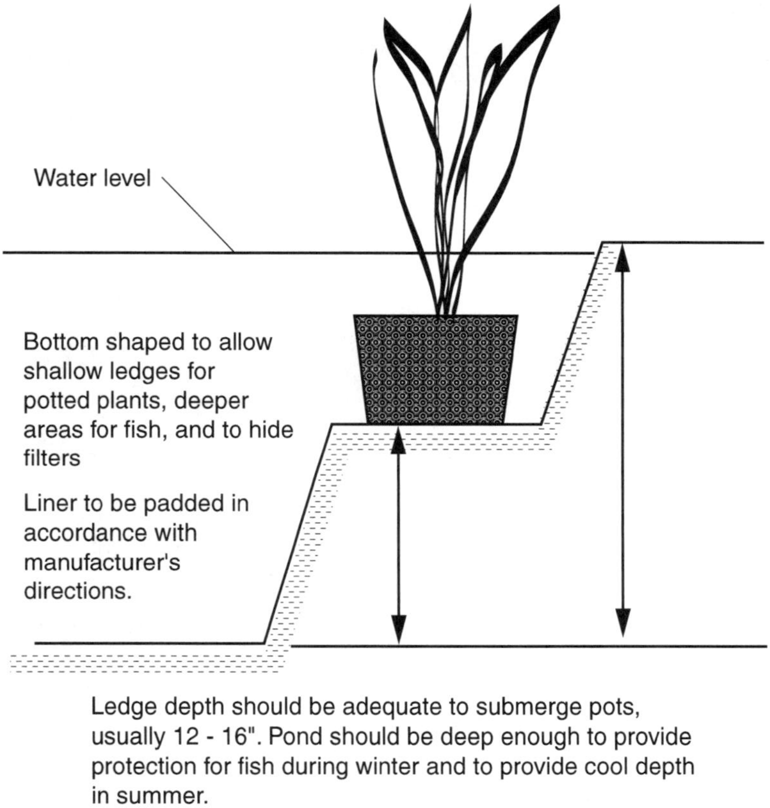

Figure 4.67 Typical ornamental pond detail.

installation. Whichever type is chosen, the installed liner should be smooth and curvilinear, and sharp corners or abrupt changes in surface aspect should always be avoided. Most pond liners are dark colored, that tends to produce a reflective surface effect. A light-colored pond surface will produce a somewhat transparent effect. In a light-colored pond, everything is visible, and so maintenance becomes a critical issue.

Pool depth is an important consideration. The minimum recommended pool depth is 16 in, a minimum requirement for operating most submersible pumps. Ponds with fish must provide an adequate depth to allow the fish to overwinter; otherwise, the fish must be removed each fall. For ponds it is important to vary the depth to allow areas for rooted plant stock as well.

Waterfalls require the addition of a pump. To select the correctly sized pump, Bazin's formula is used to estimate the flow over the falls:

$$Q = \left[0.405 + \left(\frac{0.00984}{H}\right)\right] [1 + (0.55H^2)(P + H)^2]\, LH\, (2gH)^{0.5}$$

Figure 4.68 Photograph of pond installation.

Figure 4.69 Photograph of pond.

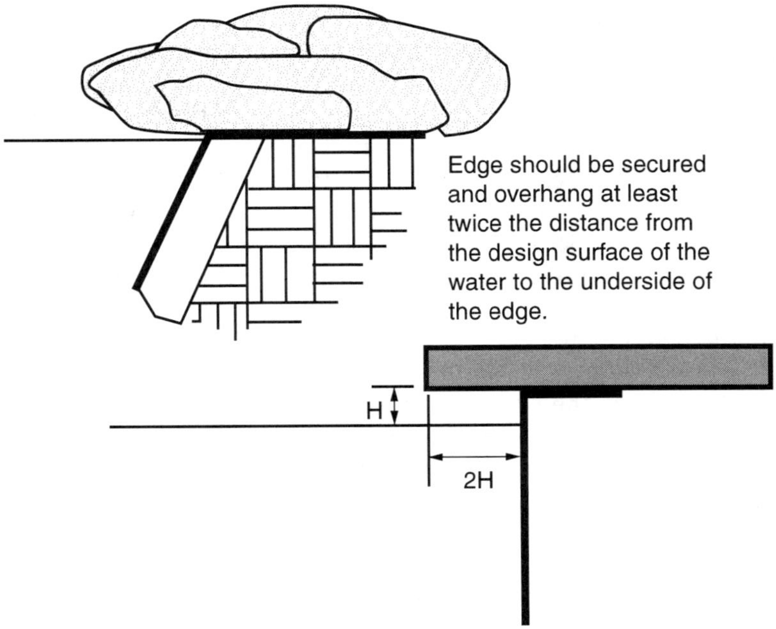

Figure 4.70 Pond edges.

where Q = volume, ft^3/s
H = head, ft (The height is taken from a position $4 \times H$ from the face of the weir.)
P = height of the water over the weir, ft
L = the length of the weir, ft
G = 32.17, universal gravity constant

Note that cubic feet per second can be converted to gallons per minute by multiplying by 448.831.

Pumps

Pond pumps come in two general types: small submersible pumps and larger centrifugal pumps located outside the pond. The centrifugal pumps are capable of moving more water, but they generate enough noise that they must be located away from the pond. They must also be protected from the weather. This requires additional plumbing. It may be necessary to use such a system to operate a large water feature. Submersible pumps are more common for smaller ponds. These pumps are located within the pond itself, often in conjunction with a filtration system. Care should be taken to locate the pump out of sight; even though it is located underwater, submersible pumps are often still visible, which can detract from the overall pond effect.

In sizing pumps for ponds or pools, it is important to remember how water is moved by pumps. The pressure of the atmosphere at sea level is approximately

14.7 lb/in^2 (or psi). This pressure is pushing on all surfaces and in all directions all of the time. Water weighs 8.34 lb/gal, and there are 7.48 gal/ft^3 of water, so there are 62.4 lb of water per cubic foot. Thus 1 ft^3 of water in a container will exert 0.433 lb/in^2 pressure at the bottom of the container (62.4 lb/144 in^2 = 0.433 lb/in^2). If the container is filled to 2 ft deep, the pressure increases to 0.866 lb/in^2 ($2 \times 62.4 = 124.8$ lb, and 124.8 lb/144 in^2 = 0.866 lb/in^2). If a tube is placed in a container of water and a vacuum of 1 lb/in^2 is drawn on it, the water will rise on the tube 2.31 ft. So, for every foot of rise, the pressure of the water increases 0.433 lb/in^2. The pressure caused by the weight of the water is called the pressure head, or simply the head. Head is measured in feet; every foot of head is equal to 0.433 lb/in^2.

If a pond were to be designed with an above-grade wall, the design would have to address the pressure that would be exerted by the water in the filled pond. Pressure on the container wall will range from zero at the water surface to the depth of the water times 0.433 lb/in^2, or 62.4 lb/ft^2 (pounds per square foot). The formula for finding the force acting on the wall is:

$$F = 31.2 \times H^2 \times L$$

where F = the force acting on the wall, lb
H = head, ft
L = the length of the wall, ft

Note that 31.2 is the constant in pounds per cubic foot based on the force at the average depth exerted at $H/3$ from the bottom of the container.

Pumps are specified usually in terms of horsepower and head. The head a pump must overcome is called the *dynamic head,* which is measured as the vertical distance the pump must lift the water, the *static head* and the *friction loss* caused by the roughness of the pipe conveying the water. Charts for various types of pipes and fittings are provided by the manufacturers of those materials. For very short runs of pipes and fittings, friction loss may be nominal. The total dynamic head is the sum of the static head and friction losses.

$$\text{Water horsepower (hp)} = \frac{(\text{F, gal/min})\ (\text{H, ft})}{3960}$$

where hp = horsepower
F = flow, gal/min
H = lift, ft

Note that 3960 is derived from:

$$8.34\ \text{lb/gal} \times \frac{\text{hp}}{3000\ \text{ft-lb/in}} = 3960$$

Pumps, however, are unable to operate at perfect efficiency. More energy must be provided to the pump than is expressed as water horsepower. Motors that drive the pumps are also not 100 percent efficient. To account for the inherent inefficiency of pumps and motors, the formula is modified by an efficiency factors E_p and E_m:

$$\text{hp} = \frac{(F, \text{gal/min})\ (H, \text{ft})}{3960 \times E_p \times E_m}$$

In most pumps and motors efficiency ranges from 50 to 85 percent and 80 to 90 percent, respectively.

Plazas and Patios

The use of plazas or patios in the site plan has become all but essential in most projects. The choices of materials and approaches differ primarily by the type of surface material. All such areas should be designed to be fairly level but with enough pitch to provide adequate drainage. Surfaces should be even and free of trip or slip hazards. The base should be sufficiently substantial to resist loads from expected traffic and to resist frost damage. Surface materials can range from poured concrete, pavers, flagstone, or brick (see Figs. 4.71 and 4.72). The base may be open-graded or impermeable. (Materials are discussed earlier in this chapter.) Bricks and pavers should always conform to the requirements of *ASTM C 902 Specification for Pedestrian and Light Traffic Paving Brick* or *ASTM C 1272 Specification for Heavy Vehicular Paving Brick,* depending on the expected volume and weight of traffic. *ASTM C 1272*– compliant brick is not necessary for most landscape and site planning functions. While the dimension tolerances and chip resistance are important, the critical elements in selecting brick or pavers for a patio are the durability and abrasion of the material.

Durability is graded as Nx, Mx, and Sx. Nx pavers or bricks should be used only for interior applications where wetting and freezing will not be issues. Mx and Sx pavers are used for exterior applications, but Sx is selected where freezing will occur. Abrasion resistance is graded as either type I, II, or III in decreasing resistance to abrasion. Type III pavers are adequate for residential or light-duty patios. Type I pavers are used for heavy traffic areas including driveways or commercial entrances. Type II pavers might be selected for restaurant entrances or similar situations.

Although concrete provides a durable and cost-competitive surface, it offers little in the way of esthetic contribution to the project. Many concrete stains and patterning methods exist to improve the appearance of poured concrete. The use of such additional steps may reduce or even eliminate the cost savings. The use of color or stains on concrete may be affected by the aggregate used in the mix. If coal ash is used, tests should be done to determine if it will affect any color applied to the concrete.

Brick paving is attractive and durable if properly specified materials are used. Brick surfaces may be either rigid or flexible depending on whether or not mortar is used to set the brick. Mortarless patios are the most common form, and they may be set over a wide range of base materials. This type of paving is at least minimally porous to allow for some infiltration of precipitation. It is

Figure 4.71 Photograph of brick plaza.

recommended that a prepared base, such as shown in Fig. 4.73, be used to reduce the amount of pumping and movement in the mortarless brick systems. Rigid paving systems are preferred where steps or exposed edges will be used (Fig. 4.74). Wherever steps, ramps, or exposed edges are used, the brick should be supported by a concrete base (Fig. 4.75). The flexible mortarless systems require support or restraint at the loose edges.

The bed for bricks and pavers usually consists of a base layer and a setting layer. The setting bed acts as a leveling course between the base and the finished surface. The base provides the strength and resistance to the finished surface. The possibilities for setting-bed materials are usually limited to sand or mortar, although sometimes asphalt is also used. There is a very broad range of choices of sand depending on the region it comes from. However, for the most part a well-graded (consistent size), washed sand with a maximum particle size of 3/16 in is acceptable. Concrete sand that complies with *ASTM C 33 Specification for Concrete Aggregates* is acceptable. Sand that meets *ASTM C144 Specification for Aggregates for Masonry Mortar,* sometimes called *mortar sand,* is also acceptable. Sand-setting beds should be between 1/2 and 2 in.

Mortar-setting beds are always used in rigid, mortared surfaces. Mortar should be prepared in accordance with *ASTM C 270 Specification for Mortar for Unit Masonry.* Type M mortar is preferred in applications where freezing is not expected, and it consists of 1 part portland cement, 1/4 part hydrated

Figure 4.72 Photograph of brick paver walkway.

lime, and $3\frac{3}{4}$ parts sand. Type S mortar may be specified where freezing is an issue. Asphalt-setting beds are usually only $\frac{3}{4}$ in thick over a concrete or asphalt base, and they consist of a mixture of about 7 percent asphalt and 93 percent sand.

Base materials are generally aggregates, concrete, or asphalt. Aggregates may be either crushed stone, gravel, or sand. The use of aggregates may be favored in places where poor drainage or frost damage is a concern. The open-graded base allows water to drain away from the patio. Aggregates should be no larger than $\frac{3}{4}$ in, but the actual size is selected as a function of the depth of the base and the type of compaction equipment to be used. Sand bases are commonly used for residential projects if the patio is to be built on undisturbed

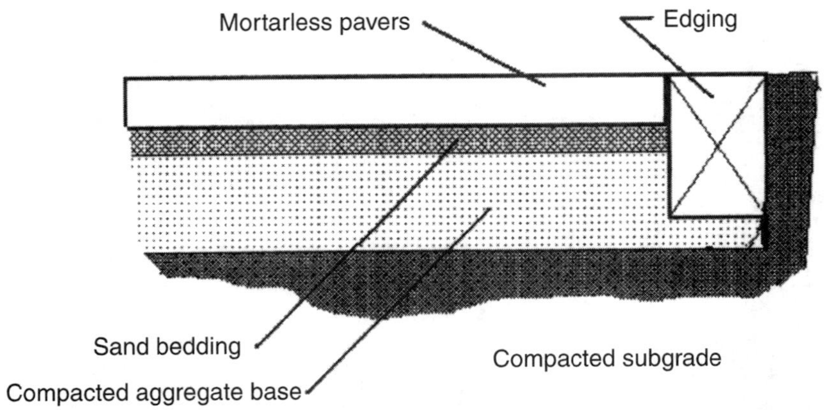

Figure 4.73 Flexible mortarless patio detail.

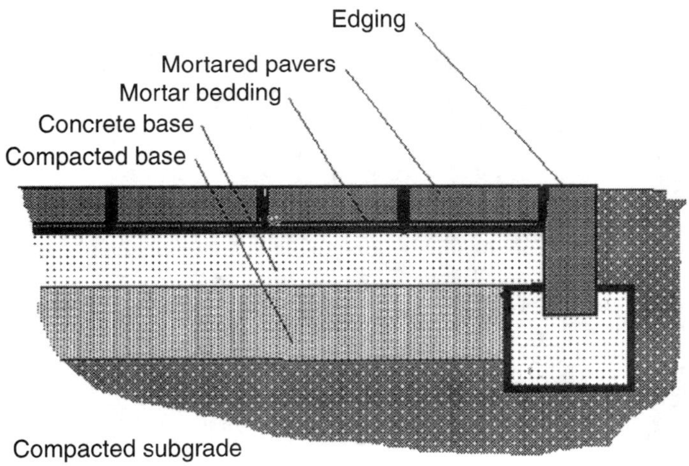

Figure 4.74 Rigid mortared paving detail.

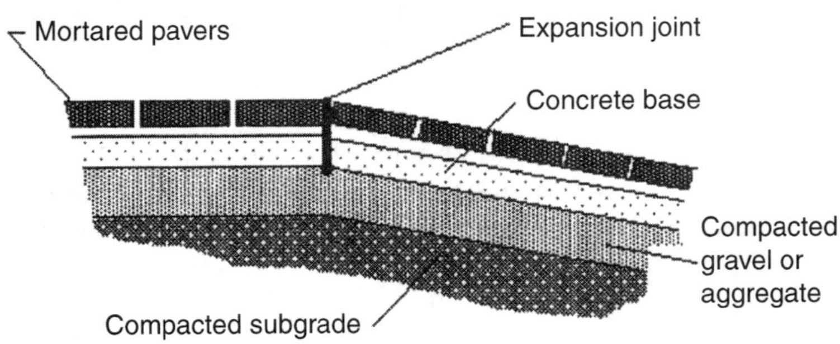

Figure 4.75 Ramp detail.

earth or material compacted sufficiently so that frost is not an issue. Sand used for the base should comply with *ASTM C 33*.

Concrete bases can be either new or existing concrete. If a mortar bed is to be used, the surface should be sufficiently roughened. If a chemical primer is to be used, the manufacturer's directions should be observed. If existing concrete is to be used, it should be carefully inspected for cracks, chips, level, and soundness. Asphalt bases should not be used for rigid paving systems.

Chapter

5

Street and Parking Lot Design

Accommodating the automobile in a site design presents many challenges, and in many instances the car may have more influence on the final design than most other considerations. This is easily observed in most contemporary development; making way for the automobile usually takes precedence over all issues of pedestrian access. Although much has been written and said about the downside of our dependence on the automobile, it does not appear likely that we will give up this means of personal transportation in the near term. Recognizing this, site designers should try to mitigate the less desirable impacts of automobile use.

The negative effects of streets and parking lots range from the obvious storm water runoff and localized microclimate changes to the isolation of pedestrians and degradation of neighborhoods. Most city dwellers are familiar with the "heat island effect" whereby pavement absorbs solar radiation and gets hot during the day and then stays warm well into the night. This effect can result in local temperature increases of 10 to 15°F above the temperature in the surrounding areas. Where heat islands exist, cooling costs are high. In climates with already-high summer temperatures, the higher temperatures can make a stressful environment actually harmful to people sensitive to heat or who have conditions that can be aggravated by heat. Warmer summers are expected to increase the number and extent of heat-related illnesses and deaths in several parts of the United States over the next 25 to 50 years. Some communities are already implementing some simple preventive strategies such as using lighter-colored paving materials that contribute less to the local heat island effects than dark-colored paving materials or incorporating more sources of shade into parking lots.

In general, it is agreed that large areas of paving are a necessary accommodation for the automobile, but such areas are at best unfriendly and at worst even stressful to people. In most cases little or no effort is made to fit the pedestrian into the design. In those applications that must accommodate both people and automobiles, a solution that favors the automobile usually wins approval. These circumstances have prevailed in the United States since 1940. Until recently even suggesting an alternative approach often brought a negative response. Thus the family's car has been accommodated at the expense of the family's environment.

The postwar suburban growth of the fifties and sixties began in the 1970s to give way to the sprawl of the 1- and 2-acre "estates." These trends reflected the growing and strong economy and an employee-friendly leisure-work lifestyle and high standard of living that were unprecedented. The environmental costs of the industrial revolution were only beginning to be recognized in the fifties. By 1970, the attention of science was turning toward our natural environment, and much of the news was not good. Today, 30 years later, we understand that we are responsible for the results of our actions, and we are beginning to adapt to the requirements of a sustainable world and economy. We also know now that the optimal solution is not necessarily to stop development but to do a better job of it.

Modern parking requirements, for example, are concerned with calculating the minimum spaces needed. Past design standards called for huge, mostly unused expanses of parking around shopping centers and malls for most of the year. Communities found, however, that although the reserve parking was useful for a few days each year, that low level of demand did not justify maintaining those parking spaces the rest of the year.

Many alternatives exist for better designs, and communities should look for good design solutions confidently. In fact, by accepting the impacts of a poor design, a community is choosing to subsidize the interests of a commercial enterprise at the expense of the community's environment. It should not be unreasonable to ask those who benefit from the development to either reimburse the community for any damage the development causes to the surrounding environment or invest more startup money to prevent the environmental damage in the first place. Pervious paving systems, even reinforced turf paving systems for overflow parking areas, may cost more initially, but the cost will most likely be offset by the enhanced value of the preserved environment in and around the development.

In many places neighborhoods were built with streets so wide that pedestrian traffic is discouraged and may actually be dangerous. In these situations, pedestrians must cross neighborhood streets more than 30 ft wide that were designed to standards for vehicle speeds in excess of 40 or even 50 mi/h, even though posted speeds may be much less. Very few of these standards allowed for any aspect of use and function beyond that of the automobile. But today the practice of street design is moving away from some of these automobile-centered standards toward a more balanced approach.

Street Design

For purposes of this discussion, streets will include local streets only; highways, collectors, and high-volume commercial streets will not be included here. The local neighborhood street is a space for which automobiles and pedestrians directly compete. Streets such as highways are constructed for large volumes of traffic and are clearly not spaces for use by pedestrians. In contrast, areas such as local streets, shopping centers, office parks, and public places must serve both pedestrians and vehicles. The concern of this discussion is with the negative impacts of local streets on environmental resources such as water and air and the social life of the neighborhood.

Preventing the negative effects of the typical street or parking lot might involve incorporating the methods of storm water management discussed in Chap. 6. By including carefully planned and larger uses of infiltration and vegetation, the physical impacts of paving and traffic can be reduced. Well-planned streets go beyond these concerns and also address the integration of both pedestrians and vehicles. In an urban neighborhood, the streetscape might represent as much as 35 percent of the total neighborhood area and all of the public or common space. In virtually all American urban neighborhoods, this common space is dedicated to the automobile, and its use by residents is incidental and at their own risk. However, the risks notwithstanding, pedestrians do try to use the streets. Neighborhood block parties and street festivals are a familiar activity in many cities. Other less obvious but more frequent uses are for recreation such as walking, bicycle riding, and playing games. Mitigating the negative environmental impacts of streets therefore includes issues of comfort, safety, access, and traffic control as well as concerns of physical impacts (Figs. 5.1 and 5.2).

Figure 5.1 Photograph of residential street.

Figure 5.2 Photograph of wide residential street.

The goals of residential street design should be to provide for reasonable vehicular and pedestrian uses (Figure 5.3). The street design should also provide access to buildings and residences in a manner that enhances the appearance, security, safety, and enjoyment of the area. Safe vehicle speeds, access for people with mobility restrictions, and a street that is friendly encourage interaction, stability, and the livability of the street. In addition to increasing the integration of pedestrian and automobile access, design solutions must be developed to overcome the problems that streets bring, such as noise, vibration, and air pollution. Streets may be intimidating to pedestrians, and they may act as a barrier to a healthy neighborhood social life if the effort required to cross the street safely is so great that it discourages residents from interacting.

Suburban street width design requirements range from 16 to 36 ft. However, although some regional differences are appropriate, the typical suburban residential cartway need not be wider than 24 ft. This width allows for either parking on both sides and one clear traffic lane or two generous traffic lanes and parking on only one side. One positive effect of the narrower street is to slow the vehicular traffic down.

Unfortunately, most design standards do not include resident satisfaction among the criteria, and most people are simply resigned to living with a less-than-perfect streetscape. Design standards tend to be prescriptive rather than performance oriented to the detriment of livability. Studies conducted by Appelyard and Lintell (1972) in San Francisco neighborhoods found a strong

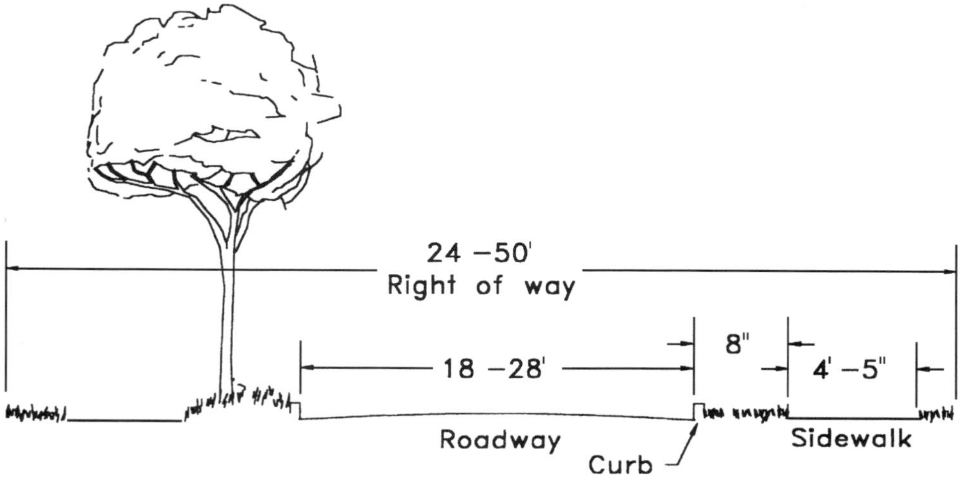

Figure 5.3 Typical street cross section.

negative correlation in residents' minds between traffic volume and values such as security, safety, neighborhood identity, comfort, privacy, and home—the greater the traffic, the less these values are perceived as being present. Simply put, on wide streets that encourage large volumes of traffic, there is less likely to be a sense of neighborhood and privacy. Interestingly, the study found that residents on all streets, regardless of actual traffic volume, are concerned with traffic and safety.

In a study published by the Institute of Traffic Engineering (1989), researchers found that the typical street design standards consist of width dimensions, grade requirements, and horizontal and vertical curves dimensions but very few, if any, performance standards. In response, the institute wrote and recommended the list of performance standards in Table 5.1. These suggestions are a good start toward developing better performance standards. However, in the study, the institute points out that street safety hazards are a result of conflicting uses of the street space. The institute suggests that the solution to these concerns is to further isolate the pedestrian from the street. Although the institute's proposal is a good strategy, it is neither the only answer nor the best answer. Observing actual neighborhood life reveals that pedestrians constantly use the streetscape for recreation and socializing. Thus imposing "design standards" that call for the segregation of pedestrians from the street will not solve the safety problems. A better solution might be to consider constructing narrower streets.

Typically the largest single-body vehicle allowed in most states is the school bus. A residential street designed for a school bus should provide for adequate turning radii and lane width. The horizontal and vertical curves should allow for adequate site distances for a school bus at an appropriate design speed. In general, it is agreed that an appropriate design speed for a residential neighborhood is about 25 mi/h. Table 5.2 compares street widths commonly used, and Table 5.3 lists street width requirements for fire vehicles.

TABLE 5.1 Performance Standards for Residential Streets

1. Adequate maneuvering and access space for the largest vehicle that will use the street
2. Adequate maneuvering space to permit an efficient level of operation speed
3. Adequate parking provisions
4. Adequate lighting and drainage and means of separating vehicles from pedestrians

SOURCE: Adapted from the Institute of Traffic Engineering.

TABLE 5.2 Comparison of Typical Street Widths

Type of street	Design speed mi/h*	Right-of-way, ft	Traffic land width, ft	Parking lane width, ft
Local, residential	20–25	30–60	9–11	8
Collector, residential	25–30	40–60	12	10
Minor arterial	30–40	100–120	12	10

SOURCE: Adapted from the Institute of Traffic Engineering and the American Association of State Highway and Transportation Officials (AASHTO).

TABLE 5.3 Street Width Requirements for Fire Vehicles

Width, ft	Source
18–20*	U.S. Fire Administration
24 (on-street parking)	Baltimore County Fire Dept., Baltimore County, Md.
16 (no on-street parking)	
18 min	Virginia State Fire Marshal
24 (no parking)	Prince Georges County Dept. of Environmental Resources, Prince Georges County, Md.
30 (parking one side)	
36 (parking both sides)	
20 (for fire truck access)	
18 (parking one side)†	Portland Office of Transportation
26 (parking both sides)	

*Represents typical "fire lane" width, which is the width necessary to accommodate a fire truck.
†Applicable to grid pattern streets and cu-de-sacs.
SOURCE: Center for Watershed Protection, Site Planning Roundtable, *Better Site Design: A Handbook for Changing Development Rules in Your Community,* 1999. (Center for Universal Watershed Protection, Ellicott City, MD). Used with permission from the Center for Watershed Protection.

The National Association of Home Builders surveyed 110 communities that allow for narrower residential streets in order to learn from their experience (Table 5.4). The majority of communities surveyed reported that the narrower streets performed as well as wider streets with regard to maintenance costs and emergency vehicle access. Parking, traffic, and access problems were rare among the experiences reported.

TABLE 5.4 Comparison of Responses from the National Association of Home Builders (NAHB) Survey

Has the implementation of reduced street widths created problems with:			
	No, %	Yes, %	No answer, %
Emergency vehicle access	82.0	8.6	8.6
Traffic congestion	81.0	14.0	5.0
Adequacy of on-street parking	63.8	29.3	6.9
Proper functioning of street	63.8	-	19.0
What specific requirements have been imposed on streets designed with reduced widths?			
	No, %	Yes, %	No answer, %
Parking on one side	79.3	19.0	1.7
No parking on street	53.4	44.8	1.7
Additional off-street parking	74.1	24.1	1.7

SOURCE: National Association of Home Builders (NAHB), "Street Standards Survey Finds Narrower Streets Perform Well," *Homebuilder,* October 1988.

Safety is the first concern of street design, and the most common and obvious means of designing safe streets has been traditionally to segregate the pedestrian and the automobile. However, experience has shown that pedestrian activities do move into the street in residential neighborhoods despite the attempt to segregate them. The street becomes a safety issue because it is used for bicycling, walking, and recreation (Fig. 5.4). Therefore, accepting that these conflicting uses will exist is a better approach to ensuring safety. Table 5.4 demonstrates that communities with more pedestrian friendly streets generally do not feel that they have sacrificed vehicular safety or access, but instead they have found ways to integrate the needs of community and vehicle (Figs. 5.5 and 5.6).

Street Layout and Engineering

Street layout and design must consider the vehicle, visual range or limitation of the operator, safety for vehicle operators and pedestrians, and the climate, as well as the geometric configuration and the character of the area in which the street will be (see Table 5.5). These factors are interrelated. Most municipalities and states have well-defined design criteria for collector roads and highways but have only general criteria for local, smaller-volume roads. In many cases the local road criteria are based on the worst-case scenario-that is, the largest anticipated vehicle. This approach has little regard for the impact of the design on the behavior of drivers or quality of neighborhood life.

As an example, one street design situation that presents quite a few of the pitfalls and possibilities just discussed is the hillside. The nature of hillside development generally constrains the standards of classic grid development.

Figure 5.4 Photograph of cul-de-sac island as play area.

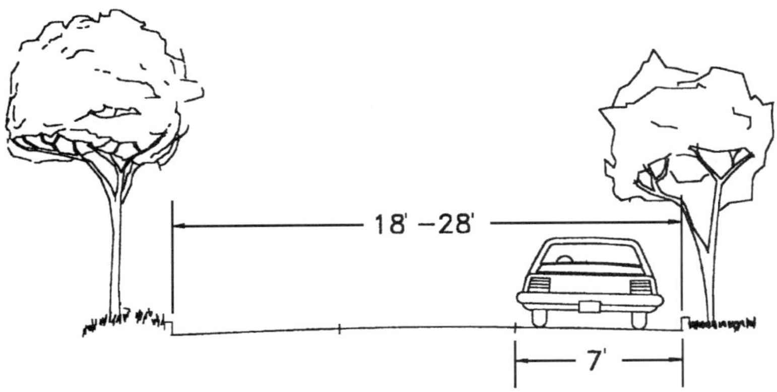

Figure 5.5 Residential street detail, parking on only one side.

Attempts to force unrealistic standards in an environment that requires special consideration inevitably creates as many problems as it solves. Furthermore, trying to fit a square peg in a round hole often increases the costs of development while not improving the living environment and actually losing the site character and some of its natural elements that made the site attractive in the first place.

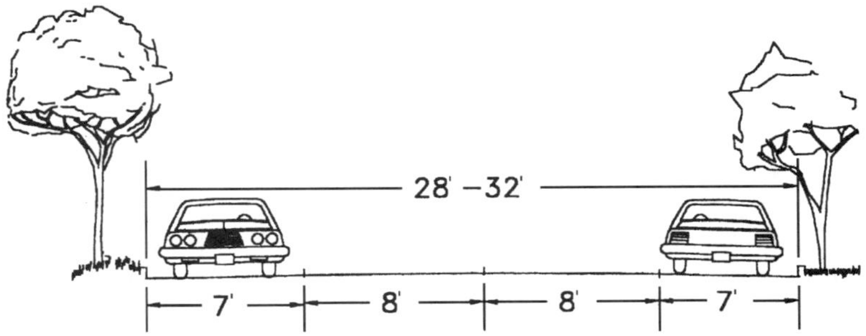

Figure 5.6 Residential street detail, parking on two sides.

TABLE 5.5 Residential Street Design Standards

	Low density, ordinary, hilly terrain	High density, ordinary, hilly terrain
Right-of-way width	40 ft	60 ft
Cartway width	22 ft	36 ft
Sidewalk width	0–6 ft	5 ft
Sidewalk distance from curb	0–6 ft	6 ft
Sight distance	20–100 ft	110–200 ft
Maximum grade	4–8%	4–15%
Maximum cul-de-sac length	1000 ft	500 ft
Design speed	30 mi/h	20 mi/h
Minimum centerline radius	250 ft	110 ft

Hillside streets generally need to be narrower and steeper to mimic the existing terrain and minimize the size and amount of cuts and fills. A rule of thumb is to use the dimensions of emergency vehicles, such as fire engines, to test design fitness. However, most design standards use the overblown requirements of vehicles built decades ago rather than the vehicles built today. Today cartway widths as narrow as 18 or 20 ft should be considered with no parking allowed. If parking is to be allowed, the designer should add a lane 8 ft wide for each side on which parking is allowed. Designers might consider varying cartway widths—narrower on slopes, wider on flat areas—to provide parking opportunities. Steep roads might be split, with a single lane in each direction separated by a wide area of steep slope. Shoulder widths might be reduced or eliminated in difficult areas. It should be noted that in most cases the split roadway may not provide a substantial savings in cost or in the amount of disturbed area because of the necessary slope lengths. As it is with all elements, the cost and benefit must be evaluated on a case-by-case basis.

Estimating traffic flow

A working estimate of traffic flow is necessary to design local roads, internal circulation, and interfaces with local collectors. The larger the project, the more important the estimate of trip generation and vehicle speeds (see Tables 5.6 and 5.7). Traffic flow is affected by a number of factors, some of which are fairly intuitive. For example, the nature of the development under consideration is important. A regional shopping center, a retirement community, a neighborhood geared to young families, and an entertainment complex will all have different traffic characteristics. In most residential cases, peak flows may be expected to occur during the rush hours between 6:00 A.M. and 9:00 A.M., and 4:00 P.M. and 6:00 P.M. Estimating peak time traffic flows from a single-family home, for example, is based on 0.8 trips per day per single-family dwelling unit during peak hours. For townhouses or multifamily units, a trip generation of 0.6 trips per day per unit is used. A single-family unit is expected to generate at least 5 round-trips (leaving and returning) each day, but the trips are not evenly loaded throughout the day. In general, there is more traffic in morning peak hours than in afternoon peak hours.

Vehicle dimensions and turning radii

Using the performance criteria given in Table 5.8, site designers should select a vehicle for design purposes that represents the largest vehicle that frequently uses the street. Different design vehicles should be selected for different hierarchies of

TABLE 5.6 Vehicle Trip Generation, Residential Areas, Number of Trips Per Day

Dwelling type	Average	Range
Single-family detached	10.1	4.3–21.9
Apartments	6.1	0.5–11.8
Condominiums	5.9	0.6–11.8
Mobile homes	4.8	2.3–10.4

Adapted from Institute of Transportation Engineers.

TABLE 5.7 Minimum Design Speeds Based on Average Daily Traffic (ADT) and Design Hourly Volume (DHV)

Terrain	ADT < 400	ADT > 400	DHV 100–200	DHV 200–400	DHV > 400
Level	40	50	50	60	60
Rolling	30	40	40	50	50
Mountainous	20	30	30	40	40

SOURCE: From *A Policy on Geometric Design of Highways and Streets.* Copyright 1994 by the American Association of State Highway and Transportation Officials (AASHTO), Washington, D.C. Used by permission.

TABLE 5.8 Vehicle Dimensions and Turning Radii

Vehicle	Length	Width	Wheel base	R*	R_1	D
Small car	15 ft, 6 in	5 ft, 10 in	9 ft, 2 in	20 ft	10 ft, 11 in	10 ft
Compact car	16 ft, 11 in	6 ft, 3 in	10 ft	21 ft, 6 in	12 ft	11 ft
Standard car	18 ft	6 ft, 10 in	10 ft, 8 in	22 ft, 6 in	12 ft, 8 in	11 ft, 6 in
Large car	19 ft	6 ft, 10 in	11 ft	23 ft	13 ft	12 ft
City bus	40 ft	8 ft, 6 in	-	53 ft, 6 in	33 ft	22 ft, 6 in
School bus	40 ft	8 ft	-	43 ft, 6 in	26 ft	19 ft, 6 in
Ambulance	20 ft, 11 in	7 ft	-	30 ft	18 ft, 9 in	13 ft, 3 in
Limousine	22 ft, 6 in	6 ft, 6 in	-	29 ft	16 ft	16 ft
Trash truck	29 ft	8 ft	-	32 ft	18 ft	16 ft
UPS truck	23 ft, 2 in	7 ft, 7 in	-	28 ft	16 ft	14 ft
Fire truck	31 ft, 6 in	8 ft, 4 in	-	48 ft	34 ft, 6 in	15 ft, 6 in

*The R value of the vehicle selected as the design vehicle should not exceed the radii of a paved circle.
SOURCE: From *A Policy on Geometric Design of Highway and Streets.* Copyright 1994 by the American Association of State Highway and Transportation Officials (AASHTO), Washington, D.C. Used by permission.

street design and for the nature of the area in which the street is situated. Collector streets should be designed for larger vehicles such as buses and tractor trailers, and residential streets should be designed for smaller vehicles. Figures 5.7 through 5.13 provide turning path dimensions for different vehicles. Also, the discussion on cul-de-sac design in this chapter includes a discussion on design vehicle selection.

Sight distance calculation

Sight distance design is concerned with providing the operator of a vehicle with safe and adequate forward visual access. *Sight distance* is the distance forward at which a driver has an unobstructed view of the road. For design purposes, minimum sight distance requirements are determined based on the assumed length of time between the driver's recognizing an object in the road and his or her being able to come to a complete stop from the design speed of the road. The factors affecting sight distance are the horizontal and vertical arrangement of the road, the height of the operator's eye, and the height of the object to be seen (see Figs. 5.14 and 5.15).

The *sight-to-stopping distance,* or *stopping distance,* is determined as a combination of the time and distance that pass from the moment of perception to reaction (PR) until the vehicle stops. PR can be expressed as follows: $\text{PR} = 1.47(t)(V)$ where PR is the stopping distance at a given speed, t is the total of the perception time and length of time braking, and V represents the speed of the vehicle.

Braking distance is calculated as $d = V^2/30f$ where d represents the braking distance, V is the velocity of the vehicle when braking begins, and f is the coefficient of friction between the tires and pavement (see Table 5.9).

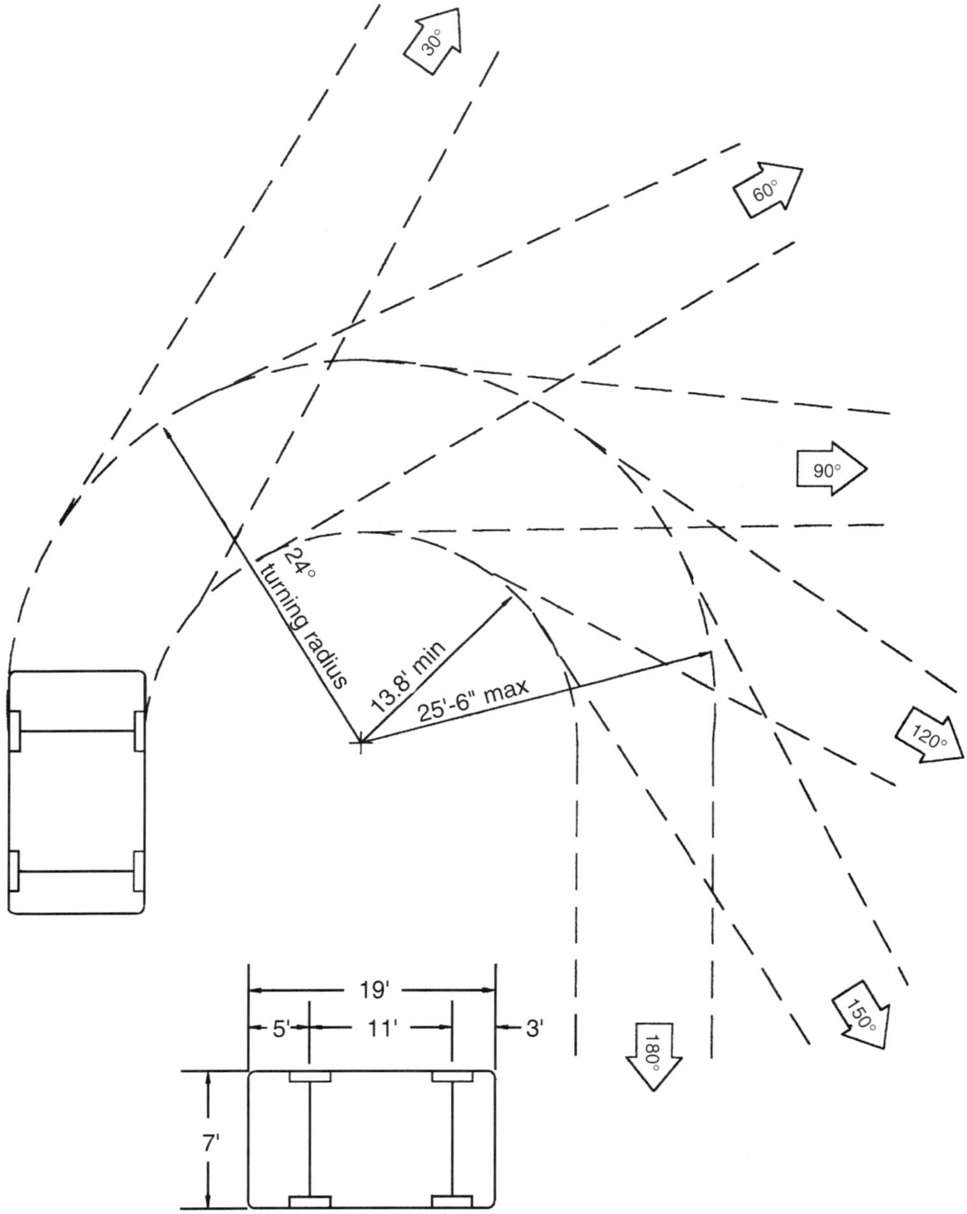

Figure 5.7 Minimum turning path for passenger car detail. (*Copyright* © 1994 *by the American Association of State Highway and Transportation Officials, Washington, D.C. Used with Permission.*)

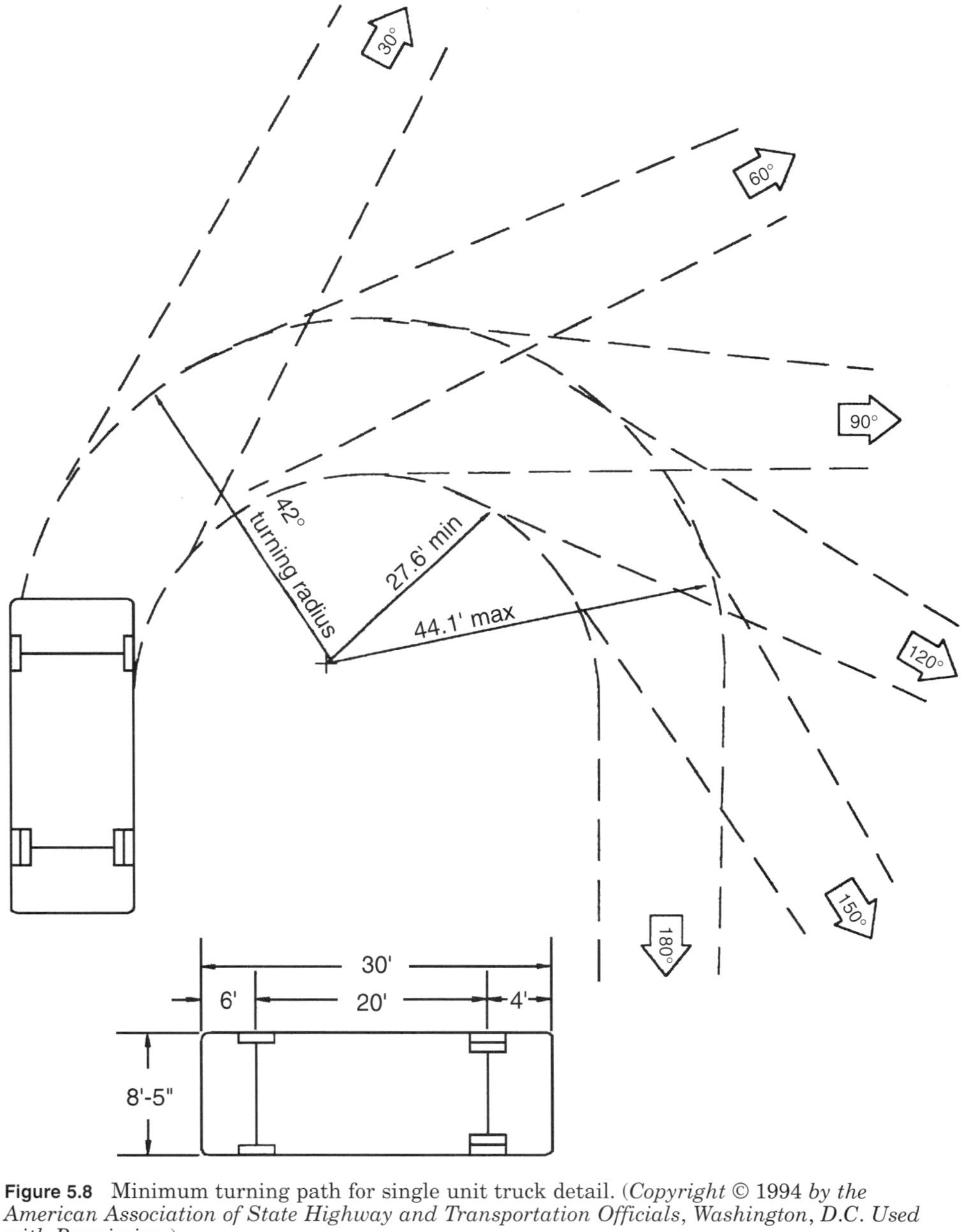

Figure 5.8 Minimum turning path for single unit truck detail. (*Copyright* © 1994 *by the American Association of State Highway and Transportation Officials, Washington, D.C. Used with Permission.*)

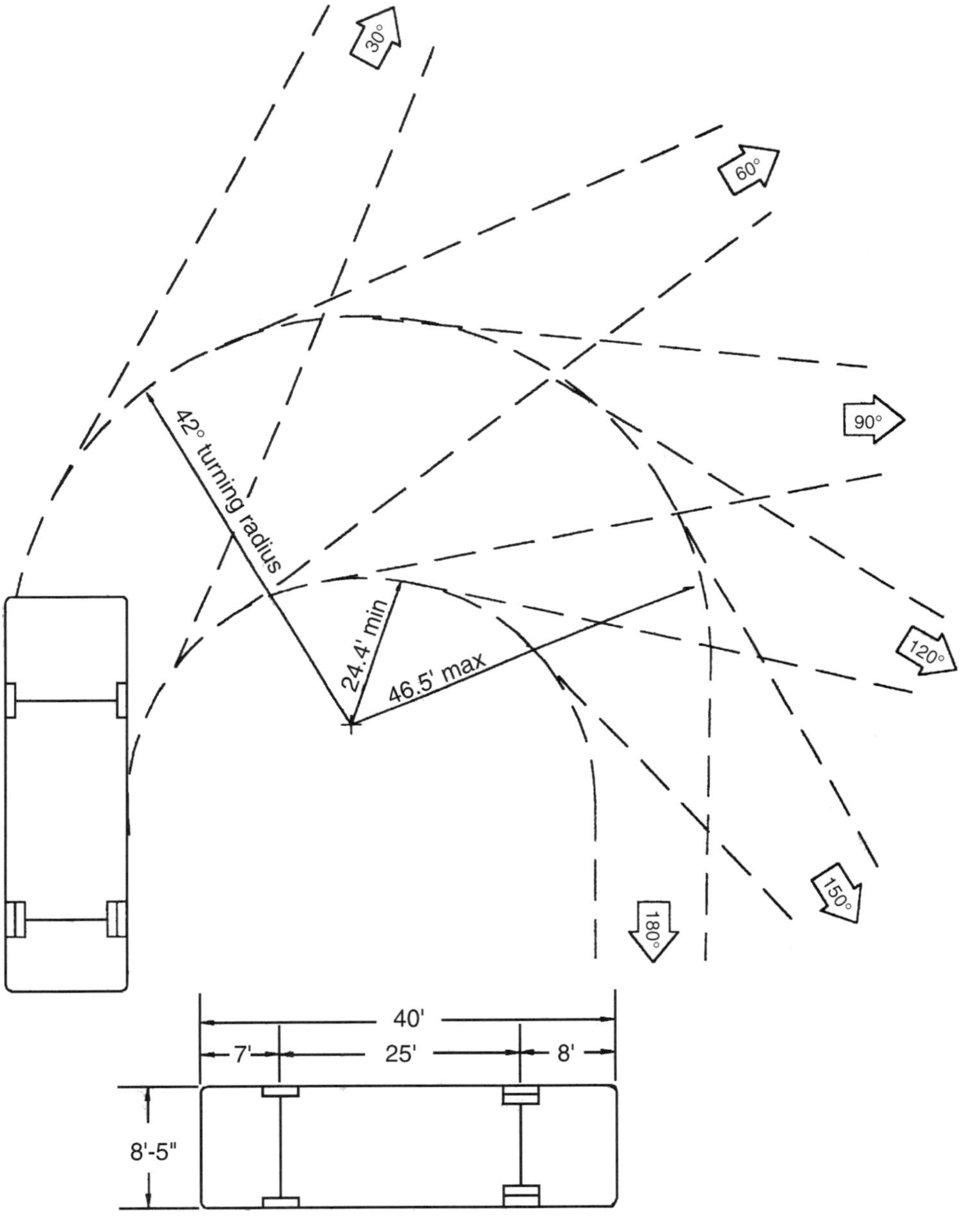

Figure 5.9 Minimum turning path for single unit bus detail. (*Copyright © 1994 by the American Association of State Highway and Transportation Officials, Washington, D.C. Used with Permission.*)

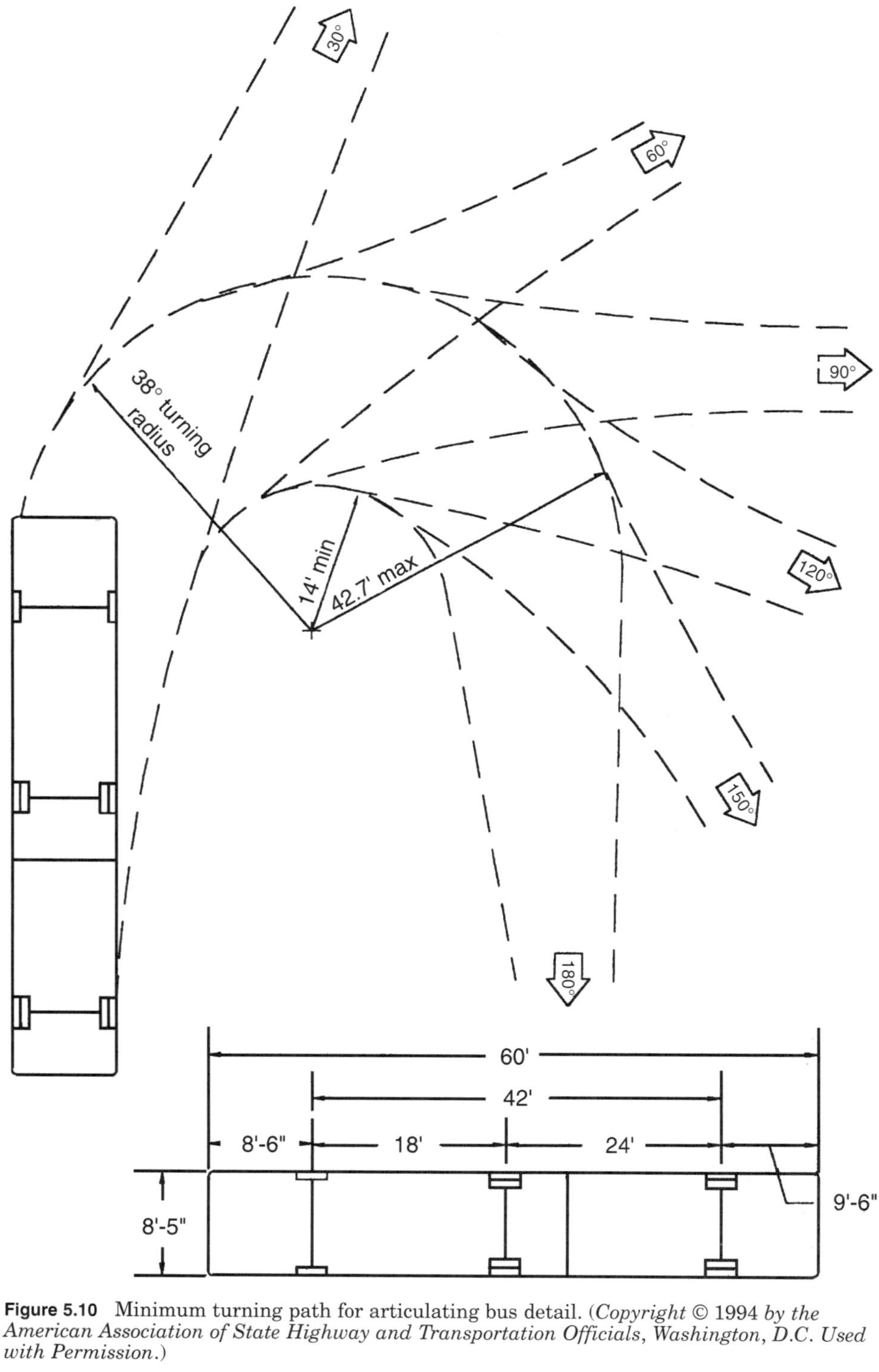

Figure 5.10 Minimum turning path for articulating bus detail. (*Copyright © 1994 by the American Association of State Highway and Transportation Officials, Washington, D.C. Used with Permission.*)

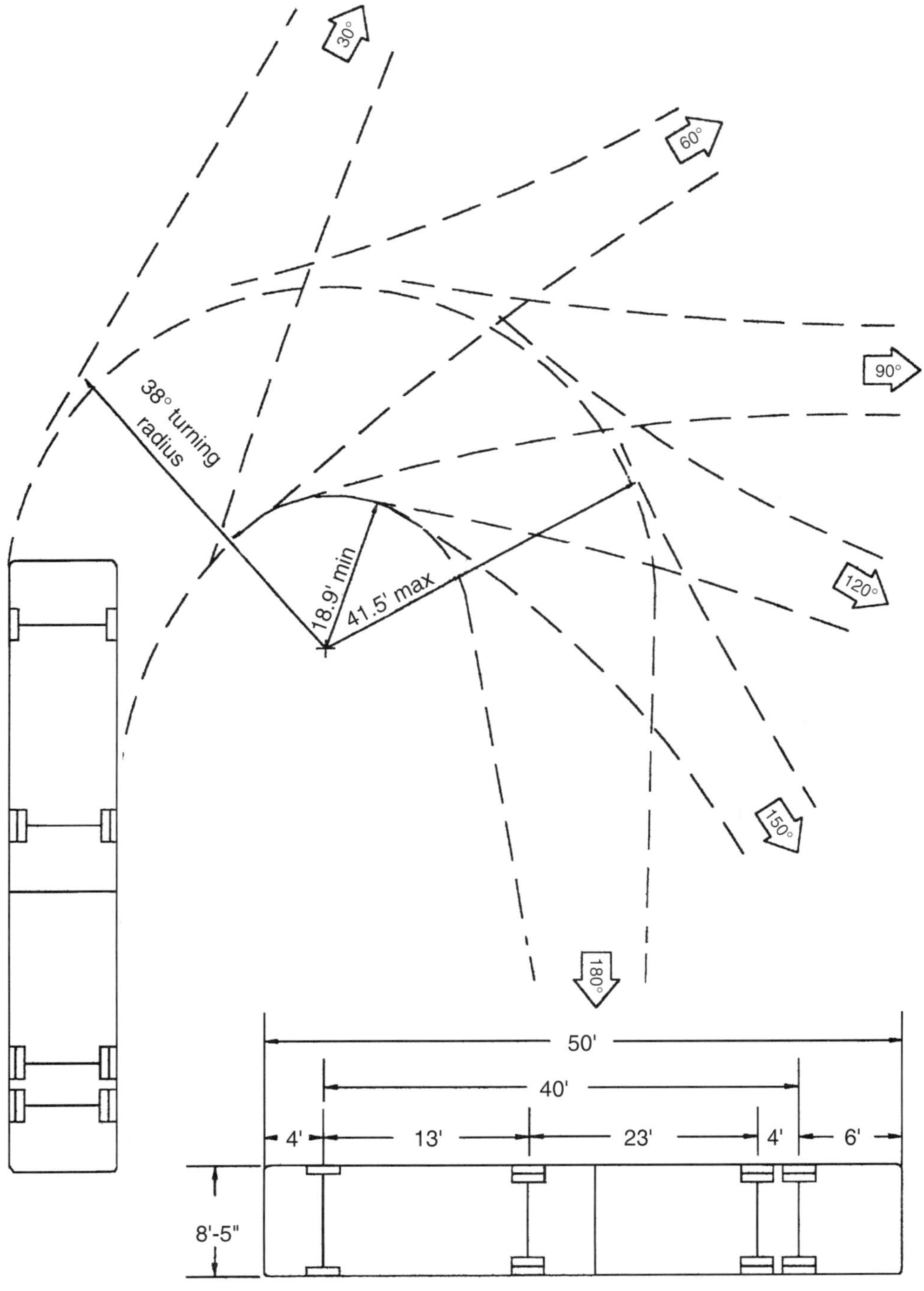

Figure 5.11 Minimum turning path for semitrailer intermediate detail. (*Copyright* © 1994 *by the American Association of State Highway and Transportation Officials, Washington, D.C. Used with Permission.*)

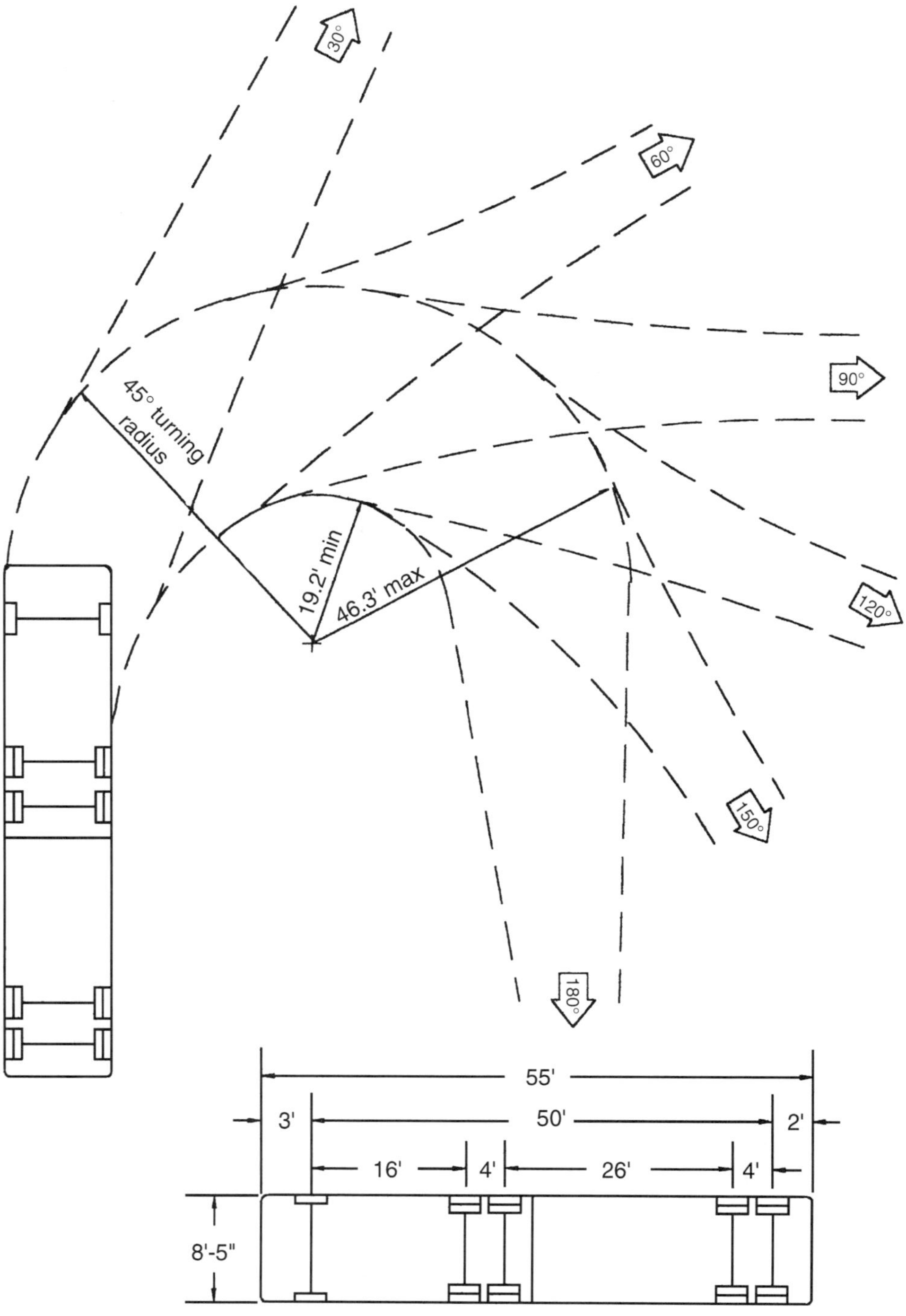

Figure 5.12 Minimum turning path for semitrailer combination detail. (*Copyright © 1994 by the American Association of State Highway and Transportation Officials, Washington, D.C. Used with Permission.*)

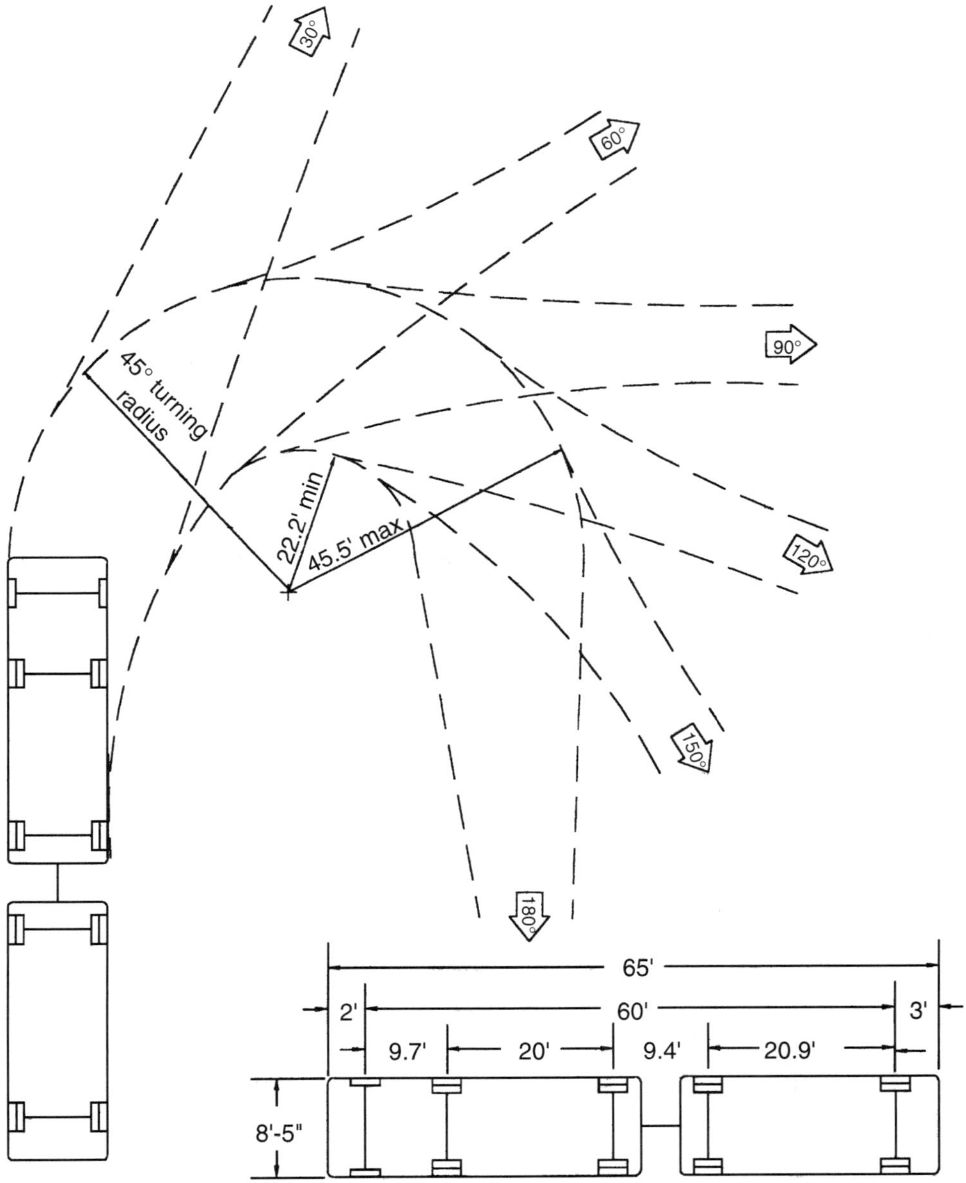

Figure 5.13 Minimum turning path for semitrailer full trailer combination. (*Copyright © 1994 by the American Association of State Highway and Transportation Officials, Washington, D.C. Used with Permission.*)

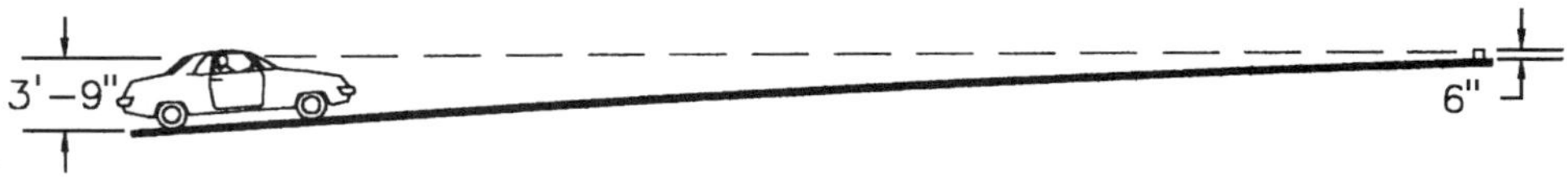

Figure 5.14 Safe stopping distance detail.

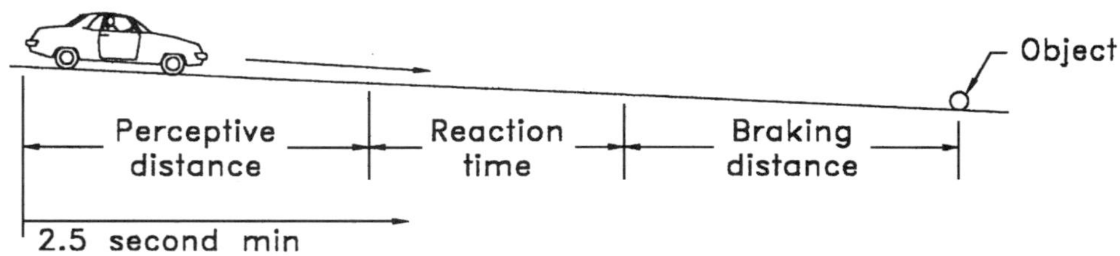

Figure 5.15 Safe sight distance design parameters.

TABLE 5.9 Coefficient of Friction *F* between Tire and Road

Design speed, mi/h	*F**
30	0.36
40	0.33
50	0.31
60	0.30
70	0.29

*Pavement assumed to be under wet conditions.
Adapted from American Association of State Highway and Transportation Officials (AASHTO), 1990.

It is an accepted practice to assume for design purposes that the driver's eye height is 3 ft, 9 in above a road surface. In general, an object 6 in high is assumed to be adequate for measuring sight and stopping distance on vertical curves (see also Table 5.10).

The weight of a specific vehicle and the grade of the road will affect stopping distance (see Table 5.11). The weight of larger vehicles is difficult to account for in a design concept; however, it is generally accepted that the greater weight of a vehicle is often offset by its increased height that allows the operator greater sight distance. Grades can be accounted for in the design. Stopping distance tends to decrease on uphill grades and increase on downhill grades. These differences are accounted for by applying the percent of grade to the coefficient of friction expressed as a decimal. Uphill grades are added (increasing the coefficient of friction), and downhill grades are subtracted (decreasing the coefficient of friction).

Sight distance on horizontal curves must also be considered. Locating visual obstructions out of the line of sight is necessary to provide safe sight distance

TABLE 5.10 Sight-to-Stopping Distances

Design speed, mi/h	Assumed speed, mi/h	Reaction time, s	Reaction distance, ft*	Coefficient of friction	Braking distance on level†	Sight-to-stopping distance (computed)†	Stopping distance rounded for design
20	20–20	2.5	73.3–73.3	0.40	333.3–33.3	106.7–106.7	125–125
25	24–25	2.5	88.0–91.7	0.38	50.5–54.8	138.5–146.5	150–150
30	28–30	2.5	102.7–110.0	0.36	74.7–85.7	177.3–195.7	200–200
35	32–35	2.5	117.3–128.3	0.34	100.4–120.1	217.7–248.4	225–250
40	36–40	2.5	132–146.7	0.32	135.0–166.7	267.0–313.3	275–325
45	40–45	2.5	146.7–165.0	0.31	172.0–217.7	318.7–382.7	325–400
50	44–50	2.5	161.3–183.3	0.30	215.1–277.8	376.4–461.1	400–475
55	48–55	2.5	176.0–201.7	0.30	256.0–336.1	432.0–537.8	450–550
60	52–60	2.5	190.7–220.0	0.29	310.8–413.8	501.5–633.8	525–650
65	55–65	2.5	201.7–238.3	0.29	347.7–485.6	549.4–724.0	550–725
70	58–70	2.5	212.7–256.7	0.28	400.5–583.3	613.1–840.0	625–850

*PR = $1.47(t)(V)$, in the table t is assumed as 2.5 s. AASHTO recommends 2.5 s as the minimum reaction time.

†$d = V^2/30f$.

SOURCE: From *A Policy on Geometric Design of Highway and Streets.* Copyright 1994 by the American Association of State Highway and Transportation Officials (AASHTO), Washington, D.C. Used by permission.

TABLE 5.11 Effect of Grade on Stopping-Sight Distance

Design speed, mi/h	Increase for downgrade, ft, %			Assumed speed, mi/h	Decrease for upgrade, ft, %		
—	3	6	9	—	3	6	9
30	10	20	30	28	—	10	20
40	20	40	70	36	10	20	30
50	30	70	—	44	30	50	—
70	70	160	—	58	40	70	—

Adapted from AASHTO, 1994.

even on relatively flat surfaces. Intersections and horizontal curves should be designed to provide drivers with a clear vision of oncoming traffic and pedestrian activity. The *sight triangle* is used to determine the clear, obstruction-free area required at an intersection (Fig. 5.16). Many local ordinances include sight triangle design requirements. Figure 5.16 illustrates the sight triangle parameters recommended by the American Association of State Highway and Transportation Officials (AASHTO). In the figure d is the distance traveled by a vehicle moving at the design speed during the time required for a stopped vehicle to get underway and cross the intersection or make a turn.

Vertical curves

The vertical design of cartways must also be considered. The vertical curve is actually a parabola as opposed to a circular curve. In most instances the vertical curve is designed from the centerline of the cartway. The following formula presumes the vertical curve is symmetrical. Designing a vertical curve begins with selecting the *point of vertical intersection* (PVI) by extending the opposing slopes to a point of intersection. The PVI is identified as a particular station along the centerline and the grades. Once the PVI and the grades for the proposed curve are known, the designer can set the curve length. The length of the vertical curve (L) is the distance of the tangent from the *point of vertical curve* (PVC) to the *point of vertical tangent* (PCT). Elevations are taken from the profile sheet. With this information the vertical curve can be calculated as follows.

Step 1. Determine the slopes for the respective tangents:

$$\text{Tangent PVC-PVI} = \frac{\text{elevation of PVI} - \text{elevation of PVC}}{L/2 \times 100 = \text{percent slope}}$$

$$\text{Tangent PVT-PVI} = \frac{\text{elevation of PVT} - \text{elevation of PVI}}{L/2 \times 100 = \text{percent slope}}$$

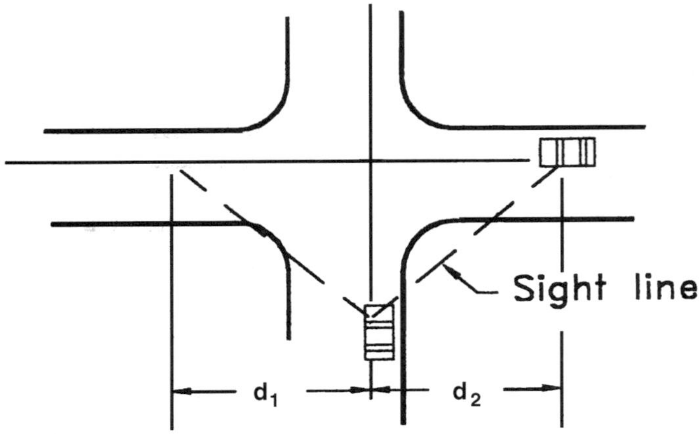

Figure 5.16 Sight triangle detail.

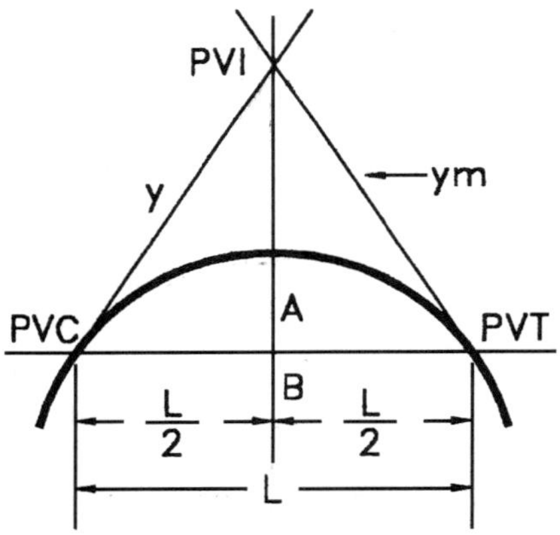

Figure 5.17 Vertical curve calculation.

Notice that the positive and negative values indicate the rise (+) or fall (−) of the slope.

Step 2. Determine *B,* the elevation of the intersection of the PVI and the secant between the PVC and PVT. This can be quickly determined by adding the elevation of the PVC and the elevation of the PVT and dividing their sum by 2.

Step 3. Calculate *ym*—that is, the vertical distance from the PVI to the vertical curve—by subtracting the elevation of *B* from the PVI and dividing by 2. This *ym* value can be used to calculate all other elevations on the curve.

Horizontal alignment

The design of horizontal curves is concerned with both the sight distance and the appropriate radius for the design speed and conditions. The horizontal curve is simply an arc of a circle connecting two tangents, which is why the circular curve formulas are relevant (Table 5.12). For most local street design, the centerline of the proposed street is used as a starting point for calculations.

The following guidelines should be observed when working with curved roadways: Whenever the horizontal direction of the street changes, a horizontal curve is used to make the transition. For most purposes horizontal alignment should be as direct as possible; however, under some conditions longer transitions might be appropriate to minimize the amount of grading or other impacts on the site. In general, abrupt or sharp curves are to be avoided as are multiple compound curves. Exceptions to these guidelines, however, are common when dealing with very low volume local roads in difficult terrain. Many local land development ordinances require specific minimum horizontal curves.

Figure 5.18, together with Table 5.12, illustrates the relationships used to calculate horizontal curves. The PC is the *point of curvature,* or the point where the curve begins. The PT is the *point of tangency,* or the point where the curve ends. Points along the curve are usually given in terms of the stationing on a given road. The *arc,* or *arc length,* is shown as L in the formulas. It refers to the length of the curve between the PC and PT. The PI is the *point of intersection,* or the point where both tangents intersect. The Greek symbol delta represents the *internal angle.* The *chord* is a straight line drawn from the PC to the PT. The *deflection angle* is the angle between the chord and the tangent. The deflection angle is always one-half of the angle subtended by the arc.

Intersections

Intersections of two or more streets should be carefully designed to allow adequate sight distance as well as smooth traffic flow. Grades at intersections should be kept to 3 percent or less. On local streets there should be a clear sight triangle of no less than 50 ft. When local roads dictate an offset

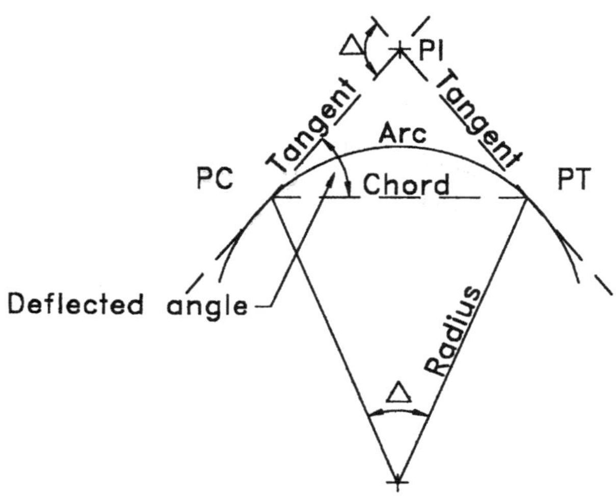

Figure 5.18 Horizontal curve calculation.

TABLE 5.12 The Circular Curve Formulas

$T = R \tan \frac{\Delta}{2}$	$T =$
$C = 2R\ \mathrm{SIN}\ \frac{\Delta}{2}$	$R =$
$R = \frac{57.3L}{\Delta}$	
$L = \frac{\Delta R}{57.3}$	
$\Delta = \frac{57.3L}{R}$	

intersection, at least 100 ft should be provided between intersections. Streets should intersect at 90° whenever possible (Fig. 5.19).

Streets for People

The streets in most residential neighborhoods are wide, overdesigned, unnecessary macadam cartways that were expensive to build and are expensive to maintain. These wide streets encourage excessive speed, increase storm water runoff, and raise the price of new construction and ongoing maintenance. In return for these overdesigned and underutilized streets, residents are exposed to an increased risk of accidents and level of noise but a decrease in neighborhood social interaction and a loss of character in the appearance of the neighborhood (see Tables 5.13 and 5.14).

Designers have made many attempts to separate traffic from pedestrian activities. An example is the "superblock" concept as constructed, for example, in 1928 in Radburn, New Jersey. Following the superblock design, 40-acre superblocks were developed at a density of four units to the acre. Access to the houses was through cul-de-sacs arranged around a central green space and pedestrian network. In this way all of the automobile traffic was kept on the outside of the superblock, and all of the pedestrian activity was focused in to the green center. The superblock concept, as it has come to be known, was proposed by architects Clarence Stien and Henry Wright. The superblock cul-de-sacs were designed to be only 350 to 450 ft long and had a cartway of only 21 ft. No curbs were used, and the right-of-way was kept to only 35 ft and included a 7-ft-wide utility corridor outside of the cartway on both sides of the street. The design resulted in 25 percent less pavement than a common grid street layout. The design also provided for reduced utility infrastructure costs. In practice, however, the superblock was associated with some observed increase in crime.

Since the development of the Radburn project, the family car has become an even greater influence on our lives. Today the most prominent image of resi-

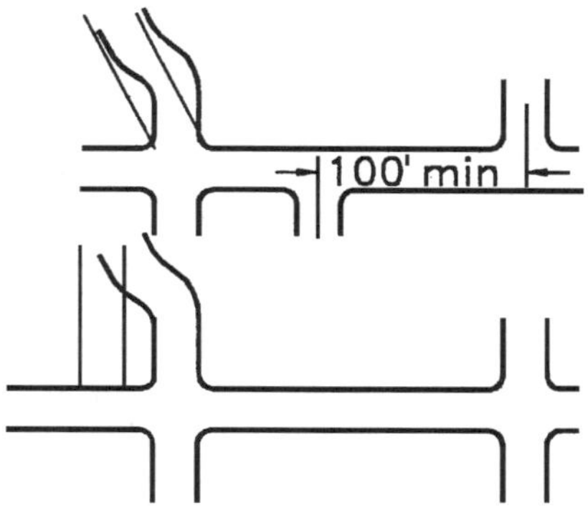

Figure 5.19 Intersection of streets detail.

TABLE 5.13 Problems with Residential Streets

Traffic accidents
Noise, vibration, pollution
Excessive traffic speed
Nonresident vehicle traffic
Diminished appearance
Maintenance
Reduction of social interaction in neighborhood
Increased storm water runoff

TABLE 5.14 Goals of Residential Street Design

1. Provide safe access to residents, including pedestrians, children, and residents with physical restrictions.
2. Reduce traffic speed and volume, and thereby reduce the number and the severity of accidents.
3. Contribute as part of the overall design to the character, stability, social interaction, and esthetics of the neighborhood.
4. Relate to the open-space, recreation, and social areas and activities of the neighborhood.
5. Provide access for emergency and delivery vehicles.
6. Provide a reasonable service life.

SOURCE: Adapted from Institute of Transportation Engineers.

dential streets in contemporary developments is of garage doors with houses attached. Nevertheless, we have been learning a great deal about how streets work and how drivers behave. Although most ordinances still require the construction of streets that encourage higher speeds and impose greater risk to pedestrians, some communities are seeking alternatives. One factor being considered is that drivers are influenced by the design of the streets they travel on just as pedestrians are influenced by the design of the streets they try to cross. Most streets are designed to be conduits or channels—straight and as level as possible. But long straight runs of streets are precisely the conditions that encourage higher speeds and less observant drivers. Posting speed limits is not an effective means of controlling speeds, especially in residential areas where conflicts between vehicles and people can be anticipated.

Typical pedestrian streets are posted at 25 to 35 mi/h, but they are designed for speeds of 45 to 50 mi/h. Under such circumstances drivers will tend to drive faster than the posted speed. Research indicates that a person struck by an automobile traveling at 20 mi/h or less is usually not seriously injured. At 20 to 30 mi/h, the person's injuries are usually serious, and at vehicle speeds over 30 mi/h, the person is often killed. The reason streets are designed for higher speeds is to protect the driver. Higher design speeds result in a longer sight distance, which makes maneuvering easier and safer for the driver. But the higher design speeds also encourage speeding, which is dangerous to the residents of the neighborhood. Perhaps a better, safer alternative would be to design streets that require drivers to be more alert and to slow down.

Nontraditional street design

Streets can be designed in ways that will result in slower and safer vehicle speeds and that will enhance the quality of neighborhood life. Design elements that tend to slow down traffic include planted islands, changes in grade, changes in street width, meandering roads, cul-de-sacs, and rotaries. The Dutch use a traffic-slowing concept called a *woon erf* to integrate traffic and neighborhoods (see Table 5.15 and Fig. 5.20).

The woon erf is a distinctly European design that reflects the high density of development and the economy of design found in Europe. The woon erf concept could be applied equally well in the United States, and U.S. designers could tap the vast experience of the Europeans in this field. The woon erf street is more expensive to build and to maintain, but developers and cities have found that the experience of residents is so positive that the higher cost is worthwhile. One interesting aspect of the woon erf is that it calls for the use of pavers instead of poured paving. The paver is used for its esthetic value as well as its role in managing storm water runoff. In some cases the use of pavers to promote infiltration may reduce or eliminate the need for other storm water management facilities, which would be a cost saving that would offset the higher cost of the pavers.

Originally the woon erf was developed for use in low-income residential areas, but the street layout proved to be so desirable that its use spread to neighborhoods of all types. Many existing streets in European cities have been converted from traditional arrangements into woon erfs. Residents report that they find the environment very desirable because of the parklike atmosphere, the visual character of the neighborhood, and the availability of social opportunities for children and adults. While it may be unlikely that every aspect of the woon erf concept would apply equally well to all of the developing resi-

TABLE 5.15 Elements of Woon Erf Design

Aspects of woon erf design that may slow down residential street traffic are the following:

1. The right-of-way are narrower and completely paved except for the planted islands and play areas.
2. Pedestrian walkways are at the same level and grade as the cartway. There is no curb separating them.
3. Vehicle traffic is permitted, though street design and activities require a reduced speed.
4. Areas of potential conflict, such as play areas and social areas, are signaled through the use of trees, planted islands, and signs.
5. Travel lanes for vehicles are narrow and change direction often to encourage lower speeds and more awareness on the part of drivers.
6. Two-way streets are encouraged because one-way streets encourage higher speeds.
7. Parking spaces are provided in clusters of six or seven and are usually at a right angle to the direction of traffic.
8. The right-of-way in a woon erf is given to the pedestrian, and the traffic speed limit is usually about 15 mi/h.
9. Signs are usually used at the entrance to a woon erf to inform drivers that they are entering a residential area in which special conditions prevail.

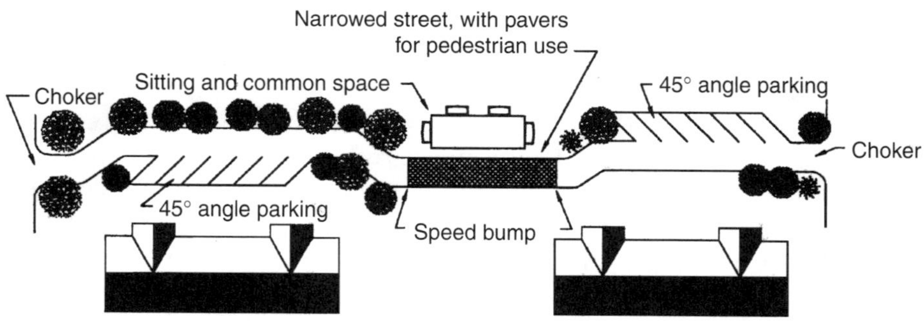

Figure 5.20 Diagram of a woon erf.

dential projects in the United States, some aspects of the design would be successful in almost any setting.

Traffic calming

Traffic calming refers to the use of design elements to increase drivers' awareness and to slow them down. As already discussed, streets designed with wide cartways and long straight runs of road tend to encourage higher speed with less attentiveness. Traffic claming devices can be used, however, to make drivers more aware of the road and of the presence of pedestrians, which reduces the number of incidents and accidents. All of their advantages notwithstanding, however, traffic calming devices should be well thought out and considered before being used in a design.

Complaints about traffic calming measures generally fall into one of two categories: driver exasperation with what is seen as interference with driving and unintended impacts to vehicles. Little can be done about the former, but some steps can be taken to prevent the latter. Most complaints are focused on speed bumps because they sometimes cause damage to snow removal equipment (and the snow removal equipment in turn sometimes damages the speed bumps) they can slow down the response time of emergency equipment, and they create accessibility problems for disabled people.

Although it is very effective, the speed bump is the least elegant of the traffic calming tools available. Other methods use the natural inclinations and behavior of drivers to reach the designer's objective (see Figs. 5.21 through 5.25). It is natural for a driver's attention to increase at changes in the road configuration or layout, and designers can incorporate this knowledge into their design to increase driver awareness of pedestrians and slow down traffic. Changes in road width, changes in grade, and changes in paving surface texture or color are all effective traffic calming methods that do not create some of the problems associated with speed bumps.

Traffic calming strategies can be adapted for use on existing streets as well as on new streets. For example, chokers can be installed at intersections to require drivers turning onto a street to turn more carefully or at midblock locations to require drivers to slow down. Drivers will automatically respond to the chokers by being more aware of pedestrians and allowing them to cross safely at crosswalks. If chokers are used at intersections with collectors, thought should be given to constructing a turning lane on the collector to allow slowed traffic to move out of the travel lanes. In place of speed bumps, grade changes can be used to slow traffic and to reduce the risk of damage to vehicles such as snow plows. Midblock chokers or street closings can also be used to reduce traffic, increase pedestrian use of the street, and foster a greater sense of neighborhood. Neighboring streets may experience an increase in traffic, however.

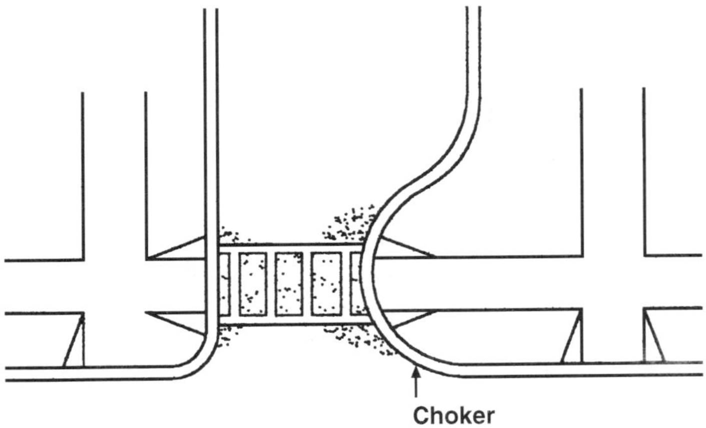

Figure 5.21 A choker used as a traffic calming device.

Figure 5.22 Photograph of a choker located midblock.

Figure 5.23 Photograph of a narrow street.

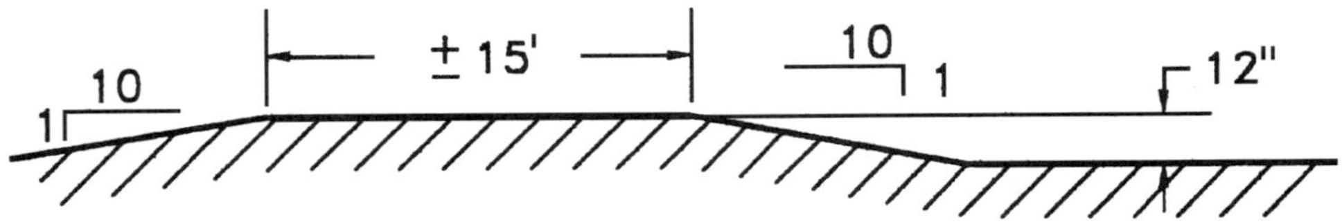

Figure 5.24 Grade change detail.

Figure 5.25 Photograph of a street closure.

Cul-de-Sac Design

The cul-de-sac was developed out of necessity but has evolved into a preferred feature of many projects. Originally a "dead end" was seen as a necessary evil or the result of poor design—necessary only to accommodate difficult topography or property shapes. But as residential development design has evolved, planners, developers, real estate professionals, and most importantly home buyers have recognized the desirability of the cul-de-sac location. The appeal of the cul-de-sac as expressed by people that live on them is the privacy, the absence of through traffic, and the sense of inclusivity and neighborhood that develops among the residents.

Along with providing the sense of a more secure environment for residents, a properly designed cul-de-sac offers more tangible benefits as well. For example, the well-designed cul-de-sac often requires less pavement for each housing unit than its equivalent in street might require because it can be built using a narrower cartway width and a landscaped island. The reduced pavement coverage has several positive results: less impermeable surface, which results in less runoff and lower street maintenance costs. The narrower cartway width is a practical response to the fact that there will be no through traffic. The limited number of residences on the cul-de-sac should be the primary guideline in determining the cartway width.

There are many ways in which to design a cul-de-sac or dead-end street, as shown in Fig. 5.26. However, the objective of the design is usually the same—that is, to develop an environment that is a pleasant place to live—safe, attractive, and desirable from the day construction is finished through to its maturity. Administrative constraints in the form of local design ordinances and requirements should serve as guidelines for the early design (see Table 5.16). These guidelines often do not account for the conditions found on a specific site or the requirements of a particular product. Generally design requirements for cul-de-sacs revolve around the number of units allowed, the length of the cul-de-sac, and the radius of the turnaround. The design of the cul-de-sac should be based on its intended use rather than prescriptive standards. A cul-de-sac designed for multifamily units should be different from a cul-de-sac designed to create a small separate community feeling. The ideal number of families on a cul-de-sac is difficult to determine. Beyond a single family, the next most basic social unit is a group of between 3 and 12 families. This is consistent with the opinions stated in informal discussions with residents of cul-de-sacs who have indicated that on cul-de-sacs with greater than 10 or 12 families, there is no special identity among the residents as there is on those with fewer families. Most ordinances limit the number of units to between 21 and 28, but there are few empirical arguments to support such a high number.

The cul-de-sac forms a cluster of residences. It is the cluster arrangement that creates a sense of privacy or exclusivity that many buyers desire. It is possible to arrange several clusters along a single cul-de-sac. In such cases, a waiver from a local guideline may be necessary. The requirement to limit cul-de-sac length seems to have developed out of a concern for traffic congestion. If these issues are adequately addressed in a design, a waiver of the requirement would seem appropriate. Cul-de-sac lengths are commonly limited to about 1000 ft, but they may run up to 1500 ft. In completing the research for this work, no empirical basis was found for determining a limit to the number of units or the length of a cul-de-sac based solely on the number of dwelling units. The most logical argument for limiting cul-de-sac length is the amount of traffic that might be generated from the single point of ingress and egress during peak traffic times. If lot sizes are larger than an acre, the guideline should be adapted appropriately. In projects with large lot sizes, the distance between houses can create the same effects as do too many units, and the sense of place and neighborhood does not develop as it would in a higher-density circumstance. In a cul-de-sac with a length that exceeds 1000 ft, an interim turnaround might be considered.

The design of the terminal end of a cul-de-sac is the source of the most discussion and concern. The choice of the design vehicle is fundamental to the design

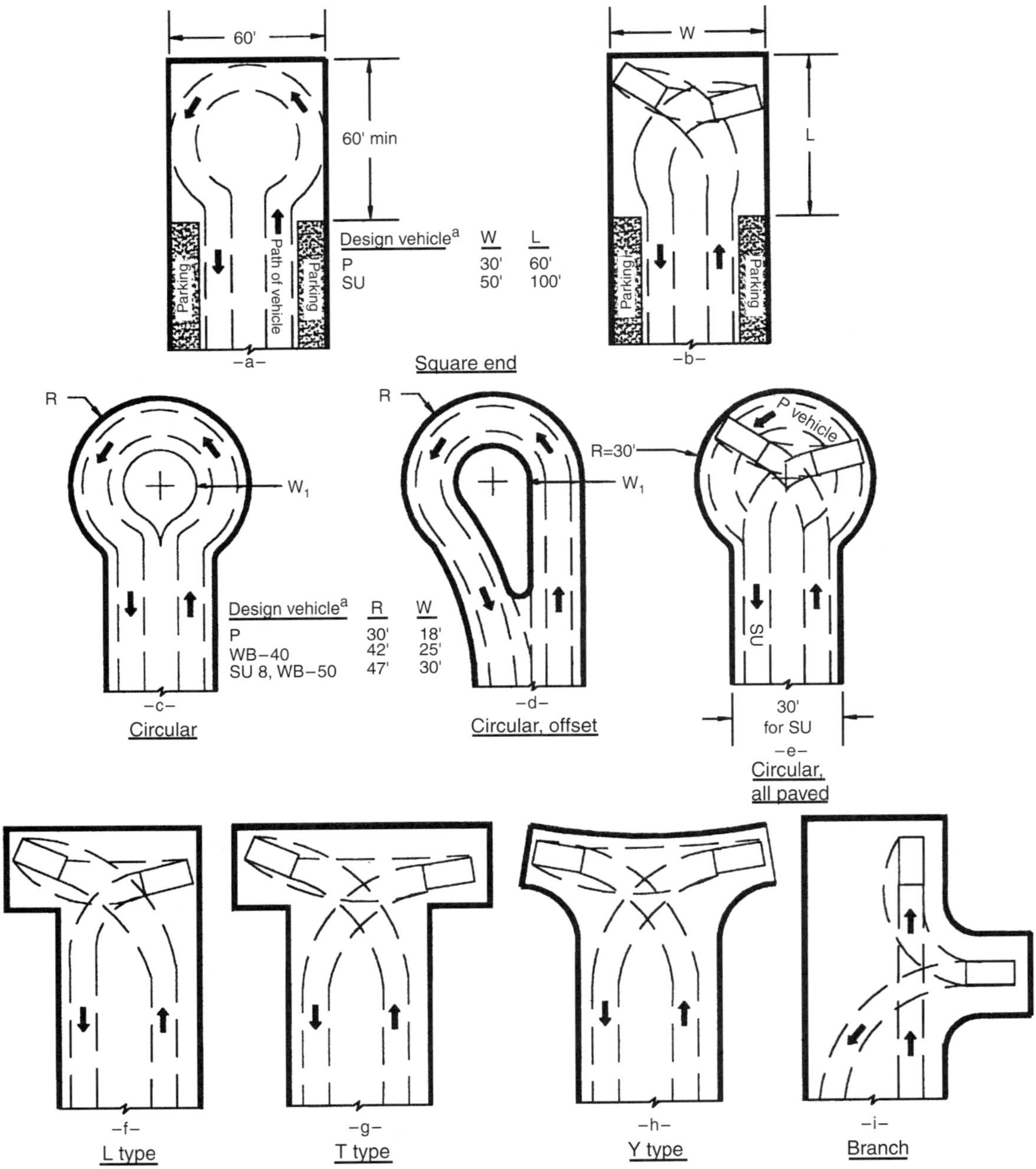

Figure 5.26 Details of cul-de-sac and dead-end street designs. (*Copyright © 1994 by the American Association of State Highway and Transportation Officials, Washington, D.C. Used with Permission.*)

TABLE 5.16 Factors Affecting Cul-de-Sac Design

Topography
The number of units on the cul-de-sac
Cartway and right-of-way requirements
Parking requirements
Cul-de-sac dimensional requirements
Drainage requirements
Landscape and lighting requirements
Site amenities such as signage, views, open-space access

of the turnaround. The *design vehicle* is the type of vehicle that is used to determine the radius and cartway width of the turnaround. Each type of vehicle has its own characteristic turning radius, which can be used to model the most desirable arrangement on a cul-de-sac. The cul-de-sac is laid out so that the design vehicle can move through the turnaround without backing up (see Table 5.17).

The design concerns on residential cul-de-sacs are how to balance free movement of vehicles and how to minimize the amount of paved surface. The recommended design vehicle for a cul-de-sac should be determined by the kind of project. An industrial park would use a large commercial vehicle as a design vehicle, but a project that is designed to attract families might be designed using a school bus, and a project in an area that receives significant snowfall might be designed around snow removal equipment. In projects designed for adult residents, the choice of design vehicle might be the family car or a UPS truck.

Designing for a full-size family car will allow for small delivery vehicles and trucks to also move through the turnaround without backing up. To design for the infrequent or occasional use by larger vehicles will result in a street that is only rarely used to its capacity. This is an unnecessary commitment of resources and money as well as an unnecessary impact on the quality of the residential environment.

The access by the infrequent larger vehicle, however, must be accounted for in the design. Consideration might be given to using a stabilized turf "shoulder" immediately outside the cartway to allow for the occasional oversteer by larger vehicles (Figs. 5.27 and 5.28). Some studies have found that the stabilized turf would actually work better if it is depressed about an inch from the edge of the road surface (Burley et al.). Since only local traffic will utilize the cul-de-sac, the cartway width can be reduced to about 20 ft and still provide year-round utility and convenience. On a short cul-de-sac or a cul-de-sac serving only a few families (four to six families), a cartway can be reduced to 16 ft. In these cases parking should be evaluated carefully since on-street parking is limited.

Ordinances often require a minimum turnaround radius of 50 ft even though most vehicles require much less. These ordinances result in large, underutilized turnarounds designed for the most infrequent uses such as fire trucks and tractor trailers. The use of a full-size car would require a minimum

TABLE 5.17 Comparison of Turning Radii of Selected Types of Vehicles

Vehicle Type	Turning Radius, ft*
Small car	19.5
Standard car	22.5
Large car	23.0
School bus	43.5
Ambulance	30.0
Trash truck	32.0
Fire truck	48.0

*The outer limits of a circular cul-de-sac.

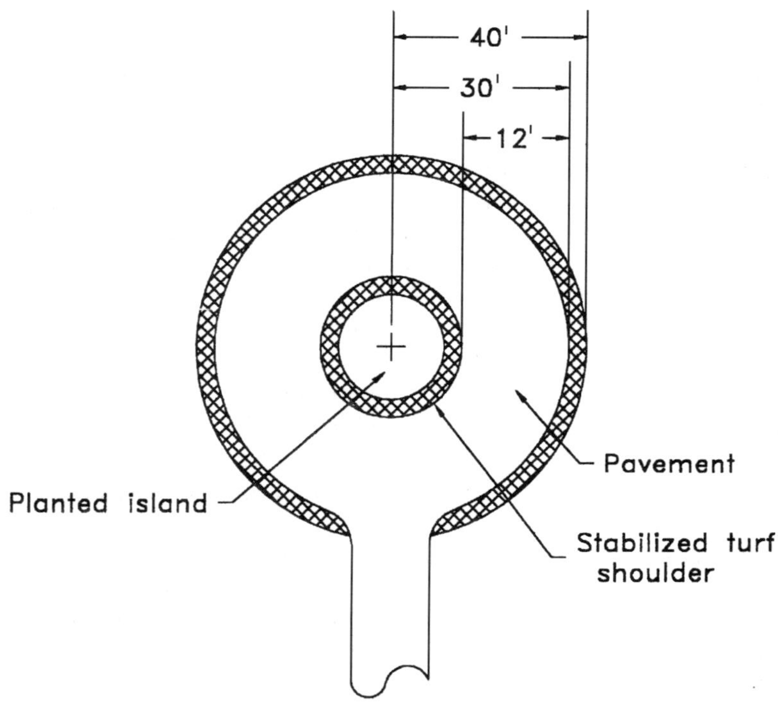

Figure 5.27 Cul-de-sac design using stabilized turf shoulder.

outside turning radius of 23 ft. This design would require larger vehicles to use at least one backing motion to move through the turnaround. The use of a slightly larger design vehicle, such as a delivery van (for example, a UPS truck) would require a 30-ft outside radius and would provide for free movement by most vehicles but would require a fire engine to make at least one backing motion.

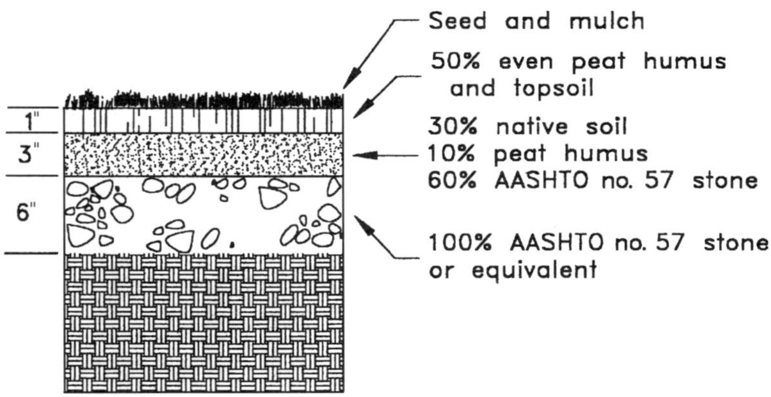

Figure 5.28 Stabilized turf shoulder.

The access of emergency vehicles is often cited as the justification of the very large and excessive turning radius requirements. In practice, however, the concern does not hold up. Fire equipment is more constrained by the presence and location of fire hydrants than it is by a restrictive turn. More importantly, in a circumstance in which an emergency vehicle is responding, it is acceptable for the vehicle to oversteer onto a stabilized shoulder or lawn or enter the cul-de-sac from the left—that is, to go the wrong way around the turn. The backing motion can be made later when the emergency is over.

Another concern in designing the turnaround is snow removal. An outside turning diameter of 32 ft is adequate for most snow removal equipment without backing up. An exception to this would be an area in which particularly heavy snowfall (>150 in/year) occurs. In these areas the additional turnaround width may be used for storage of snow and maneuvering of larger equipment. In areas where snow storage is needed, a center island is often used for this purpose. In these cases turnarounds of up to 85 ft have been used and a larger center island built. The larger radius may allow for additional or larger lots as well.

As there are for any other design feature, there are positive and negative aspects to using a cul-de-sac. The positive aspects such as security and attractiveness can be difficult to quantify, unless the popularity of homes located on "courts" is an accurate measure. In some areas homes on cul-de-sacs are valued higher and increase in value faster than similar homes on adjacent streets. The other positive aspects of the cul-de-sac include less pavement required per unit, more groundwater recharge, and usually more open space. The criticisms of cul-de-sacs are related primarily to providing for access by large vehicles, but, as discussed earlier, large vehicles can be accounted for in a well-thought-out design.

Traditional Street Design

Traditional neighborhood streets have been narrower and more pedestrian friendly than many modern subdivision ordinances will permit. Many of the most desirable communities in the United States feature relatively narrow and curvilinear street networks not found in new developments. The most common reasons given for building wide streets are adequate traffic flow, parking, and access for service or emergency vehicles; however, communities with narrower streets must have similar service requirements per dwelling unit. Parking presents challenges in communities designed and constructed before the increase in automobile use, but for the most part these problems are solved or at least accommodated by the residents. It is therefore reasonable to assume that new projects can easily be designed to address both the use of automobiles and the interests of pedestrians.

Parking Area Design

Parking lots can be massive seas of asphalt contributing to a degradation of local water quality and an increase in urban heat. In addition to the environmental consequences, a parking lot is, by function if not design, a place where people and vehicles mix fairly freely, a contest in which the vehicle is better suited. However, despite the problems with parking lots, we will probably continue to build them and use them in the foreseeable future. As we move toward incorporating principles of sustainability into site design, we need to consider the options available for making parking lots environmental friendly.

How much parking is enough?

Parking requirements are usually set by local municipalities as a ratio of so many spaces per dwelling unit, square feet of retail space, or seats in a theater. The ratio is usually based on what is thought to be the minimum number of spaces needed to accommodate the maximum amount of parking demand. This method of calculating parking requirements creates conditions in some instances in which most of the parking area is rarely used. Note in Table 5.18 the range of minimum requirements as opposed to the actual demand. It is clear that in many cases there is a requirement for parking that is far beyond the actual average demand. Parking demand for homes and industrial applications are more easily calculated than the ephemeral demands for a shopping center. (Table 5.19 gives the number of accessible parking spaces required per total parking spaces in the lot. Table 5.20 gives

TABLE 5.18 Comparison of Minimum Parking Requirements and Average Use

Use	Typical min requirements	Actual average demand
Industrial parks	0.5–2.0/1000 ft^2	1.48/1000 ft^2
Single-family homes	1.5–2.5	1.11
Convenience stores	2.0–10.0	-
Shopping centers	4.0–6.5/1000 ft^2	3.97/1000 ft^2
Medical or dental offices	4.5–10.0/1000 ft^2	4.11/1000 ft^2

TABLE 5.19 Accessible Parking Space Requirements

Total parking spaces in lot	Required accessible spaces
1–25	1
26–50	2
51–75	3
76–100	4
101–150	5
151–200	6
201–300	7
301–400	8
401–500	9
501–1000	2% total spaces
1001–over	20 plus 1 for each 100 over 1000 spaces

NOTE: Access aisles adjacent to accessible spaces should be 60 in wide minimum except that 1 of every 8 accessible spaces, but not less than 1, should be served by an access aisle 96 in wide minimum and should be designated "van accessible." This requirement is not necessary if all required accessible parking spaces conform with "universal parking design" specifications.

the number of parking spaces required for various land uses.) The response to uncertainty has usually been to presume the worse case and plan accordingly.

Planning for an average condition may have unacceptable repercussions, and in some cases the risk of lost sales because of insufficient customer parking is the most common criticism. Retail businesses tend to make much of their gross income in relatively narrow windows; most retail stores make most of their money in the weeks preceding the end-of-the-year holidays. To limit parking is seen as having a cost in the form of lost sales. In addition to the retailer's interests, commercial loans for retail development often have parking requirements that exceed the requirements of the already substantial local ordinance. The upshot is that most often the retailer's needs for parking have prevailed. This approach has worked well for retailers, but it has been shown to have undesirable environmental impacts, and those costs have been borne by communities. The net effect of this approach to planning for parking is that many parking lots are used to capacity for only a few days each year. But the impacts of the excess paved area continue every day regardless of whether the parking is used or not. Even if the shopping center is unsuccessful, the negative impacts on the community continue. Mitigating the negative impacts of parking lots is difficult to do in the context of a single land development project, unless of course the project is very large.

However, some strategies have emerged to address the impacts and issues associated with parking lots, and any reductions in the size of paved parking areas will benefit all concerned. Developers will benefit from having to provide fewer spaces because parking lots contribute significantly to the development

TABLE 5.20 Required Parking for Various Land Uses

Type of land use	No. of spaces
Residential	
Single-family homes	2.0/dwelling unit
Multifamily homes	
Efficiencies	1.0/dwelling unit
1–2 bedrooms	1.5/dwelling unit
3 or more apartments	2.0/dwelling unit
Dormitories	0.5/dwelling unit
Hotels or motels	1.0/dwelling unit
Commercial	
Offices, banks	3.0/1000 ft^2 GFA*
Businesses and professional services	3.3/1000 ft^2 GFA
Commercial recreational facilities	8.0/1000 ft2 GFA
Bowling alleys	4.0/lane
Regional shopping centers	4.5/1000 ft2 GFA
Community shopping centers	5.0/1000 ft2 GFA
Neighborhood centers	6.0/1000 ft2 GFA
Restaurants	0.3/seat
Educational	
Elementary and junior high schools	1.0/teacher and staff
High schools and colleges	1.0/2–5 students
Medical	
Medical and dental offices	1.0/200 ft^2 GFA
Hospitals	1.0/2–3 beds
Convalescent and nursing homes	1.0/3 beds
Public Buildings	
Auditoriums, theaters, stadiums	1.0/4 seats
Museums and libraries	1.0/300 ft^2 GFA
Public utilities and offices	1.0/2 employees
Recreation	
Beaches	1.0/1000 ft^2
Swimming pools	1.0/30 ft^2
Athletic fields and courts	1/3000 ft^2
Golf courses	1/acre
Industrial	
Industrial manufacturing	1.0/2–5 employees
Churches	
Churches	1.0/4 seats

*GFA is gross floor area.

SOURCE: Charles W. Harris and Nicholas T. Dines, *Time-Saver Standards for Landscape Architecture* (New York: McGraw-Hill, 1988), pp. 210–225. Used with permission of McGraw-Hill.

cost of a site. Increased runoff from additional spaces will also increase the size, and thus the cost, of storm water management facilities. In addition, the additional spaces and storm water management facilities require the use of expensive commercially zoned land. Costs of parking spaces range from $1200 to $1500 per space, which make them a considerable factor in the cost of development (Markowitz 1995). Some communities have revised their ordinances as a solution to their problems and the developer's problems. Instead of requiring a minimum number of spaces, these revised ordinances have a requirement for a maximum number of spaces. Developers wishing to have more spaces must demonstrate a need for them during the land development approval process and must address how to offset the expected environmental impacts. In this way parking is developed that will meet the needs of the store operator as well as those of the community.

Shared parking arrangements are also becoming more widely used and accepted. In shared parking arrangements, land uses with complementary parking demands cooperate to lower costs and derive the maximum use from a facility. Businesses with daytime peak demand may cooperate with a business that has night-time peak demand. Communities seeking to reduce the environmental impacts of parking may elect to provide incentives in the ordinance for such arrangements.

Design strategies are directed toward minimizing the amount of impervious surface and maintaining the predeveloped rate of infiltration. These objectives are most often accomplished in three ways: (1) minimizing the overall parking space area, (2) requiring smaller spaces dedicated to compact cars, and (3) designing spillover parking with pervious surfaces. Figures 5.29 through 5.38

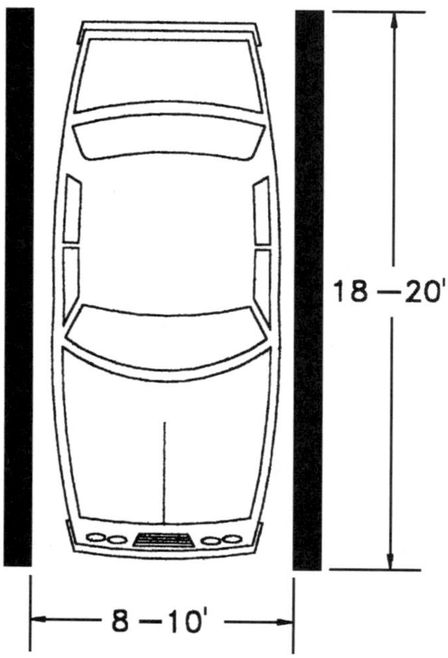

Figure 5.29 Typical parking space detail.

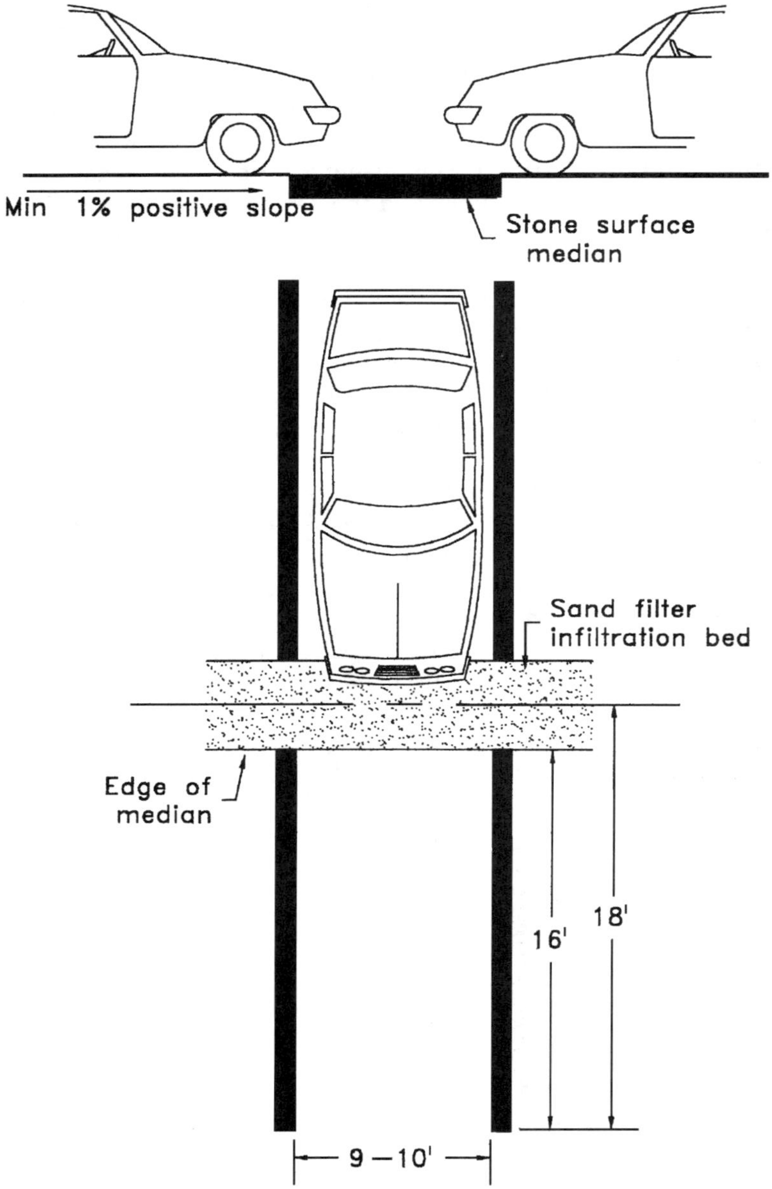

Figure 5.30 Revised parking space detail.

provide parking space specifications. Arguments against smaller spaces are usually centered on the recent proliferation of larger cars, minivans, and sports utility vehicles. In fact, however, except for the largest SUVs, none of the larger cars have a footprint larger than a car. Even the larger vehicles are not more than 7 ft wide, and many are actually smaller than a full-size car. As SUVs and light trucks are required to meet the same fuel efficiency and pollution

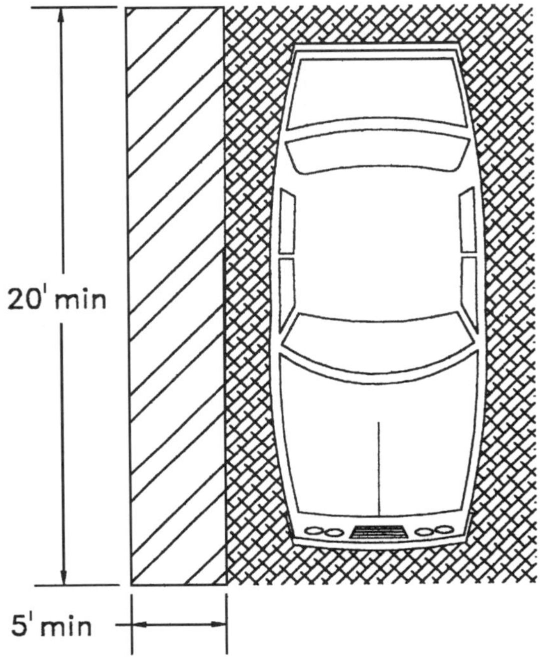

Figure 5.31 Accessible parking space detail 1.

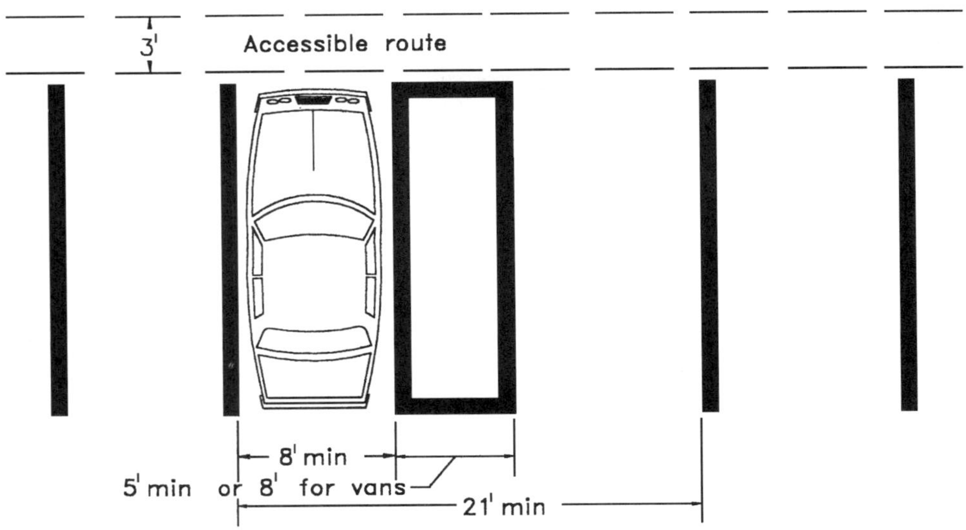

Figure 5.32 Accessible parking space detail 2.

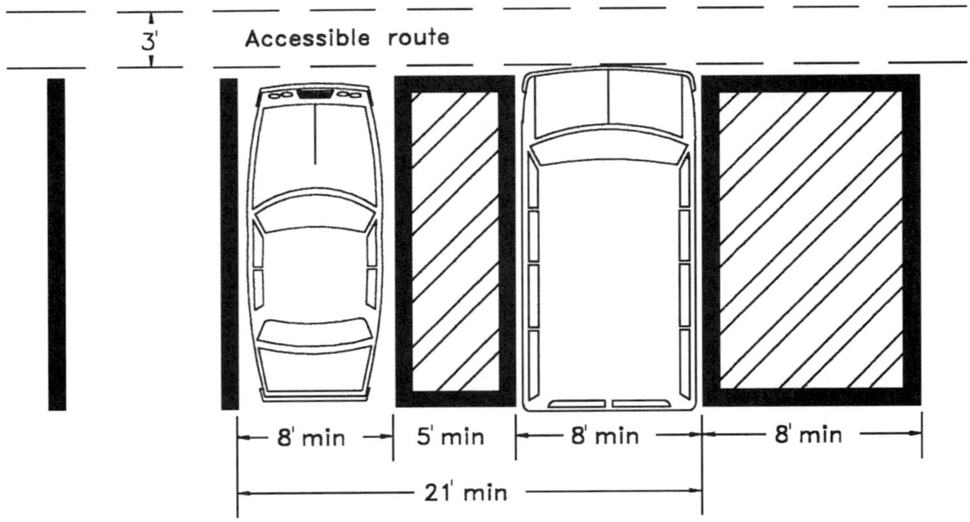

Figure 5.33 Accessible parking space detail 3.

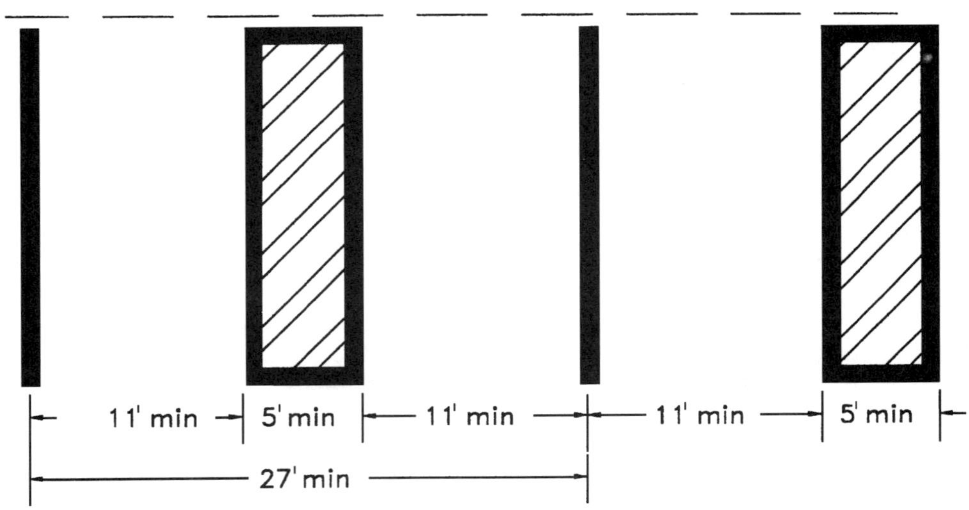

Figure 5.34 Universal parking space detail.

requirements of automobiles over the next few years, we may expect to see some moderation in their size.

Way finding

Finding the way through a parking lot can be a challenge. To help people navigate within a parking lot, designs should include subtle way-finding aids as well as directional signs. Way-finding cues may be of particular importance in

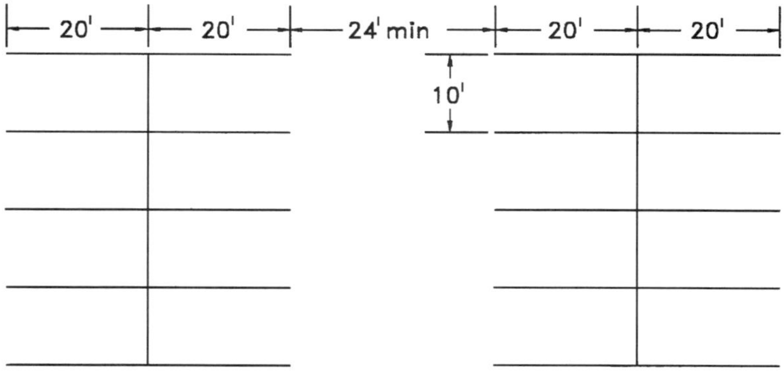

Figure 5.35 90° parking, two-way aisles detail.

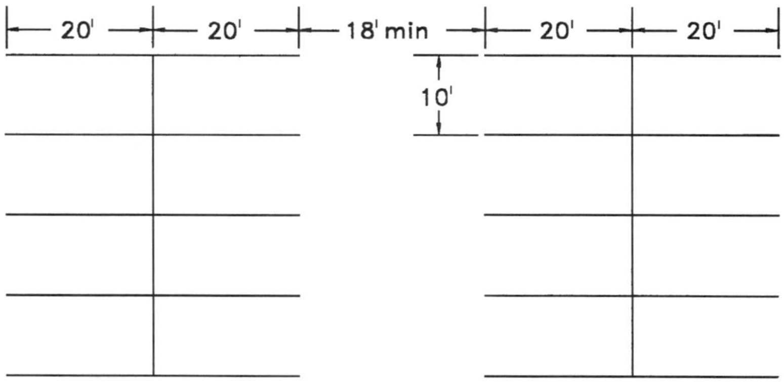

Figure 5.36 90° parking detail, one-way aisle.

very large parking areas or in areas that are used often at night or in parking lots with unique or tortuous layouts dictated by site constraints (Figs. 5.39 through 5.42).

Pavement design

The choice of material for paving surfaces is usually dictated by local ordinances, and there are two types in general use: *asphalt,* which is a dark bituminous concrete, or *portland cement,* which is a clay and limestone (or a substance similar to limestone) concrete. The functional difference between the choices lies in how load is transferred through the material to the subbase. If heavier loads from trucks or buses are expected, a more substantial pavement is warranted. In every case the conditions of the subbase are critical as well. The pavement design must take into account local soil and geotechnical conditions. In general, rigid pavements spread the load over a larger area than flexible pavements. Designs with a thick supporting subbase will also provide significant support. Designers will select a combination of subbase and pavement type to support the expected

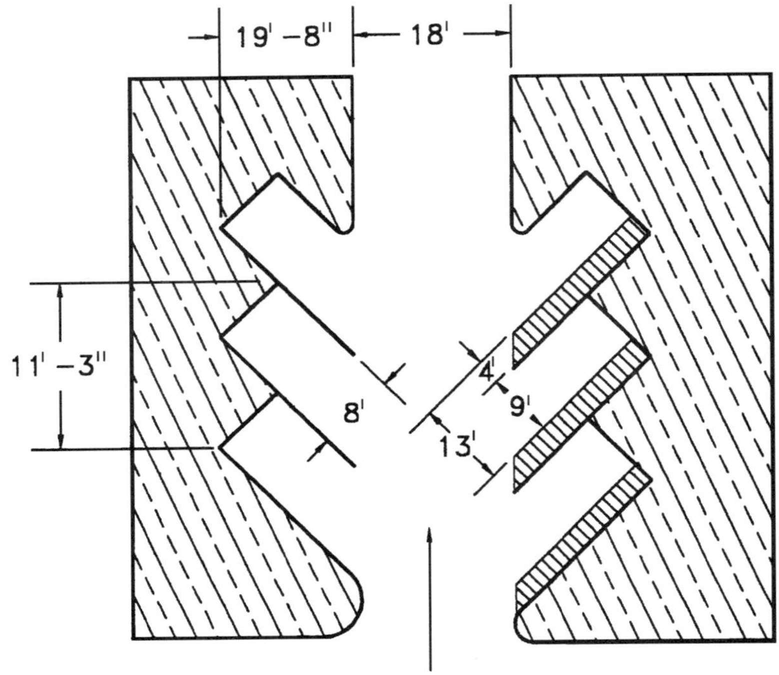

Figure 5.37 45° parking detail, one-way aisle.

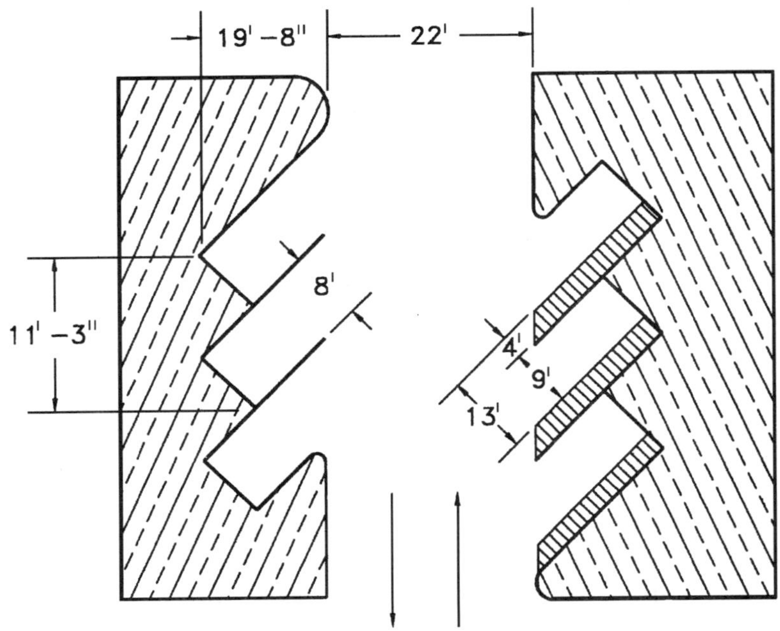

Figure 5.38 45° parking detail, two-way aisle.

Figure 5.39 Photograph of parking lot signs and designated paths.

traffic loads on the subgrade conditions found at a site. The design of pavement for most streets is also specified by local ordinances, which prescribe required pavement thickness and materials, and cross section. *Cross section,* in this context, refers to a standard or required use of materials of a certain thickness and arrangement. Designers should review local standards prior to designing paved areas. Pavement thickness should be based on or consistent with AASHTO or appropriate ASTM standards in addition to local specifications (see Table 5.21). Figures 5.43 through 5.47 provide diagrams of pavement structures.

Porous paving to reduce runoff

Table 5.22 summarizes the issues related to the various types of alternative pavement materials. The increase in runoff from paved surfaces, and the

Figure 5.40 Photograph of parking lot paving details.

necessary construction costs and land used for detention basins to offset the increase, can be reduced through the use of porous paving. Porous paving allows runoff to pass through the paved surface and infiltrate back into the soil and into the groundwater. Studies have found that the permeability of some porous pavements can be as high as 56 in/h [Bay Area Stormwater Management Agencies Association (BASMAA) 1997]. A great deal of work and study have gone into developing porous paving systems and methods, but the use of this approach is still very localized. Objections to porous paving are often unfounded, but the "facts" can be hard to find and the objections go unchallenged.

Porous paving is installed and constructed using the same equipment as typical bituminous concrete paving. As a result of the studies and the experience of the pioneers in this field, only a practiced eye would be able to detect the difference between porous paving and the typical impervious paving. The concern that porous paving leaves voids that could catch a narrow heel or cane is unfounded; the voids are too small. The advantages to using porous paving designs include a recharge of groundwater, a reduction in the amount of particulate from runoff into streams and ponds, the preservation of open space, an improved site appearance, and the reduction or elimination of land dedicated to surface storm water facilities. Objections to the porous paving approach usually include a concern that the pavement will become clogged and no longer function. Studies show that properly constructed surfaces do not

Figure 5.41 Photograph of markings in parking lots.

clog. Early experience with clogged surfaces was traced to the construction and improper design of the paved area. Some designs incorporate an edge drain system in case clogging does occur many years in the future.

The structural integrity of porous paving is often criticized as well, and it must be acknowledged that the materials and methods do have limitations. The porous paving designs do not hold up well under truck traffic or heavy loads. These concerns can be addressed by limiting the use of the porous paving to parking areas or roads designated for automobile and light truck traffic only.

Maintenance of the porous surface is limited. Some experience with porous systems has shown that in some cases the use of ice-melting chemicals and snow plowing can be reduced because the underlying stone retains heat so that ice and snow melt away.

Figure 5.42 Photograph of planted islands in a parking lot.

TABLE 5.21 Asphalt Pavement Thicknesses for Parking Areas, Passenger Cars

Subgrade	Bituminous surface, in	Bar course, in	Subbase, in
Gravelly or sandy soils, well drained	1–3	4	—
Fine-grained soils, slight to nonplastic	2–3	4	—
Fine-grained soils, plastic	2–3	4	4

NOTE: The base course is assumed to have a CBR of at least 70. The subbase or bank run aggregate is assumed to have a CBR of 40–60. Also note that a 5-in concrete slab would accommodate a wheel load of up to 6000 lb on all three soils. In fine-grained plastic soils subject to frost, a granular base of 4 in would be required for drainage. In addition, note that a 4-in concrete slab would accommodate a wheel load of up to 4000 lb on all three soils and would require a granular base in plastic soils.

SOURCE: Charles W. Harris and Nicholas T. Dines, *Time-Saver Standards for Landscape Architecture* (New York: McGraw-Hill, 1988), pp. 440–443. Used with permission of McGraw-Hill.

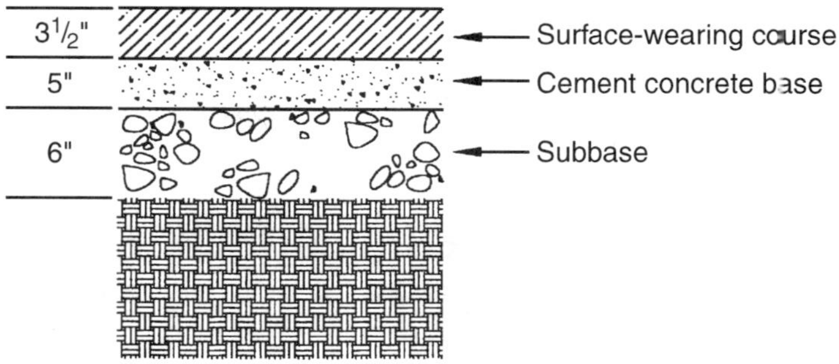

Figure 5.43 Typical parking lot paving detail A.

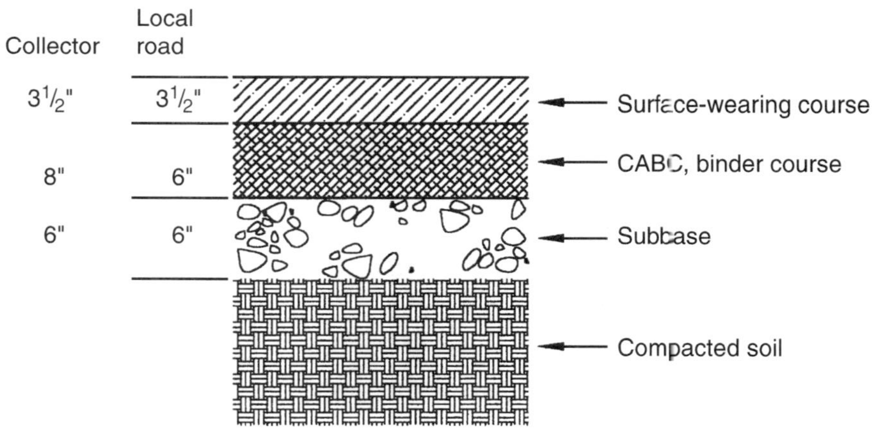

Figure 5.44 Typical parking lot paving detail B.

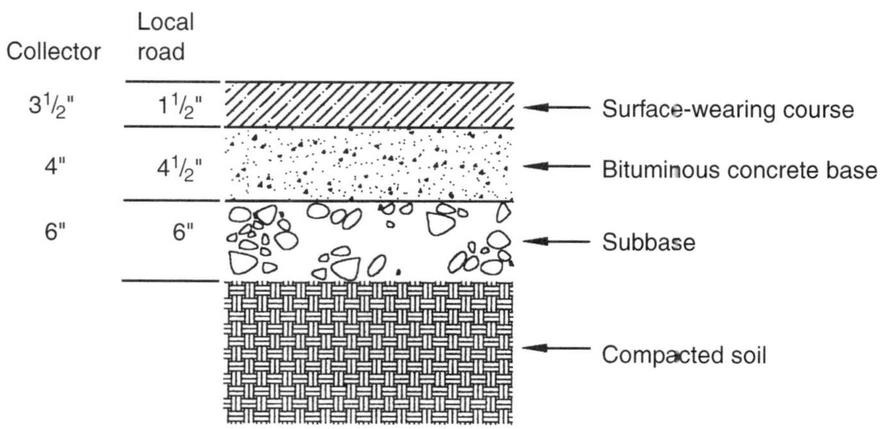

Figure 5.45 Typical parking lot paving detail C.

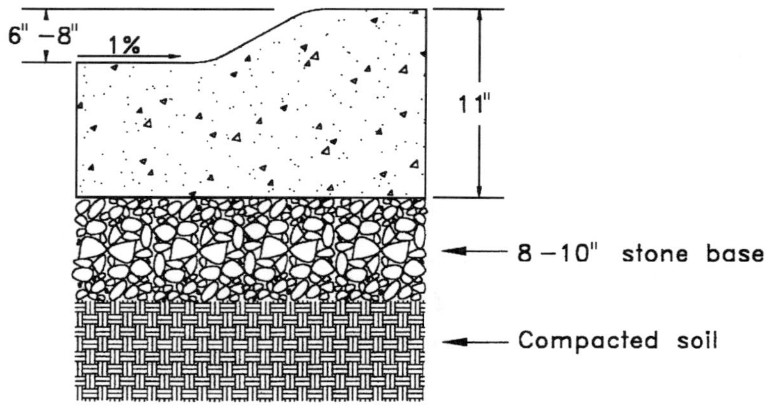

Figure 5.46 Mountable curb detail.

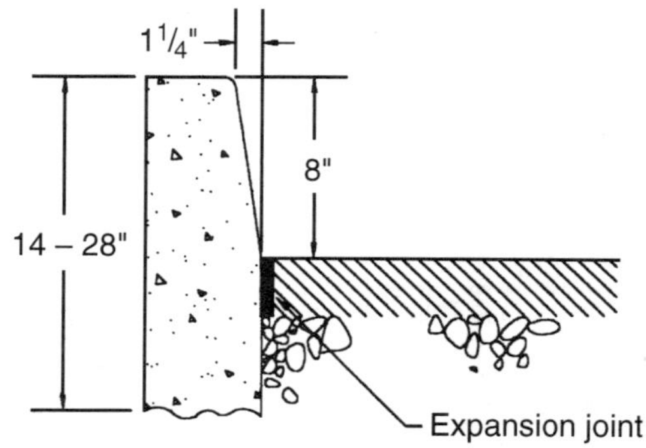

Figure 5.47 Curb detail.

The cost of porous paving must be considered on a project-by-project basis. In comparing paving costs, the overriding factor is the cost of the aggregate. Any higher costs, however, for porous paving need to be balanced against the costs reduced or eliminated by not having to construct a basin and by the additional land area available to the project. The higher the land cost, the greater the feasibility of the porous paving approach.

Reducing the negative environmental impacts of parking lots

The negative environmental impacts of paved parking areas are substantial. Generally speaking, the impacts of paved parking areas are the same as they are for streets. Increased storm water runoff, excessive heat retention, and high risks to pedestrians are found in degrees at least equal to if not greater than those on streets. There is a number of strategies that can reduce these impacts, and these strategies should be incorporated into the design. For example, parking lots should have ample shading, ways to intercept runoff, and breaks in the long expanses of pavement.

By building islands across the slope (parallel to the contours), storm water can be intercepted and redirected into the soil of a planted island. Planting the

TABLE 5.22 Summary of Issues Related to Various Types of Pavement Materials

Material	Initial cost	Maintenance cost	Water quality effectiveness*
Conventional asphalt and concrete	Medium	Low	Low
Pervious concrete	High	High	High
Porous asphalt	High	High	High
Turf block	Medium	High	High
Brick	High	Medium	Medium
Natural stone	High	Medium	Medium
Concrete unit pavers	Medium	Medum	Medium
Gravel	Low	Medium	High
Wood mulch	Low	Medium	High
Cobbles	Low	Medium	Medium

*Relative effectiveness in meeting storm water quality goals.

SOURCE: Center for Watershed Protection, Site Planning Roundtable, *Better Site Design: A Handbook for Changing Development Rules in Your Community* (Site Planning Roundtable, 1998), p. 77.

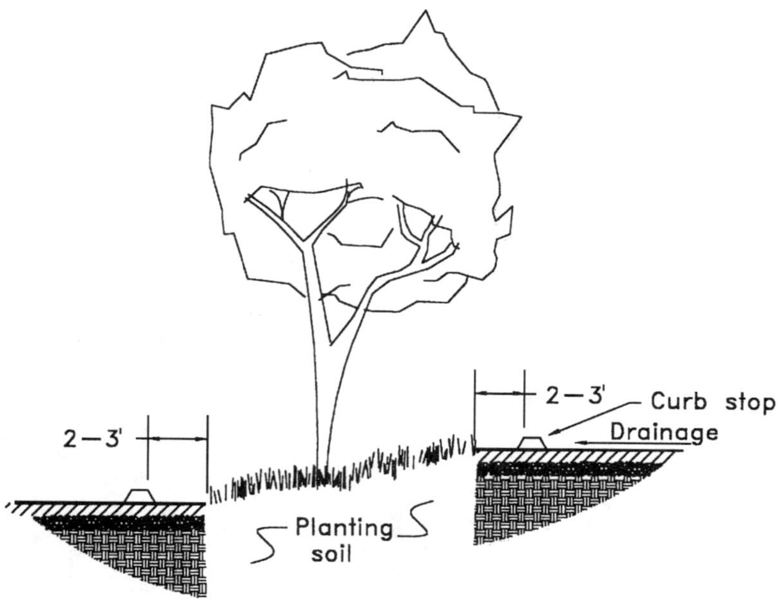

Figure 5.48 Planted islands in parking lot detail.

islands with trees and ground covers will help to shade and cool the parking lots and reduce maintenance costs. The planted island also helps break up the paving visually by making it a more appealing site. A rule of thumb for determining the minimum amount of planted surface areas in a parking lot is to allow about 5 percent of the parking area to be planted (Figs. 5.48 through 5.51).

In areas where deicing chemicals are used, consideration might be given to protecting the plantings with stone mulch or other material if the deicing mate-

Figure 5.49 Photograph of planted islands.

Figure 5.50 Photograph of planted landscaped buffer.

Figure 5.51 Photograph of berm and trees.

rials will harm the plants. The use of landscape fabric will contribute to keeping maintenance costs down where stone or other mulches are used. The landscape fabrics are woven and so allow water and air to pass through the material, but they block light and therefore reduce weed growth. Islands should be designed for snow storage in those areas where large snowfalls are common. In such areas it may be appropriate to space the islands out and to make them deeper or wider to allow for greater efficiency in plowing and snow removal.

Chapter

6

Infrastructure

Storm Water Management

Storm water runoff—rainwater that simply rolls off and away from pavement and buildings rather than seeping into the ground below them—has become the focus of many modern water protection efforts. As storm water runoff moves across the top of the earth's surface rather than penetrating the surface immediately upon falling, it acts according to the laws of nature and seeks the lowest point, where it naturally meets other rainwater and quickly forms rivulets and streams. This newly created waterway system collects and concentrates contaminants on its way through developed areas. When the runoff finally enters the groundwater, it is often very polluted. Of course, the more built up the area, the more polluted is the runoff it generates. National and local environmental protection policies are beginning to reflect the public concern about this problem, insisting that storm water runoff be channeled safely throughout the life cycle of a development project.

To prevent water pollution, site designers must now address storm water management practices not only for the finished development but for the construction site as well, from the time the site is first disturbed. As site work begins and progresses and the natural characteristics and irregularities of the site are graded and removed, the volume and velocity of storm water runoff increase. Once the vegetation is gone and the soil disturbed, rainfall and runoff are no longer deflected off the surfaces of the vegetation, and they are unable to infiltrate into the compacted soil. So the water runs off the surface to collect in the lower areas of the site. In the recent past to keep the pooled runoff from compromising the site work, rainwater was concentrated into pipes and conveyed to the most convenient point of discharge.

The increase in impervious area that results from development has important consequences for environmental quality. Studies have shown consistently that habitat quality in streams is poor once a watershed reaches from 10 to 15 percent imperviousness (Booth and Reinelt 1993). To prevent such damage, the methods and practices of design and construction must evolve to address the causes of it. Storm water management is a critical element in this evolution toward sustainable site design practices.

Until recently, storm water has been viewed primarily as a problem, to be collected and disposed of as quickly and as efficiently as possible. Today, with growing concern for our environment and recognition of the need for more sustainable development methods, simple collection, conveyance, and discharge strategies have become less acceptable. Falling groundwater tables, dry streams, and degraded surface quality have convinced us that storm water must begin to be seen differently. As our understanding of the environment and sustainability has improved, we have realized that storm water is an important resource.

As the runoff moves across the developed surfaces of lawn and pavement, it washes particles and pollutants into the system and ultimately into the groundwater. Pollutants from such sites include nutrients, sediment, bacteria, oil and grease, heavy metals, chemicals, and pesticides. These are known as *nonpoint source pollutants* (NSPs) because they do not originate from a single pipe or discharge point. Nonpoint source pollutants are regulated under the National Pollution Discharge Elimination System (NPDES) as prescribed by the Clean Water Act. Under federal regulations construction sites over an acre in size must address discharge from a site, and large developments and municipalities must acquire an NPDES permit to discharge storm water. For the most part, these regulations are enforced by the individual states.

In many urban environments the pollution from storm water runoff from parking lots and streets is much greater than the pollution from factories and sewage treatment plants. Storm water that is directed across paved surfaces and collected into gutters and pipes conveys runoff at a velocity that scours the surface and washes the pollutants along with it. The most obvious method for reducing the negative impact of development on water quality is straightforward: If the amount of paving and roof surface is reduced, the amount of increased runoff is reduced. The problem with this solution is obvious; without paving and roof, and without being able to discharge at least some of the storm water, there probably is no project. In this case the environment is preserved, but there is no construction or building. Alternatively, the more water is retained on the site, the less negative impact there will be on water quality from pollutants. Development and water quality solutions are a matter of effective design. New development can be designed, and even existing development can be refit, to slow runoff velocities and volumes and to encourage infiltration. Instead of addressing storm water as a problem, it is becoming more important and necessary to see it as a resource. Thus development and environmental protection can coexist as long as sites are designed using effective and sustainable strategies. Tables 6.1 and 6.2 compare the effectiveness of various storm water management strategies. Looking at the tables, it is clear that many of the effective strategies are very familiar and cost effective.

TABLE 6.1 Pollutant Removal Effectiveness of Storm Water Management Practices for Parking Lots

Management practices	Suspended solids, %	Phosphorus, %	Nitrogen, %	Metals, %
Dry swales	91	67	92	80–90
Grass channels	65	25	15	20–50
Roadside ditches	30	10	0	
Sand filters	85	55	35	Lead 60
Filter strips	70	10	30	40–50

NOTE: Bioretention Facilities assumed to be the same as dry swales.
SOURCE: Adapted from the Center for Watershed Protection, *Better Site Design: A Handbook for Changing Development Rules in Your Community,* prepared for the Site Planning Roundtable, Ellicott City, MD, 1998.

TABLE 6.2 Comparison of Costs for Storm Water Management Facilities

Practice	Construction cost	Annual O&M	Useful life, years
Infiltration trenches	$0.20–$1.20/ft^3	3–13% of capital cost	25
Vegetated swales	$4.5–$8.5 per linear foot	$0.50–$1.0 per linear foot	50
Vegetative	Existing vegetation $50.00–$200.00 acre	-	50
filter strips	From seed, $200.00–$1000.00/acre	$800 per acre	50
	From seed with mulch, $800.00–$3500.00/acre	-	50
	From sod, $4500.00–$48,000.00/acre	-	50
Sand filters	$1.00–$11.00/ft^3	7% of construction cost	25
Wet ponds	$0.50–$1.00/ft^3	0.1%–1% of capital cost	50
Bioswales	Not available	Not available	Not available

SOURCE: Adapted from U.S. Environmental Protection Agency (EPA), EPA-840-B-92-002, January 1993.

Estimating peak runoff with the Rational Method

The *Rational Method* is a common approach to calculating peak discharge. The primary strengths of the Rational Method lie in its relative simplicity and accuracy when applied to watersheds or relatively small basins. It can easily be adapted for watersheds that divide into smaller subsheds and have different surface characteristics. The Rational Method uses the area of the basin, a runoff coefficient, and the intensity of a selected design storm to determine the peak discharge:

$$Q = CiA$$

where Q = peak discharge, ft^3/s
C = runoff coefficient (a ratio of the amount of surface runoff to rainfall)
i = rainfall intensity for a storm duration equal to the time of concentration
A = area of the basin (or subshed)

In general, the Rational Method proceeds in the following order: (1) Determine the time of concentration in order to determine (2) I, the rainfall intensity. (3) Determine a runoff coefficient and (4) the area of the drainage area. The time of concentration T_C is the length of time required for a drop of water to travel from the farthest hydrologic point through a completely saturated drainage area to a point of discharge. The T_C is a function of the slope of the land and the surface roughness and whether the flow occurs in sheet or concentrated form. To determine T_C, determine the longest hydrologic flow path; if appropriate, subdivide the path into sections of different conditions (forest, paved surfaces, lawns, and so on). Calculate the velocity along the path using the formula

$$t = L/v$$

where t = travel time, h
L = length of the flow path, ft
v = velocity for Manning's formula or from a prepared chart

Travel time for sheet flow is calculated as follows:

$$t = \frac{0.007\,(nL)^{0.8}}{(P)^{0.5}s^{0.4}}$$

where t = travel time, h
n = Manning's coefficient of roughness
L = flow length, ft
P = design storm, 24-h rainfall, in
s = average slope of surface

Design storms are selected on a project-by-project basis, and the 24-h storm rainfall will differ from place to place. Up-to-date rainfall and rainfall-intensity charts for specific locations are usually available from state transportation agencies and local planning agencies. It should be noted that rainfall-intensity charts are not the same as total precipitation or storm duration charts. *Rainfall intensity* specifically refers to a storm with a duration equal to the Rational Method Time of Concentration, or storms with a duration of less than one hour. Rainfall intensity is a regional calculation determined using the Steel formula:

$$i = \frac{K}{t + b}$$

where i = rainfall intensity
k = rainfall coefficient
b = rainfall coefficient
t = travel time

The rainfall coefficients k and b are statistical constants developed by region for storms of different frequencies. In recent years the Steel formula rainfall coefficients have been shown to be somewhat less reliable in western states (see Fig. 6.1 and Table 6.3).

A runoff coefficient C is selected from a chart like the one shown in Table 6.4 based on the conditions found at the site. Composite or weighted C values are calculated by multiplying the area of each type of ground cover by the appropriate C factor, adding up all of the results, and dividing the sum by the total land area A. Other pertinent formulas are given in Table 6.5.

In general, the designs of open channels and pipes are similar in that both may be determined using Manning's formula:

$$Q = \frac{(1.49/n)\,(a/p)^{2/3}\,(s)^{2/2}}{a}$$

The essential difference between them is that while open, the hydraulic properties of open channels continue to increase as the channel fills; pipes reach their greatest discharge at about 93 percent of total depth. (See also Table 6.6 and Figs. 6.2 and 6.3.)

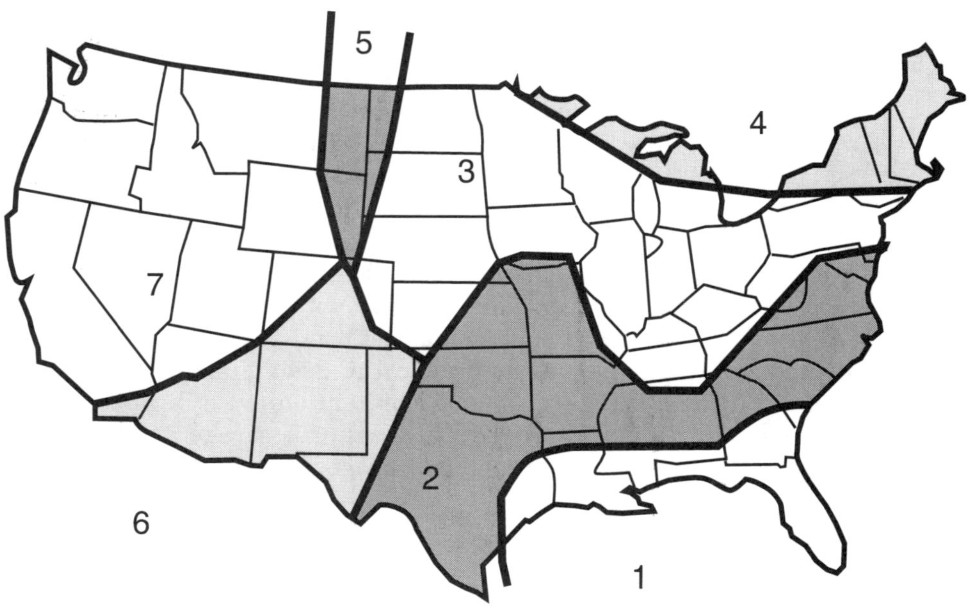

Figure 6.1 Map of rainfall regions for Steel formula.

TABLE 6.3 Values of k and b for the Steel Formula

		Region						
Design storm	Rainfall constant	1	2	3	4	5	6	7
2	*k*	206	140	106	70	70	68	32
	b	30	21	17	13	16	14	11
5	*k*	247	190	131	97	81	75	48
	b	29	25	19	16	13	12	12
10	*k*	300	230	170	111	111	122	60
	b	36	29	23	16	17	23	13
25	*k*	327	260	230	170	130	155	67
	b	33	32	30	27	17	26	10
50	*k*	315	350	250	187	187	160	65
	b	28	38	27	24	25	21	8
100	*k*	367	375	290	220	240	210	77
	b	33	36	31	28	29	26	10

SOURCE: From Harlow C. Landphair and Fred Klatt, Jr., *Landscape Architecture Construction,* 2d ed. (New York: Elsevier Science Publishing, 1988).

Strategies for storm water management in arid areas

Much of the United States receives less than 35 in of rain each year, some even less than 15 in. Strategies for storm water management are appreciably different for these areas. Even though rainfall depths are much smaller, arid (less than 15 in of rain) and semi-arid (from 15 to 35 in of rain) areas have a much greater pollutant load for each storm event and may experience substantially greater sediment loads due to the lack of stabilizing vegetation. The drier areas are also more concerned with groundwater quality due to the high pollution loads and permeability of some western soils. Therefore, pollution prevention is a critical part of storm water runoff control strategies. Such strategies include street sweeping and drain cleaning.

Many of the strategies and practices recommended for areas where precipitation exceeds evaporation rates are not practical for drier regions where dry ponds are favored. For example, wet ponds are desirable in areas with surplus moisture, but they are impractical in drier climates. Other practices such as sand filters, filter strips, and bioretention are still important tools. Key elements in storm water design in these areas are to minimize groundwater pollution, channel erosion, and encourage infiltration (Fig. 6.4).

Swales

Vegetated swales are important tools in implementing a sustainable storm water management strategy (Table 6.7). Unlike pipes, vegetated swales

TABLE 6.4 Rational Method Runoff Coefficients

Type of Terrain	Steep, >7%	Rolling, 2–7%	Flat, 0.2%
Wooded			
Heavily	.21	.18	.15
Moderately	.25	.21	.18
Lightly	.29	.25	.21
Lawns	.35	.30	.26
Uncompacted bare soil	.60	.60	.50
Impervious	.98	.95	.95
Residential			
25,000-ft^2 lots	.40	.36	.32
15,000-ft^2 lots	.50	.45	.40
12,000-ft^2 lots	.50	.45	.40
Townhomes (45% impervious)	.65	.60	.55
Apartments (75% impervious)	.82	.79	.74
Pasture			
Good condition	.25	.21	.18
Average condition	.45	.40	.36
Poor condition	.55	.50	.45
Farmland			
Nongrowing season	.50	.46	.41

encourage infiltration, act to filter the water by providing many surfaces for deposition, and reduce the velocity of water. In addition, using a swale infiltration system increases the infiltration capacity of the typical swale to allow infiltration of low-flow, frequent storm events while providing the capacity to convey runoff from large infrequent storms that cannot be easily infiltrated through the soil. Swales, rather than pipes, are the choice of conveyance systems. One advantage is that by using grass-lined swales instead of pipes, site costs can be reduced by as much as $12.00 a linear foot. Adding a swale infiltration trap, the cost of the swale may be increased $2.30 a linear foot, but because downstream piping and swale systems are smaller, a net savings is possible. Swales are generally calculated to function with at least 20 percent of the depth as freeboard. This allows for some retardance in the design flow due to vegetation or debris.

The design of vegetated swales usually assumes an acceptable velocity of water, which for a turf channel is from 2 to 4 ft/s. The design of channels is accomplished using Manning's equation. The choice of the coefficient of roughness is the critical step in the use of the equation. Generally a freeboard of at least 20 percent of the design depth or 6 in (15 cm) is used to protect against underestimates of the n value and roughening of the channel by vegetative

TABLE 6.5 Formulas for Storm Water Calculations

Manning's Formula

$$V = \left(\frac{1.49}{n}\right)\left(\frac{a}{p}\right)^{2/3}(s)^{1/2}$$

where n = coefficient of roughness
a = area cross section, ft^2
p = wetted perimeter, ft
s = slope, %
V = velocity, ft/s

Flow through a Grate

$$Q = 0.66CA(64.4h)^{0.5}$$

where Q = discharge, ft^2
C = orifice coefficient (0.06 of square edge opening, 0.8 for round edge opening)
A = area of opening, ft^2
h = depth of flow over opening, ft

Volume of Flow in a Channel

$$Q = \frac{V}{a}$$

where V = velocity, ft/2
Q = discharge rate, ft^3/s
a = cross-sectional area of the channel, ft^2

Emergency Spillway

$$Q = CLH^{3/2}$$

where Q = discharge over spillway, ft^3/s
C = coefficient for spillway surface (3.1 for grass)
H = height over invert of spillway, ft
L = length of spillway, ft

Water Flow through Stone Medium

Control of the water through stone medium is a function of slope and medium size. The rate of travel through the medium may be determined by:

$$Q = 0.04\left(\frac{h^{1.57}}{L^{0.57}}\right)W$$

where Q = volume of water, ft^3/s
h = depth of water, ft
L = flow path length, ft
W = width of channel, ft
0.4 = hydraulic conductivity, K

SOURCE: Gert Aron and Charles McIntyre, "Permeability of Gabions Used as Outlet Control Structures in the Design of Detention Basins" (Pennsylvania State University, 1990).

TABLE 6.6 Coefficient of Roughness

Very smooth, like glass or plastic	.010
Smooth pipe (PVC, concrete, vit. Clay, etc.)	.013–.015
Concrete pipe 24 in and under	.013
Concrete pipe over 24 in	.012
Galvanized corrugated pipe	.024
Straight unlined earth channels in good condition	.020
Open channels lined with asphalt	.013–.017
Open channels lined with brick	.012–.018
Open channels lined with concrete	.011–.02
Open channels lined with rip rap	.02–.035
Channels lined with vegetation, 11–12 in	.09–.15
Channels lined with vegetation, 6–10 in	.055–.08
Channels lined with vegetation, 2–3 in	.045–.06
Natural channels,* regular section	.03–.5
Natural channels, dense vegetation	.05–.7
Natural channels, irregular with pools	.04–.10
Rivers, some growth	.025
Winding natural streams in poor† condition	.035
Mountain streams with rocky beds, some vegetation along banks	.040–.050

*Refers to minor streams with a top width of less than 100 ft at flood stage.
†Refers to very rough condition, erosion, and so on.
SOURCE: Data from Chow 1987, Brewer and Alter 1988, Ferguson and Debo 1990.

growth and incidental obstruction or debris. Design of unlined channels is usually limited by either the velocity a channel will allow without suffering damage or the tractive force. Trials of channel design should be tested using the formula $V = Q/a$ where velocity is a function of the quantity in cubic feet per second over the wetted area.

In cases in which velocity may exceed the recommended rates, it may be necessary to reinforce the channel using geotextile fabric (Fig. 6.5). A variety of companies manufacture geotextile fabrics for an even greater variety of applications. Permanent fabrics designed to reinforce vegetated channels extend the designer's choices significantly. Some geotextiles have demonstrated the ability to maintain the channel integrity and hold the vegetation in place even at velocities as great as 13 ft/s. The ASTM has developed standards for testing geotextiles to provide users with a great degree of certainty when specifying materials.

Geotextiles should be selected on the basis of their ability to resist the flow of moving water, to protect the channel surface, and to hold the vegetation in place (Table 6.8). Some geotextiles are designed to act as *armor*—that is, they

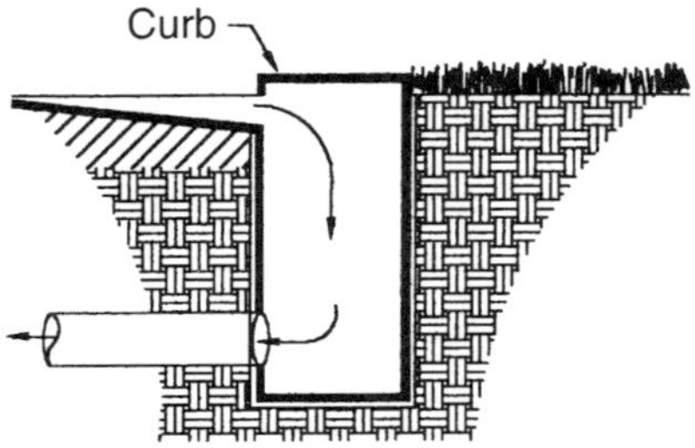

$$Q = 0.7L(a+y)^{1.5}$$

where:

Q = Inlet rate, ft^3/s
L = Length of curb opening, ft
a = Depression of curb inlet below existing gutter line, ft
y = Depth of flow of inlet, ft

Figure 6.2 Curb inlet flow calculation.

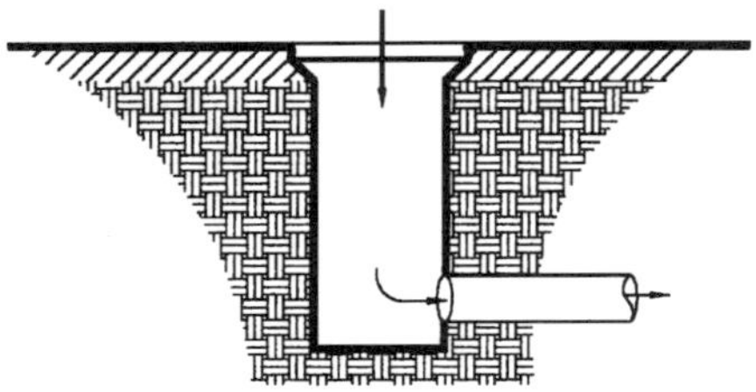

$$Q = 3Py^{1.5}$$

where:

Q = Flow through grate, ft^3/s
P = Perimeter of grate surface, no allowance for bars in grate, ft
y = Depth of flow over grate, estimate not to exceed 0.4 ft

Figure 6.3 Grate inlet flow calculation.

Figure 6.4 Photograph of infiltration bed in New Mexico.

TABLE 6.7 Advantages of Swales

Economical; costs less than pipes
Increases infiltration and groundwater recharge
Provides surfaces for deposition of particulate
Reduces runoff velocity
Easy to maintain
More natural appearance than curbs

Figure 6.5 Trapezoidal swale detail.

cover the channel surface and protect it from the erosive force of flowing water in much the same way a concrete lining might be used. Other materials are three dimensional and are incorporated into the soil to act in concert with the roots of pants to mechanically resist erosion. The permanence of the geotextile material is very important. Some materials are designed to biodegrade or photodegrade while still others are intended to remain in place permanently. In most cases swales must be installed to be in service immediately, even before the vegetation is established. This may require the use of temporary geotextiles, which would remain in place until the vegetation had become sufficient to stabilize the channel. The proper installation of the appropriate geotextile is critical. Many failures of installed swales occur because the geotextile was not installed properly.

Shape and sinuosity are critical factors in the hydraulic and environmental functions of a channel. *Shape* refers to the cross-sectional configuration of the channel, and *sinuosity* refers to the length of a channel over a given distance. Increasing the length of a channel within a given distance requires increasing the number and amplitude of curves within the distance. This increase in length allows a flatter slope over the same distance, which in turn results in slower velocities, less erosion, and more infiltration. The channel can be designed to contain more water in high-flow conditions and incorporate special high-flow channels that operate in flood conditions. The incorporation of vegetation and pools in the channels will increase the natural capacity for retention and treatment of water-borne contaminants. Careful selection of channel bottom media and plants will further enhance the environmental functions of storm water channels.

TABLE 6.8 Limiting Velocities for Channel Design

Material	n	Velocity for clear water, ft/s	Velocity for water with sediment, ft/s
Fine sand	.02	1.5	2.5
Sandy loam	.02	1.75	2.5
Silt loam	.02	2.0	3.0
Firm loam	.02	2.5	3.5
Stiff clay	.025	3.75	5.0
Shales, hardpan	.025	3.75	5.0
Fine gravel	.02	2.5	5.0
Coarse gravel	.025	4.0	6.0

The swale infiltrator may also be used to introduce biological treatment to the storm water in circumstances in which this would be desirable (Figs. 6.6 through 6.8). The media for the swale may serve as surfaces to which bacteria and other microorganisms may attach, and media may also act as biofilters of the storm water as it passes through. The velocity of water through the media must be controlled if contact time between the biological agents and the water is to be adequate. The actual time will be a function of the toxins and the ability of the specific biological agents to consume or act on it. Determination of this time factor will require some bench testing by a microbiologist. Once provided with the requirements, the site designer can design the swale accordingly.

Infiltration and Recharge

The preferred method of storm water quality management is to reintroduce the runoff into the soil as quickly as possible to provide the opportunity for groundwater recharge. Some sources point to infiltration as a method of pollutant removal, but there is in fact limited cleansing value to infiltration systems. In any case, infiltration affects or removes only the particulate matter and pollutants that might attach to soil particles. Water-soluble pollutants, such as nutrients, pesticides, or salts, will travel through the soil medium because they are dissolved in the water. In circumstances in which such water-soluble pollutants are a particular risk, the design must provide for biological treatment such as algae treatment in wet ponds or microorganisms in wetlands or in bioretention beds (also called *rain gardens*).

Another good reason to consider infiltration is the loss of groundwater recharge that accompanies a typical detention basin development. The Chester County Planning Commission in Pennsylvania has created some conceptual models of development for its planning purposes. A study they commissioned found that the typical developed square mile in the study area lost about 10 in of recharge water (storm water runoff) each year. This 10 in would

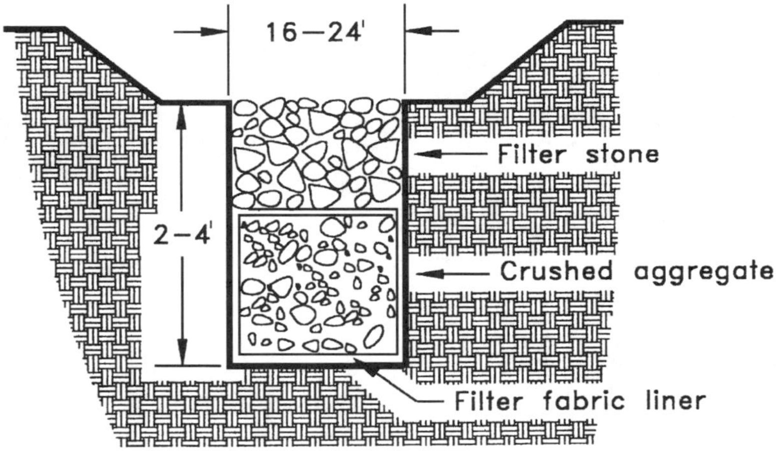

Figure 6.6 Swale infiltrator detail.

Figure 6.7 Photograph of parking lot island infiltrator.

represent 40,785,879 gal of water each year! By using infiltration systems where it is possible, this kind of loss can be significantly reduced. About 70 percent of the homes in the United States use groundwater, so protecting and managing our groundwater resources is an important and necessary undertaking.

A significant body of knowledge exists on the use of soil as a filter medium in such situations as ground sewage disposal systems. The Environmental Protection Agency recommends a 2- to 4-ft vertical separation from the bottom

Figure 6.8 Photograph of a swale infiltrator.

of the infiltration facility to the seasonal top of the water table or bedrock. The feasibility of infiltration is determined in a four-point test (see also Table 6.9):

1. The soil texture is in a class with an infiltration rate that will permit adequate percolation of collected water through the soil.
2. There would be an available ponding or dewatering time of 3 but no more than 7 days.
3. There is adequate depth to provide a vertical depth of from 2 to 4 ft, minimum, between the infiltration bed and bedrock or the seasonal high water table.
4. The site topography (slope), nature of the soil (fill, stability), and the location of foundations, utilities, wells, and similar site features are appropriate to the proposed infiltration system. The four points were developed from EPA, 840-B-92-002 (January 1993) as well as information from several state programs and standards.

Soil texture is an important element in determining infiltration rates. Infiltration rates that are too slow will not allow the ponded water to drain within the desired time. Soils with an infiltration rate of 0.17 in/h or less or with a clay content of 30 percent or more may be unsuitable for infiltration. Infiltration feasibility may be determined based on the allowable ponding time T_p or storage time T_s. Ponding and storage times should be kept to a reasonable

TABLE 6.9 General Infiltration Properties of Soils by Texture

Texture class	Effective water capacity, *Cw*	Minimum infiltration rate, *f*, in/h	Hydrologic soil group
Sand	.35	8.27	A
Loamy sand	.31	2.41	A
Sandy loam	.25	1.02	B
Loam	.19	0.52	B
Silt loam	.17	0.27	C
Sandy clay loam	.14	0.17	C
Clay loam	.14	0.09	D
Silty clay loam	.11	0.06	D
Sandy clay	.09	0.05	D
Silty clay	.09	0.04	D
Clay	.08	0.02	D

minimum, and 72 h is frequently used for a reasonable period to drain an infiltration structure.

The depth of an infiltration structure can be determined using the ponding time and the infiltration rate of a soil, as follows:

$$d = fT_p$$

where d = maximum allowable design depth, ft
f = minimum infiltration rate, in/h
T_p = maximum allowable ponding time (surface storage) , days

The maximum depth of an infiltration trench or dry well in which the storage is within a porous medium such as stone ballast can be calculated as follows:

$$d = \frac{fT_s}{V_r}$$

where d = maximum allowable design depth, ft
f = minimum infiltration rate, in/h
T_s = maximum allowable storage within subsurface storage, ft^3
V_r = void ratio of aggregate reservoir, expressed as a percent

Construction of the infiltration system requires special care. Equipment should be kept from being driven over the bottom of the system. The weight and motion of construction vehicles compress the soil, closing its pores and limiting its infiltration capacity. The infiltration surface must be protected. Once the site is located, the infiltration area should be staked out and identified. Equipment should work from the side of the area while excavating the trench.

Methods adapted for infiltration include dry wells, swale traps, catch basin traps, infiltration trenches, and basins and rain gardens. These facilities are designed to collect and trap the storm water in the earliest stages of the runoff process. Typically these types of facilities are small in size and located throughout a site. The cumulative effect of small collection facilities offsets the increase in runoff due to the development.

Dry wells

Dry wells are small excavated pits that are backfilled with aggregate in the same manner as infiltration trenches. The primary difference between the dry well and infiltration trench is the means in which the water is collected into the system. Trenches are located parallel to the contours and extend along a certain point in the site to intercept the runoff. In contrast, dry wells are usually designed to collect runoff directly from a roof drain or outfall (Fig. 6.9).

Dry wells are used primarily to collect the runoff from small areas such as roofs or sections of roofs. In design, the dry well must meet the same tests as the infiltration basin or trench. In most applications the dry well is situated in a visible location near a structure, and the appearance of the dry well at the surface is important. A soil filter is used in most cases to make the dry well disappear from view. The soil filter consists of the top 1 ft of the dry well, which has been backfilled with top soil.

In some cases the dry well concept can be incorporated into a catch basin (Fig. 6.10). The sizing of the catch basin infiltrator can be geared toward the more frequent storms. This may allow the rest of the conveyance system to be sized differently, which will reduce construction costs. For example, if a 1-year storm can be retained and infiltrated, downstream conveyances may be designed for storms less the 1-year storm Q.

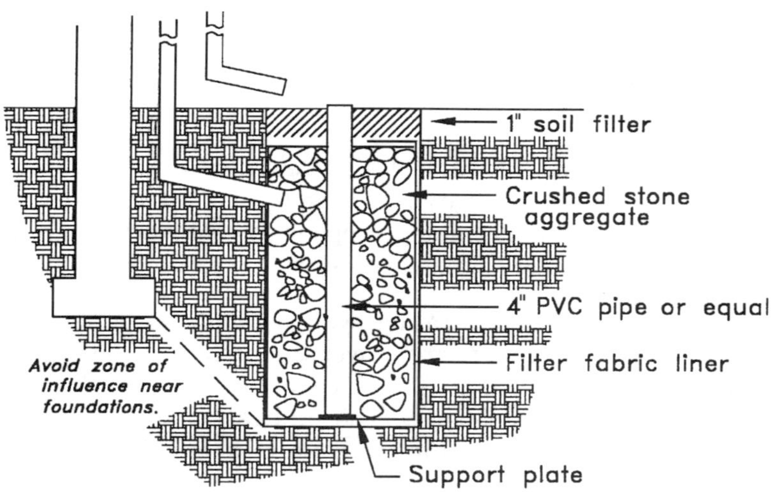

Figure 6.9 Roof drop dry well detail.

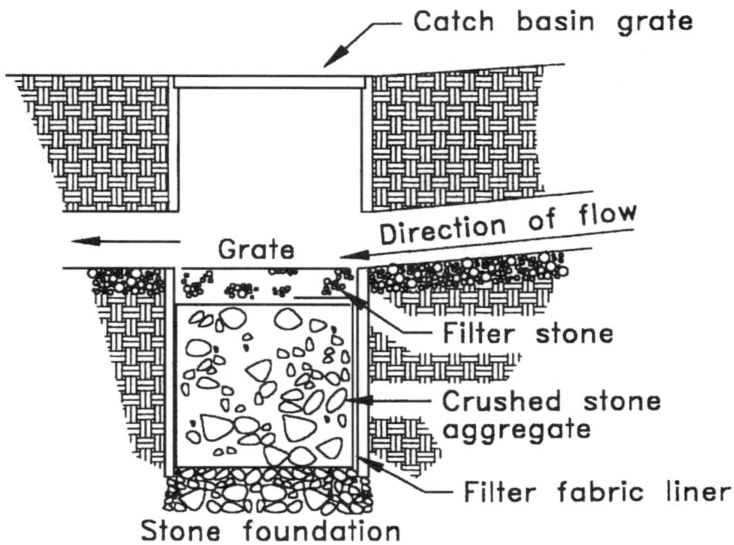

Figure 6.10 Catch basin infiltrator detail.

Filter strips

The vegetated filter strip removes particulates such as metals and phosphorus by filtration through the surfaces of the vegetation and promotes some infiltration as runoff is slowed. The surfaces of the plants also act as surfaces for the deposition of contaminants that might exist as films in the runoff such as hydrocarbons. The presence of a healthy soil medium and plant community provides some inherent microbial action on the contaminants present in the runoff. The microbial action will continue even after the surge of storm water has passed. Properly designed and constructed filter strips may have a particulate trapping efficiency of up to 95 percent (Tourbier et al. 1989). Contact time—that is, the time the water is in contact with the vegetation—should be maximized by slowing the velocity of the runoff and by designing the strip to be as wide as possible. Velocities should be no more than 1 ft/s (0.3 m/s). Filter strips should be designed with a minimum of 2 percent slope and should not exceed 4 percent (Fig. 6.11). If the slope of the filter strip is less than 2 percent, an infiltration underdrain may be required. The vegetation selected for filter strips is usually grass. The filter should be a minimum of 15 ft (4.4 m) wide. It may be necessary to use sod in order to allow the strip an opportunity to secure and establish itself. Native sod-forming grasses are recommended; tall fescue, western wheatgrass, ryegrass, and Kentucky bluegrass are all recommended for filter strips. The filter strip should be designed to receive a perpendicular sheet of runoff as a concentrated flow will limit the effectiveness of the filter strip and may damage it.

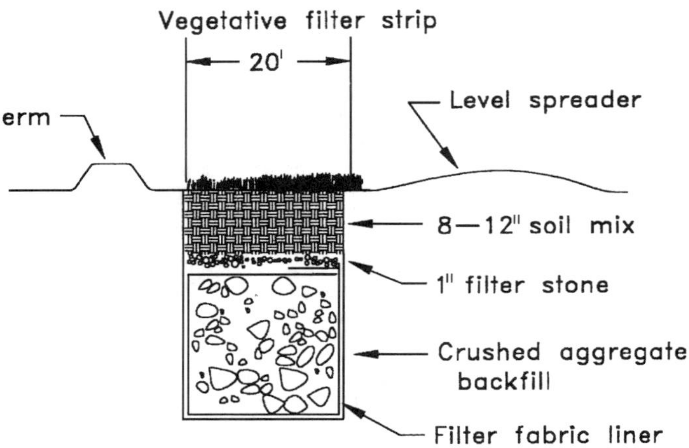

Figure 6.11 Filter strip detail.

Sand filters

Sand filters were among the first water treatment systems devised, and they are still used in many systems today. The sand filter is effective in removing suspended solids, but it has no designed biological treatment capacity and it cannot remove soluble pollutants. In general, the sand filter is at least 1.5 ft deep, and it should be used in conjunction with other media such as peat or other systems to address soluble contaminants (Fig. 6.12). Sand filters have been designed using a layer of peat to increase the efficiency to 90 percent of suspended solids, 70 percent of total phosphorus, 50 percent of total nitrogen, and 80 percent of trace metals (Pitt). The peat/sand filter is usually planted with a cover of grass to increase the removal of nutrients and provide filter surfaces for the deposition of films. The combination of peat and sand makes a very effective filter, although effectiveness can be increased still further when the filter is used in conjunction with a presettling facility.

Filters of peat or composted materials are a relatively new technology and have a much greater removal efficiency—for example, up to 90 percent removal of soluble metals, 95 percent of suspended solids, and 87 percent of hydrocarbons. The peat compost filter has only moderate rates of nutrient removal (40 percent of total phosphorus and 56 percent of Kjeldahl nitrogen) (Pitt). Peat is an effective filter material by itself, but it releases effluent with a greater turbidity. The compost used in such filters should be of deciduous leaves. The peat compost filter requires maintenance and a change of filter material approximately once every 2 years depending on loading.

The principles of the sand filter may be applied to parking lots (Fig. 6.13). In such situations, the parking lot edges would be designed with sand filters to improve the quality of infiltrating runoff.

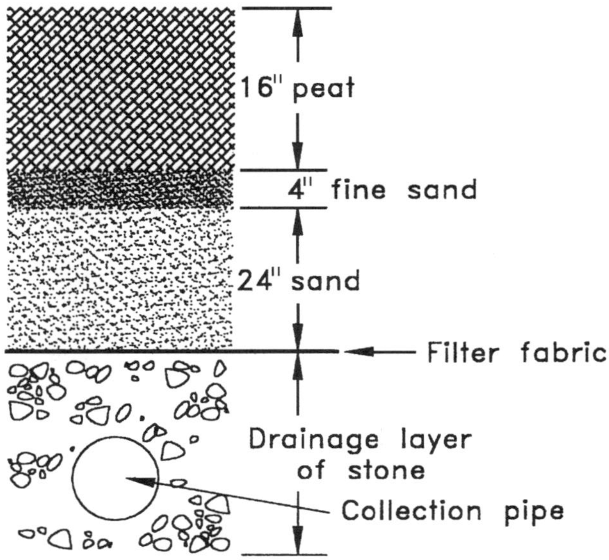

Figure 6.12 Peat/sand filter detail.

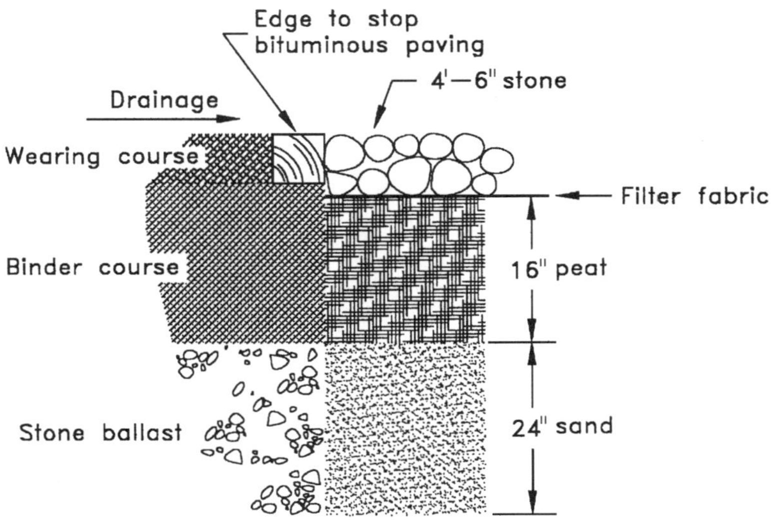

Figure 6.13 Parking lot sand filter strips detail.

Infiltration trenches

An infiltration trench is another method of capturing water and allowing for recharge (Figs. 6.14 through 6.16). The infiltration trench is generally 2 to 10 ft deep (0.6 to 3 m). The depth is constrained by the same criteria as the basin (that is, depth to bedrock or the seasonal high water table). The trench is lined with filter fabric and filled with stone. The spaces between the stone provide

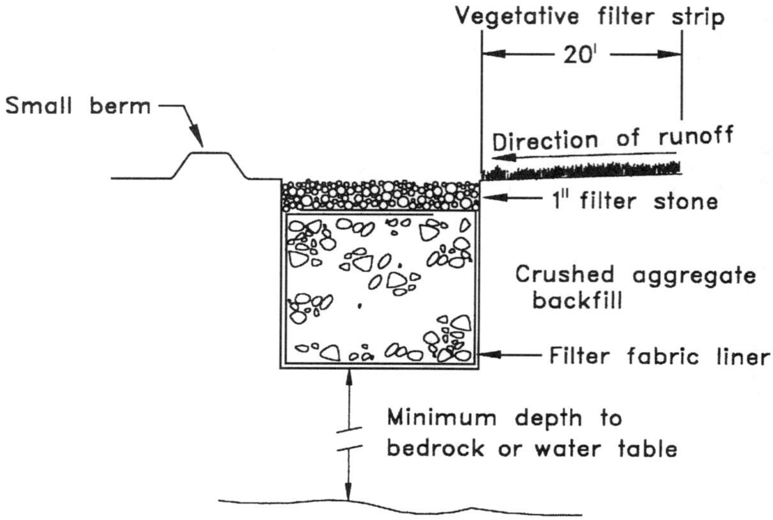

Figure 6.14 Infiltration recharge basin detail.

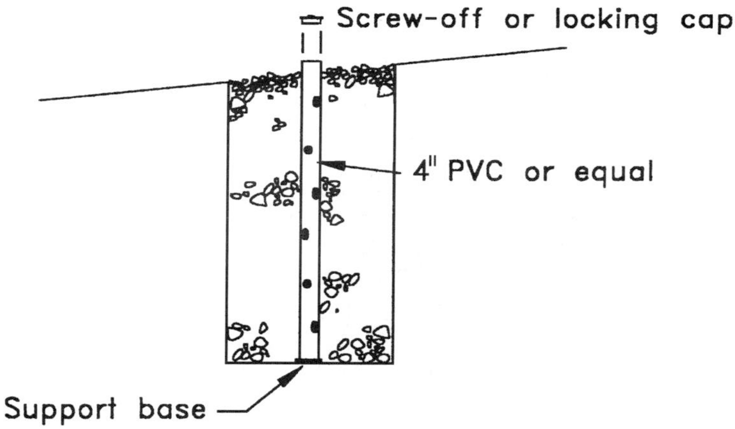

Figure 6.15 Inspection pipe detail.

the storage area for the runoff. Void spaces of backfill are assumed to be in the range of 30 to 40 percent for aggregate of 1.5 to 3 in (4 to 8 cm). The term *void spaces* refers to the spaces between the solid particles of the fill. Although an emerging spillway is usually not designed for an infiltration trench, the design and construction should consider and address the circumstance of an overflow. An observation well should be installed in the infiltration trench to monitor the sediment level in the trench and to monitor the dewatering time (Maryland State Department of Education 1999).

The amount of void space—referred to in most standards as the "percentage of voids"—varies with the type of material. The National Stone Foundation suggests that 35 percent voids is a good rule of thumb. Trenches designed using

Figure 6.16 Photograph of infiltration trench in median.

this rule have functioned well within expected performance. A more accurate percentage of voids necessary can be calculated as follows:

$$n = 1 - \left(\frac{d}{G \times 62.4} \right)$$

where n = voids, percentage
d = dry density of stone
G = specific gravity of stone

The dry density of a particular stone is usually available from the quarry where stone is graded to specifications (see also Table 6.10). If more specific data are not available, the mean specific gravity of 2.6 is used. To determine the cubic feet of stone necessary to store a given volume of water, the following equation may be used:

$$k = \frac{9}{N} \quad ft^3$$

where k = volume of stone required, ft^3
9 = storage volume revenue, ft^3
n = percent of void

TABLE 6.10 Dry Density Per Cubic Foot of Typical Stone

Stone size	d/ft^3	n, %
2A modified	108.92	0.328
2B	96.08	0.408
3A	96.00	0.408
3A modified	112.0	0.320

NOTE: d is the dry density per cubic foot and n is the percentage of voids.

A study at Pennsylvania State University determined a formula for determining the rate at which water will move through a gabion. This same formula has been used to determine the rate at which water will move through the stone ballast of the infiltration trench, or to design an outlet structure of stone. The formula is as follows:

$$Q = 0.40\left(\frac{h^{1.57}}{L^{0.57}}\right)W$$

where h = the ponding depth or head, ft
L = flow path length, ft
W = width of the structure, ft
0.4 = hydraulic conductivity (a constant)

This formula can also be used to determine the control of an infiltration and detention system with an outlet structure—that is, the formula can determine whether the outlet structure will control the peak flow or the time through the ballast.

Infiltration basins

For the purposes of design, the infiltration basin serves the same function as the detention basin—that is, to offset the increase in runoff from the developed site. It is designed according to the parameters described for an infiltration trench. The infiltration basin allows water out through the pore space in the soil rather than through a surface outlet structure. This contributes to some recharge of the aquifer and minimizes the pollutant impact on the receiving surface water. In general, infiltration basins have a large surface area and can provide the maximum possible soil surface contact for the collected runoff. The greater the surface area, the faster the volume can infiltrate into the soil. It is important to remember that oil and grease, floating organic material, and fast-settling solids need to be filtered from the infiltration basin. This may be accomplished through the use of a vegetative filter strip, which will act as a filter by slowing the velocity of the surface runoff and providing many surfaces with which to filter grease and oil.

Construction inspection should pay particular attention to the level of the infiltration system. It is very important that water enter the infiltration system as a sheet flow. Concentrated flows should be spread out using a level spreader or other device. It is also very important that sediment-laden runoff be diverted away from the infiltration trench during construction.

Rain gardens

Rain gardens, also called *bioretention basins,* are simply shallow areas designed to collect storm water. They are usually designed to drain fairly quickly to allow typical landscape plants to be used. Rain gardens have become more common in landscape designs, and they are usually described as landscape features without highlighting their hydrologic role. Rain gardens provide an elegant opportunity for designers to incorporate the esthetic and functional elements of a landscape into a single feature. Rain gardens have been found to have excellent pollution-removal capabilities, removing 60 to 80 percent of nutrients and as much as 99 percent of heavy metals.

Rain gardens are designed after natural upland areas. The use of native plants reduces the maintenance costs and the need for supplementary water supplies in most cases. The rain garden is sized to the area that is contributing runoff. For the most part, rain gardens are designed to serve drainage areas up to an acre. The volume of the rain garden is based on the needed level of control—for example, the first half inch of rainfall might need to be controlled. The rain garden must be able to collect the desired volume and allow the excess to drain away or bypass the rain garden. In general, multiple small rain gardens have been found to operate better than a single very large garden. Larger rain gardens tend to become and remain saturated. Rain gardens are ideal for use with planted islands in parking lots as well as with other types of planted features within a landscape.

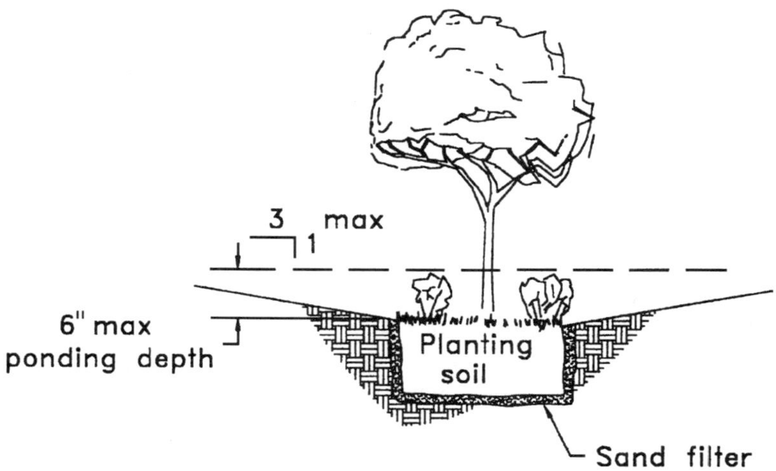

Figure 6.17 Rain garden detail. The shallow depression retains water after storm events and encourages infiltration.

Maximum ponding depth in the rain garden should be limited to about 6 in, and the design should allow the pond to drain away within 3 days to protect the native plants and to discourage insects from breeding. A minimum of 2 ft of planting medium is required. To assure appropriate permeability, the clay content of the soil should be no greater than 10 percent. Studies have found that a soil pH of between 5.5 and 6.5 is ideal for maximum absorbtion of pollutants and associated microbial activity.

Rain gardens increase the treatment capability of retention by adding the biological elements of the filter strip to the infiltration trench. Quite a lot of work has been done with bioretention strips and basins, but, in spite of a good deal of variability, all systems are a combination of filtration and biological action by soil and plant communities and infiltration. These have been effective innovations for the treatment of urban runoff and can be adapted to some brownfield sites. The bioswale uses a grass strip as a filter to reduce runoff velocity and to remove particulates. The swale employs a medium of sand or other material with a topsoil cover to further filter the runoff, to stimulate microbial growth, and to encourage rooting of the swale plantings.

The bioretention swale has received high marks for treatment of the first flush of runoff. Hydrocarbons are degraded, and metals are bound to organic constituents in the topsoil layer. As a system, the bioswale has very high removal efficiencies. For example, the bioswale has been known to remove up to 92 percent of suspended solids, 67 percent of lead, from 30 to 80 percent of phosphorus, and 75 percent of petroleum hydrocarbons.

Detention and Retention Basins

Although detention basins are designed to mimic the storm water flow that existed before development, they do not account for the many other aspects and functions of the natural drainage basin. As a result, the downstream water bodies may suffer from degraded water quality, loss of habitat, and reduced recreation value. Before development, the site provided a buffer for flows, retaining water and detaining runoff, so that flooding was delayed and reduced. Other incidental values included filtering of fine particles as well as increased infiltration and diversion created by the vegetated and irregular natural surfaces. Comprehensive design is able to identify and imitate some of these characteristics of the existing natural storm water management system. The natural system is nearly always superior to the "designed" system, and so efforts should be made to retain and enhance the existing drainage system whenever possible.

The most familiar methods of managing storm water runoff are detention or retention basins. *Detention basins* are usually "dry" basins that fill with water only during a rain. They work by delaying the storm water so that it is released at a rate that mimics the predevelopment flow. A *retention basin* holds the water in a pool. The only outlet is through emergency spillways that allow the basin to overflow in a controlled manner if it should become too full. The retention basin loses water through infiltration and evaporation. These basins are often part of a larger plan that incorporates storm water management into water features.

Detention basin effectiveness is a function of the drainage area in which it operates and the location of the basin in the watershed. The lower the basin is in a watershed, the less effective it is. In fact, a properly constructed basin could actually make a downstream flooding problem worse. The basin functions by detaining the water it collects and then releasing the water at a rate that is calculated to be equal to the predeveloped rate. The development of a site results in more runoff, so that at an equal rate of discharge, it will take longer for the runoff to be discharged.

When it rains, it takes a period of time for the runoff to collect and run to the low points. In a watershed this "lag time" can be hours or days, depending on the size of the drainage area. Before a site is developed, areas low in the watershed may collect and empty before the main portion of the flood travels down to that point in the watershed. The runoff from the lower end is discharged before the "flood" arrives. After development and the installation of the detention basin, this increased runoff is stored and its discharge is delayed. In some cases the delay may be long enough to coincide with the "flood." In such cases the basin may actually make the flood worse by contributing more water to the peak flow. The result may be that the project, and certainly the downstream landowners, may be better off without the basin. The impact of a basin can be predicted quantitatively but not without some expense.

In terms of storm water management, not all sites are created equally. If a project site is high in the drainage area, it can be difficult to collect enough water in the basin to offset the increase in runoff due to development, and detaining floods nearer the bottom of a watershed may create more problems than it solves. Clearly the best design solution is to locate the basin with careful thought and analysis if it is to serve its purpose within a watershed.

Detention basins may be designed to meet a predeveloped storm rate of discharge, but the nature of a basin is to concentrate the flow, usually from a single outlet, and to extend the time of discharge to account for the increased volume from the developed site. Site designers must study existing drainage patterns and pathways to identify opportunities that exist on the site. Much consideration should be given to using existing drainage paths. Where increases in flows and velocities will occur, it may be appropriate to enhance existing drainage ways to account for the increases rather than obliterating them in favor of a new path. The drainage patterns on a site that have developed as part of natural landscape processes can often be converted into effective drainage ways for the new development. In some cases these drainage pathways, left in place through the site, may double as a greenway and walkway for pedestrians, combining the use of some landscape features for drainage and open space.

Existing channels may require some attention to stabilization and alignment because of the change to the hydrologic and hydraulic character of the project site, but these projects should be undertaken with care so as to imitate the natural appearance and function of the drainage ways. Changes in the volume and concentration of runoff will have an impact on streams. Consideration should be given to creating pools and ponding areas that would collect and trap water in high flows. The use of flat areas to intercept runoff and encourage infiltration might be developed to function also as wetlands. Developed wetland pockets could prove to be important storm water management features, as well as visually interesting habitat areas.

Designers and developers should encourage the local governments to create storm water management authorities that encompass the entire watershed. Although watersheds may include any number of municipalities, the watershed-based approach is a more accurate and effective means of managing storm water runoff. Watershed authorities have been created in a number of places, most notably in Florida. In principle, the storm water authority is able to develop a more efficient and less expensive management approach. Public facilities are more likely to be located where they will provide the greatest benefit and avoid the expense of many uncoordinated and often poorly maintained private facilities. Developers can contribute to the authority based on runoff quantities and avoid the expense of dealing with storm water on a site-by-site basis.

Designers must find ways to put the landscape to work. Storm water detention basins should be used to improve a site by finding new ways to offset the storm water increase and provide benefits that the simple detention basin does not offer. These benefits might include recreation or esthetic focal points, perhaps even wildlife habitat, or a water quality enhancement. With careful design and consideration, perhaps all of these.

An alternative to the infiltration and recharge methods is the development of a wet pond or a retention basin (Fig. 6.18). The design of wet ponds and wetlands is usually a matter for design professionals who are trained and experienced with balancing the site constraints with a pond. It is generally a balancing act between cost, site issues (such as slope or drainage area), appearance, and the pond function. Wet ponds can be used advantageously as part of an effective process by which certain urban pollutants are removed through settling in the permanent pool. The geometry of the pool is an important aspect of its capability to remove or reduce pollutants. Ponds are sized with regard to the flow of water through the pond, pond volume, pond depth, and the expected particle sizes to be encountered in order to allow for the necessary settling time. The activity of plants and microorganisms necessary for the reduction of pollutants occurs primarily at the bottom of the basin. The shape of the basin (geometry) must be designed so as to minimize the currents within the basin and maximize the travel time from the point where storm water enters the pond to any point of outlet or overflow (Washington Council of Governments).

The surface area is usually designed also in relation to the depth of the pond to avoid *dead storage*—that is, areas that do not get mixed into the rest of the pond. Pond depths will vary according to the purpose. The *marsh,* or *littoral, zone* is usually 6 in to 2 ft deep and provides the most effective removal of nutrients and some other pollutants. The design of the basin should also

Figure 6.18 Photograph of wet pond. Note the open water beyond the shallow littoral area and the natural appearance of the pond system.

include an area equal to 33 percent of the pond surface area that is from 3 to 6 ft deep for fish. An additional 25 percent of the pond should be at least 3 ft deep and within 6 ft of the shore. This combination of shallow and deep areas will help the pond function well.

The minimum drainage area to be considered for the wet pond should be about 10 acres. The drainage area size should be adjusted according to the rainfall characteristics of an area, the amount of anticipated runoff, the type of land use, the pond geometry and depth, and the settling rate of the expected particulates. The drainage area should be large enough to contribute adequate supplies of water to the pond. The first parameter to consider is the ratio of the drainage area to the pond surface area. The recommended range of the ratio is from 10 to 50. This range could represent a 1-acre pond in a 10-acre watershed or a 10-acre pond in a 500-acre watershed.

If the volume of a pond is much greater than the volume of runoff coming into it, there will be a longer residence time. The residence time is important because the settlement of pollutants will occur primarily when the water is not moving in the pond. Studies have shown that two-thirds of the incoming sediments will settle out in the first 24 h. Significant reduction of phosphorus, however, can take up to 2 weeks. Phosphorus is a pollutant with serious water quality consequences. The volume required for a 2-week storage period is very large. Volumes this large will affect the pond's ability to function as a detention basin and meet the peak discharge control requirements. Planning for the combination of activities—storm water detention and water purification—must be carefully balanced.

The ratio of the wet pond volume to the mean runoff volume is another key guide to pond design. In general, the larger the surface area of a pond, the greater its pollution removal efficiency. The smaller the area ratio (larger pond surface), the greater the efficiency of the pond in removing pollutants. A pond can be made deeper to achieve water quality, but increased depth is not as effective as increased surface area. A volume ratio of 2.5 is suggested in order to achieve 70 percent removal of sediment loads, or a residence time of about 9 days. The 9-day residence time is generally recognized as a middle ground, providing water quality improvements but avoiding the large volume required for the 2-week residence time.

Other Storm Water Management Methods

Other ways of dealing with storm water might be described as avoidance strategies—simply not creating the runoff eliminates the need to deal with it. Runoff can be minimized by reducing paved surfaces or by using more permeable surface materials. Increased permeability of paving surfaces offers the greatest opportunities to minimize runoff (see Table 6.11). A key concern of the development community, however, is that this strategy will equate to a limit on development as to the amount of discharge allowed on a parcel. To zone for such a low density, however, would probably cause development to sprawl even more, increasing the amount of roads and infrastructure necessary and resulting in other undesirable impacts. In fact, watershed managers have come to

TABLE 6.11 Costs of Various Types of Permeable Pavements

Product	Manufacturer	Cost per square foot*
Asphalt	Various	\$0.50–\$1.00
Geoweb	Presto Products	\$1.00–\$2.00
Grasspave	Invisible Structures	\$1.00–\$2.00
GRASSY PAVERS	RK Manufacturing	\$1.00–\$2.00
Geoblock	Presto Products	\$2.00–\$3.00
Checkerblock	Hastings Pavement	\$3.00–\$4.00
Grasscrete	Bomanite Co.	\$3.00–\$4.00
Turfstone	Westcon Pavers	\$2.00–\$3.00
UNI Eco-stone	Concrete Paving Stones	\$2.00–\$3.00

*Includes material cost, typical shipping cost, and installation cost on a fully prepared base course. Does not include cost of gravel or soil and grass fill or labor. These costs are approximately \$0.10–\$0.25 per square foot.

SOURCE: This table was adopted from the Center for Watershed Protection, *Better Site Design: A Handbook for Changing Development Rules in Your Community,* prepared for the Site Planning Roundtable, 1998.

realize that the best way to reduce the unwanted stream impacts of development may be to concentrate development into areas with 80 or even 100 percent impervious coverage. Such dense development will have a negative impact on some streams or portions of streams, but the impact on others will be avoided or minimized.

Using open-joint pavers instead of other impermeable paving materials can allow some infiltration through the joints. Figure 6.19 illustrates the effect of an open-joint paver design utilizing 1/4-in joints filled with sharp sand. The 4- by 8-in design with a 1/4-in open joint provides an opening equivalent to 60.8 in^2 in an area of 32 by 24 in.

Reducing cartway widths where possible, as discussed in Chap. 5, may contribute to the quality of development in a variety of ways. For example, it may benefit the development immensely to encourage the use of smaller paved areas in cul-de-sacs by using smaller radii and designing centers with rain gardens or other infiltration features such as grass pavers (Figs. 6.20 and 6.21). By reducing parking lot size requirements and space sizes, or encouraging shared parking arrangements, the amount of area required for parking can be reduced.

We should remember that when a site is developed, the small, intermittent and ephemeral streams are replaced by curb and gutter flow, but these replace only the conveyance of runoff and none of the filtering and delay of natural channels. To counteract the loss of filtering, lawns can be graded to include subtle channels or collection areas that will delay runoff and increase infiltration. Likewise the edges of parking lots or driveways can be designed to collect runoff and encourage infiltration. Where swales are used, vegetation should

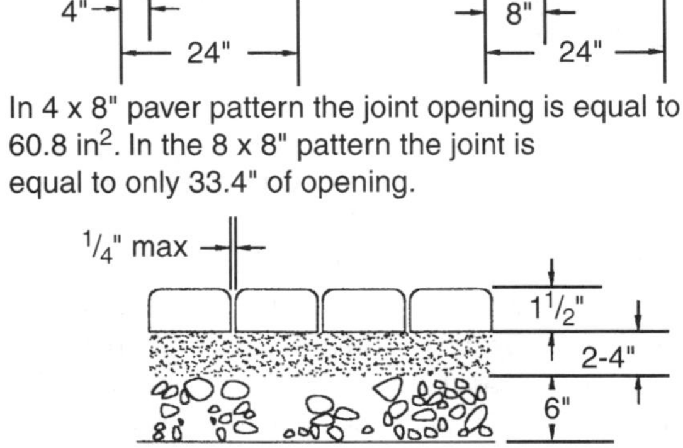

Figure 6.19 Open-joint paver detail.

Figure 6.20 Photograph of grass pavers.

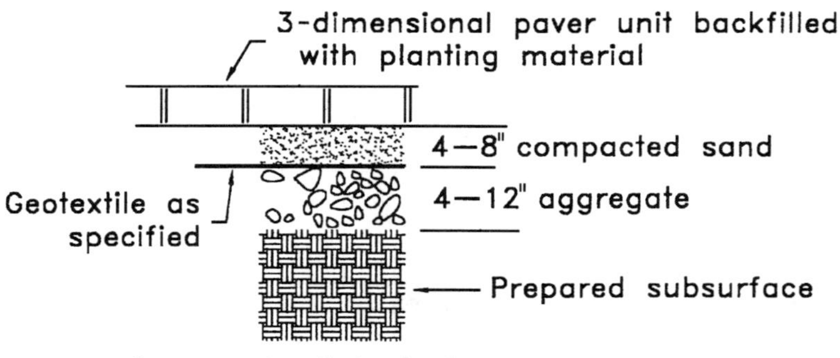

Figure 6.21 Grass paver installation detail.

be chosen that will provide a relatively high Manning's coefficient of roughness and delay runoff. By increasing the travel distance and decreasing the rate of runoff, more water is retained and may infiltrate into the soil.

There are other strategies as well such as rooftop rain storage systems, as shown in Fig. 6.22. Such rooftop installations have been used for some time in Europe and have been successfully introduced to the United States in recent years. Even common flat roofs can be designed to reduce runoff by increasing the roughness of the roof surface or by restricting roof drains. These installations can significantly reduce runoff volumes and in turn reduce

Figure 6.22 Rooftop rain garden photograph. (*Used with permission of Charles Miller, P.E., of Roofscapes, Inc.*)

development costs. In older cities the installation of rooftop systems can contribute to reducing the costs of rehabilitating infrastructure (Miller 1998).

Sanitary Sewer Collection Systems

The most common types of sanitary sewer collection systems are gravity flow systems. Gravity collection systems are designed to use as few pumps as possible by taking advantage, to the extent possible, of the natural lay of the land. Gravity systems in areas with little topographic relief can be limited by the practical depth limits of installation and have difficulty with infiltration and with sedimentation problems associated with low velocity. In places with very steep topography high-flow velocities may result in problems of odors (caused by turbulence) or separation of liquids and solids.

Although most municipalities establish fairly explicit design standards for sanitary sewer design and construction, sewer flows are generally calculated using a peak hourly flow rather than an average flow since the distribution of flow is not constant. Peak flow times tends to be about 1.5 times greater than the maximum daily flow or about 3.0 times the average flow. Peak periods of flow are coincidental with some lag time, with the peak water demand. Flows are lowest in the early morning hours, and they increase as people rise and prepare for their day. After a midday dip, flows increase again in the evening hours (Tables 6.12 and 6.13).

Gravity sewers are usually designed to provide a minimum flow velocity of at least 2 ft/s, known as the *scouring velocity.* At velocities less than 2 ft/s solids may settle out of the flow and result in sedimentation in the pipe, reducing the capacity of the pipe and eventually causing a blocked pipe. Velocities are also best kept below 10 ft/s to minimize turbulence and splashing. Higher velocities may also increase the wear and decrease the life of concrete pipes and facilities. Municipal design standards may require the use of abrasion-resistant pipe materials when velocities exceed 10 ft/s. As a rule of thumb, a scouring velocity should be reached and exceeded during peak flow periods but not necessarily during low-flow periods.

In general, gravity sewers are not constructed with pipe diameters less than 8 in. Also, manholes should be placed at regular intervals to provide future access to maintenance workers and equipment. Manholes should be placed at all changes in grade, pipe size, and direction (Table 6.14).

The depth of utility trenches is a concern for several reasons. The safety of workers is the first concern. Designers should be aware of the risks involved in working in deep trenches, particularly in unconsolidated or unstable soils. Whether or not working drawings should include shoring instructions or warnings is a practice that varies from location to location and even from firm to firm. Including safety instructions on working drawings may increase the designer's liability for construction conditions; on the other hand, noting the installation should be in compliance with federal or state safety standards, and regulations may protect the designer from liability. Trench depth also influences the costs of installation and maintenance.

From a performance standpoint, trench depth is a concern in terms of the earth loads on the pipe. The ability of a pipe to resist deformation under a load is a function of the depth and the width of the trench. As trenches are dug wider, the sides of the trench offer less and less support to the pipe. As the

TABLE 6.12 Relationships between Dry-Weather Domestic Wastewater Flows

Minimum daily flow = 0.66 average daily flow
Minimum hourly flow = 0.5 minimum daily flow, or 0.33 average daily flow
Maximum daily flow = 2 × average daily flow
Maximum hourly flow = 1.5 maximum daily flow, or 3 × average daily flow

TABLE 6.13 Wastewater Flows

Source	Gallons per person per day
Airports (per passenger)	5
Camps	
With central comfort stations	35
With flush toilets, no showers	25
Construction camps, semipermanent	50
Day camps (no meals)	35
Resort camps (day and night)	100
Seasonal cottages	50
Country clubs (per resident member)	100
Country clubs (per nonresident member present)	25
Churches (per seat)	6
Dwellings	
Boarding houses	50
Luxury residences	150
Multiple-family homes	60
Single-family homes	75
Factories (gallons per person per shift)	35*
Hospitals (per bed space)	250
Hotels (2 persons per room)	80
Hotels without private bath	50
Institutions (other than hospitals)	125
Laundries (self-service, per machine)	300
Mobile-home parks (per space)	250
Motels (per bed, with kitchen)	50
Motels (per bed, no kitchen)	40
Picnic parks (per visitor)	5
Restaurants (toilet and kitchen waste, per patron)	10
Restaurants (per meal served)	3
Restaurants (additional for bar/lounge)	2
Schools	
Boarding	100
Day (no gym, showers, or cafeteria)	15
Day (with gym, showers, and cafeteria)	25
Day (with cafeteria only)	20
Service stations (per vehicle served)	10
Theaters (per seat)	5
Travel trailer or RV parks with hookups (per space)	100
Office workers (workers per shift)	25

Adapted from American Society of Civil Engineers, *Design and Construction of Sanitary and Storm Sewers*, 1969. (*Used with permission of the ASCE.*)

TABLE 6.14 Suggested Manhole Spacing

Pipe size, in	Min spacing, ft	Max spacing, ft
8–15	400	600
18–30	600	800
36–60	800	1200
Greater than 60	1200	1300

depth of the trench increases, the backfill and surface loads increase. The selection of pipe material should include the consideration of the loads that will be developed in buried pipes. Most pipe manufacturers provide tables for the designer to determine loads on pipes under various field conditions.

Bedding materials also contribute to the performance of pipes by assisting the pipe in handling the backfill and surface loads. Pipes are tested in laboratories for their ability to support loads and are given a strength rating usually expressed in pounds per foot. Since bedding materials vary from place to place and will perform under field rather than laboratory conditions, they are identified by standard class designations. The standard class designation refers to a bedding material's ability to support the "load factor" and is based on a "three-edge bearing test." The ability to support the pipe to the three-edge bearing load is rated as 1. The most common bedding materials are concrete, crushed stone, or suitable local materials.

Relatively low strength portland cement concrete (2000 lb/in^2) is used to support sewer pipe in the class A bedding and will develop a load factor of from 2 to more than 3 depending on the degree of steel or wire reinforcement used. Class A bedding is expensive and is usually used only in very deep trenches or where there are anticipated high surface loads. When class A bedding is used, pipes are usually blocked in place and grade while the concrete is poured in place. Class B and C bedding are commonly constructed of crushed stone ranging in size from $^1/_4$ to $^3/_4$ in. In most cases, stone bedding is superior to compacted sand or gravel, except for some plastic pipes. Class B and C bedding are the most common types used due to their fairly high load factors, which range from 1.5 to 1.9, and their reasonable cost. Class D and sometimes class C bedding may be constructed from some local or native materials by carefully excavating a trench bed to use the compacted native material as bedding. Class D bedding is usually not recommended. See Figs. 6.23 through 6.25.

Onsite Sewage Disposal Systems

Onsite sewage disposal systems, also called *soil absorbtion systems,* are commonly used in areas where public sewage collection and disposal are not available. In general, these systems are composed of a holding tank and a drain field. The tank provides for settlement and some digestion of the solids. Liquids rise to the surface and are conducted either by gravity or by a pump to a drain field.

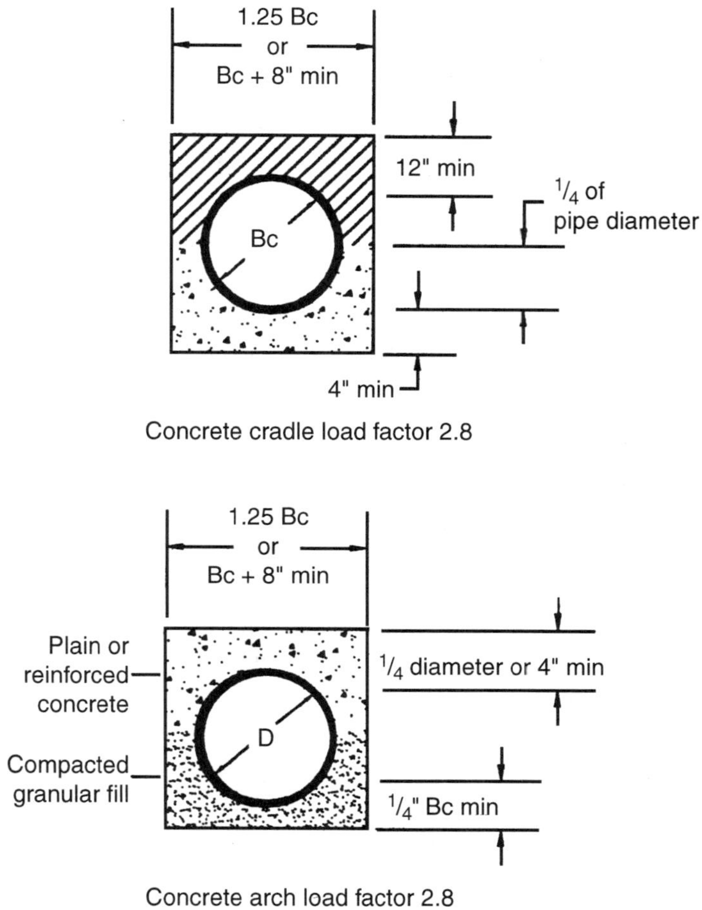

Figure 6.23 Class A pipe bedding detail. (American Society of Civil Engineers, *Design and Construction of Sanitary and Storm Sewers*, 1969. Used with permission of ASCE.)

The drain field is essentially a manifold system of perforated pipe that distributes the effluent from the tank to a prescribed area. The design of onsite sewage disposal systems is usually conducted under state or local guidelines and requirements, but there are similarities found in the regulations.

Onsite sewage disposal systems utilize the ability of soil to absorb and treat the effluent as it percolates through the soil matrix. The soil must have a fairly high degree of permeability but not so high as to present a danger of contamination to groundwater supplies. Onsite design proceeds on the basis of field tests that measure the permeability of the soil—that is, the rate at which waste will percolate through the undisturbed soil. The *perc rate* is determined by filling a series of small holes or pits. After the hole has drained thoroughly once, the hole is refilled and the rate at which the water drains is recorded. An acceptable perc rate is one that is neither too fast nor too slow according to

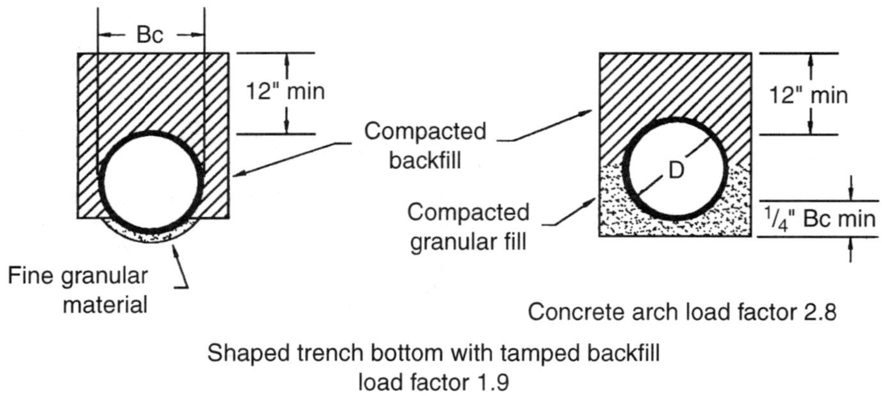

Figure 6.24 Class B pipe bedding detail. (American Society of Civil Engineers, *Design and Construction of Sanitary and Storm Sewers*, 1969. Used with permission of ASCE.)

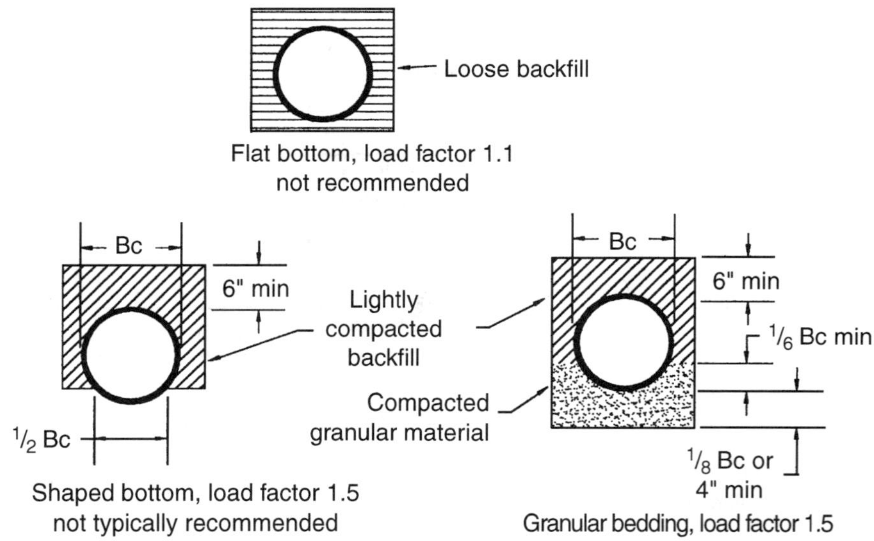

Figure 6.25 Class C pipe bedding detail. (American Society of Civil Engineers, *Design and Construction of Sanitary and Storm Sewers*, 1969. Used with permission of ASCE.)

local regulations. Other design concerns include the slope of the proposed drain field and distance to the water table and bedrock. The actual size of the drain field is calculated as a function of the perc rate and the volume of loading (see Table 6.13). The drain field must be large enough to allow for complete absorbtion of the effluent.

Onsite system failures are common. The proper installation of the distribution manifold is critical. The pipe system must be installed flat to provide for even distribution over the entire drain field. If low spots should occur, the waste effluent will tend to collect in the low spots and overload those portions of the drain field. Likewise, if the holding tank becomes filled with solids, the quality of the effluent will degrade and solids may block the distribution system, causing overloading.

In areas where perc rates or soil depth is inadequate for onsite disposal, sand mound systems are sometimes used. *Sand mound systems* are raised areas constructed of materials, usually sandy soils, that are designated by local authorities as having acceptable percolation rates. The sand mound is designed to meet the percolation and loading requirements in much the same way as a typical in ground system except that the mound is raised, installed above grade usually on top of the native soils. In some applications the sand mound system is coupled with a small aerobic treatment system. In any case the sand mound is always loaded under pressure using a pump to assure even distribution of the effluent throughout the drain field.

Chapter

7

Landscape Restoration

Landscape restoration encompasses a broad range of activities and concerns. Although there is not an established or formal distinction within the professions, as a matter of practice, *restoration* could refer to rehabilitation, reclamation, or remediation efforts. *Rehabilitation* refers to actions taken to restore the environmental functioning and the vitality of a landscape. In some rehabilitation projects the salient underlying features of the landscape are still present, but because of urbanization or other landscape disturbances, the quality and functioning of the landscape have been negatively impacted or degraded. Stream and wetland restoration and landscape revegetation projects are examples of rehabilitation. *Reclamation* projects are usually undertaken on landscapes where features have been obliterated by development or agricultural or mining operations. Reclamation projects usually require the construction of new landscape features in order to replace what was lost in the process of exploiting the land area previously. Reclamation projects might include constructing wetlands or infiltration features such as rain gardens or eliminating invasive exotic vegetation and encouraging the return of native species. *Remediation activities* are concerned with mitigating a pollution condition that has resulted from activities conducted on the site previously. Dealing with acid mine drainage or contaminated runoff from a brownfield site could be examples of remediation. A given landscape restoration project may involve all three.

Landscape restoration as an area of professional practice is not new, but it has grown dramatically in recent years. As site development practices expand to include concerns with environmental impact and sustainability, many of the innovative practices used in landscape restoration may become more common.

Restoring Vegetative Cover

Soil structure is the arrangement of soil particles into aggregates of the mineral soil material, organic material, and microorganisms. The capacity of soil to agglomerate into aggregates is an important characteristic of a soil ecosystem. On the disturbed site, the ability of the soil to form aggregates is destroyed through the operations of grading and compaction. Soils with a granular structure naturally allow for infiltration and resist erosion, and the loss of that soil structure results in a decrease in soil permeability and an increase in erodability and runoff.

In addition to being desirable esthetically, vegetation also prevents erosion and runoff and is key to the long-term maintenance of the soil structure. For revegetation to succeed, stabilizing the site as soon as possible is critical (Darmer 1992). Plans for revegetation must be in place in the project design phase so that revegetation can begin while the site is under construction, which is especially necessary if vegetation is part of the plan to mitigate construction site runoff. The aspects of the site design that pertain to revegetation include soil preparation, the appropriate selection of materials, and the maintenance of the plants and soil.

Site evaluation and plant selection

Soil analysis. The first step that should be taken in restoring vegetation is a soil analysis. Soil tests will provide the fundamental data for determining the characteristics of the soil and the cultural requirements and amendments necessary for a successful revegetation effort (Sobek et al. 1976).

In most circumstances, a soil analysis should be conducted during construction. Typically on a disturbed site, soil is often dry, compacted, and infertile and bears little resemblance to the original native soils. On many sites fill has been brought from off site, sometimes from a myriad of sources, and on other sites new "made-land" conditions exist. The result is unpredictable if not unproductive soil. In some cases it is necessary to consider reworking the soil up to a depth of 30 in. Attempting to establish plant growth without knowing the characteristics of the soil could yield higher uneven results.

The procedure chosen for collecting and analyzing soil samples will depend on the objectives of the site plan, as well as the homogeneity of the soil, the ease of collection, and the construction of the sampling equipment. (See also Table 7.1.) For sites where contamination might be an issue, the required decontamination procedures might also affect the collection of soil samples. Sample planning can be conducted using aerial photography, USGS, or site maps. The character of the sample is a function of the objective. Different types of samples are required for different analyses: For example, a bulk-density soil analysis requires an intact sample core. A nutrient analysis requires a well-balanced composite sample that consists of about 15 to 20 cores taken at random locations throughout one field or area. The area should be no more than about 20 acres, and each sample should represent only one general soil type or condition. If the sampling is being conducted for engineering classification, it might be beneficial to sample areas separately that are clearly unique.

TABLE 7.1 Planning and Collecting Soil Samples

1. Subdivide the area into homogenous units, and, if necessary, also subdivide these into uniformly sized areas.
2. Establish a grid to locate sampling points. Composite samples of each area should be composed of between 10 and 20 samples. Care must be taken to use uniformly sized cores and/or slices of equal volume and of equal depth to develop the composite.
3. Test for the following:

   Standard water pH and/or buffer pH

   % organic matter

   Cation exchange capacity

   Particle size distribution

   Salinity

   Available nutrients

SOURCE: Adapted from the U.S. Environmental Protection Agency, *Process Design Manual for Land Treatment,* Section III (Washington, D.C.: Government Printing Office, 1977).

Sampling for nutrient analysis is usually conducted with clean stainless steel, chrome, or plastic buckets and tools. Brass, bronze, or galvanized tools should be avoided. If the sampling is for engineering analysis, the samples are usually collected as intact cores taken with split spoons on drilling rigs or direct push machines. When hand samples are needed, they are taken with hand augers or coring equipment. Samples collected for chemical analysis might require special nonreactive equipment and special sample preservation activities.

Soil sampling may also be conducted in other ways and for other purposes. *Geophysical techniques* are either profiling or sounding methods used to identify the presence of buried metal objects or to map the subsurface features. *Profiling* is used to define the lateral extent of a feature such as an area of buried wastes. The result is a contour map of the area and/or object. *Sounding* is a radar technique used to determine the depth of an object at a specific location. Soundings are taken and superimposed on a grid pattern to allow interpolation of the depth and area of objects. Methods of geophysical testing include ground-penetrating radar, electromagnetic exploration, seismic refraction, and magnetometer surveys.

Ground-penetrating radar (GPR) provides a shallow cross section of subsurface objects. GPR can penetrate up to 40 ft in sandy soils, but only the first 4 ft in clay soils or soils containing conductive wastes. *Data prom GPR* must be used in conjunction with supporting data from bore hole logs and resistivity or conductivity tests. The GPR capability is affected by terrain and site vegetation. The GPR antenna is dragged behind a vehicle along a cleared path that is at least 3 or 4 ft wide. The distance between paths varies by the type of equipment used. A typical day's survey, including the interpretation of data and preparation of a report, costs between $5000 and $20,000.

Electromagnetic exploration (EM) includes several techniques, each of which requires the contrasting of the conductivity of the materials being screened. These readings are useful in mapping metallic plumes and locating buried objects such as tanks, pipes, utilities, or drums.

Seismic refraction is a geological tool generally used in exploratory work, but it has proven useful in some ESA work. It is used for mapping bedrock surface areas and groundwater or environmental pathways where influences over contamination plumes are being sought. The *magnetometer survey,* like the electromagnetic exploration, is used to determine magnetic anomalies on a particular site.

Other soil testing methods are *soil gas studies,* which are used to identify the presence of *volatile organic compounds* (VOCs), which could include solvents, oils, gas, and cleaning fluids. Samples of the air in the soil are collected at predetermined depths and locations and analyzed. Areas are usually selected for study because they are known or suspected to be hazardous waste dumping or disposal areas. The presence of VOCs in the soil indicates contamination near the monitoring point and possibly in the groundwater. Soil gas surveys are used to determine the placement of borings and monitoring points and/or wells so as to more precisely define the area of contamination. Test results are mapped to illustrate the area of contamination and to track a contaminant that may be mobile. Samples are sometimes taken as *grab samples,* for which a probe is inserted into the vadose zone, which is the area above the groundwater table. Air is drawn with a vacuum pump into a sample container. In another soil gas test method, called *static sampling,* the air to be tested is drawn into a tube containing activated charcoal, which absorbs the gases. A third soil gas testing method does not involve collecting air samples but rather takes an above-ground reading with a *photoionization detector* (PID) or a *flame-ionization detector* (FID).

Plant selection. In addition to examining the soil, a visit to the area surrounding the site to identify local vegetation can provide important information on native plants. As shown in Table 7.2, it is important to know the climate and precipitation requirements for the native species if they are included in the proposed vegetation. A visit to the area will also allow the designer to notice differences that occur at various elevations, slopes, exposures, and aspects. This information will be invaluable in creating a design that can be sustained through the seasons and through climate extremes such as droughty or wet years, or heavy winds or snows (Brown et al. 1986).

The introduction of acrylic polymers in recent years has improved the success of stabilization and revegetation projects particularly in droughty soils. Acrylic polymers are added and mixed with soils to create a film that allows air and water to penetrate but still binds the soil particles together. It is nontoxic, and runoff does not stain concrete.

The barren and compacted soil surface of a disturbed site is a difficult place to reestablish plants. There is little protection for seedlings and young plants

TABLE 7.2 Criteria for Selecting Plants for Restoration Projects

Grasses
Availability of seed
Resistance to erosion and traffic stresses at the site
Adaptability to critical conditions such as pH, soil texture, drainage, salinity, and wind erosion
Adaptability to climate of site such as sunlight, exposure, temperatures, wind, and rainfall
Resistance to Insects and Disease
Compatibility with other plants selected
Ability to propagate
Consistent with long-term maintenance and succession plans
Shrubs and Trees
Availability in required quantity
Capability to produce root systems as encouraged or constrained by the site characteristics
Ability to become quickly established
Tolerance of site conditions—acid, saline, wet, droughty, or compacted soils
Compatible with principals of secondary succession
Ability for vigorous growth after relief of moisture stress; regrowth after damage
Ability to reproduce
Value to wildlife
Ability to create islands of fertility by being a point of accumulation for organic matter, detritus, and nutrients
Ability to withstand traffic stresses
Resistance to insects, diseases, and other pests
Compatibility with other plants selected for the project
Relative maintenance requirements and/or costs
Tolerance for site-specific stresses

even with mulches that can protect plants only during the earliest stages of their growth. Young seedlings are particularly susceptible to damage from wind and rain and heat or cold stress. The negative effects of climatic extremes in an area are often even worse on the disturbed site. Without cover, the wind will dry the unprotected surface more quickly and erode the unprotected soils. Rainfall impact on unprotected soils may result in significant erosion and downstream sedimentation.

To survive in this harsh environment, the plant material selected to stabilize and revegetate the site must be able to establish quickly. In general, native species are a good choice because they have a predictable performance and growth habit in the geographic area. They have adapted to the general soils and climatic conditions and are best suited to the extremes as well as the average conditions that can be expected.

Introduced species should be carefully evaluated before they are used. The EPA has established criteria for selecting plant materials, differentiated for grasses and forbs and for shrubs and trees (Sobek et al. 1976). In addition, the regional offices of the Natural Resources Conservation Service (NRCS) and the various state universities and agricultural colleges also provide recommendations for seed mixtures to be used for effective erosion control. These offices are an excellent source of information. The designer and developer need to consider the various mixtures of plants in light of specific site characteristics. Some plant materials that work well in a commercial or industrial application may not be appropriate or desirable in a residential application. In addition, it is necessary for the designer to have accurate topography for the site as well as climactic information and data about site hydrology.

The site characteristics must be evaluated against the cultural requirements of selected plant materials. The constraints on establishing and maintaining the plants that must be considered are the growth habit, rooting depth, and rate of establishment. The time of year for seeding is also an important consideration in conjunction with the rate of maturation from germination and the expected temperature and precipitation. Some cool season grasses will not germinate in high temperatures, or once germinated, they will suffer from the extreme temperatures. Warm season grasses require these same higher temperatures to germinate. In addition, the plants selected for the revegetation plan must be compatible with each other as well as resistant to insect damage and diseases. The long-term permanent stabilization plan should include grasses, legumes, shrubs, and trees. Generally speaking, perennials are best planted in the fall, and annuals should be seeded in the spring.

Cultural operations

Seedbed preparation generally occurs after "finish grading" is complete. Generally the surfaces left to be seeded are hard and smooth and not ready to be seeded by any means. Graded slopes that are to be seeded should be 2:1 or flatter. Steeper slopes may require special treatment if vegetation alone is to be used to stabilize them. The interim condition between finish grading and stable vegetated slope is a fragile one. On slopes steeper than 3:1, *steeping the slope* is sometimes used to help vegetation become established (Fig. 7.1). Slopes should be left in a rough condition. A smooth slope is a more difficult surface on which to establish vegetation than a slope left with clods and imperfections. Another version of the stair-stepped slope is the *tracked slope* for which a serrated blade is drawn across the slope, parallel with the contours (Fig. 7.2). The surface is left with many locations in which seed may become established. Seeds blown or washed from one location are likely to be deposited in microsites created by the "steps" in the slopes. Gradually, over time, the edges of the stairs wear down, and debris from above fills in the trough, leaving the desired smooth surface (Rogoshewski, Bryson, and Wagner 1983; Maryland Water Resources Administration 1983).

Compaction is a problem that can extend deep into the soil. Simply "scratching up" the seedbed will not provide for infiltration or air but will create only a thin, weak layer of loose soil at the surface. Although an initial stand of veg-

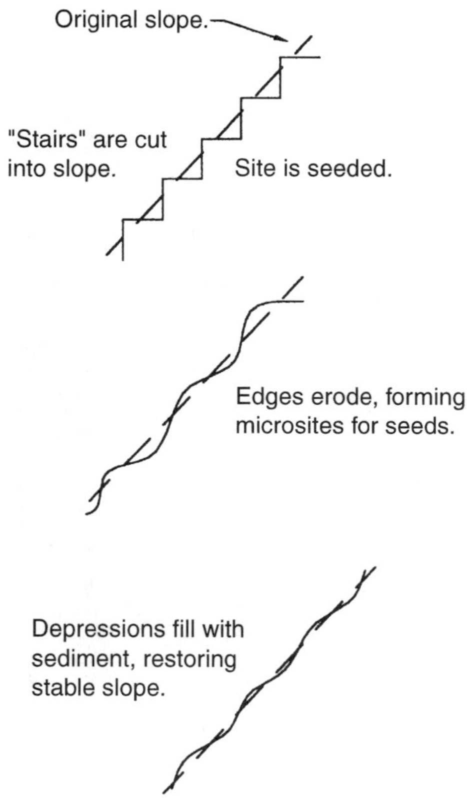

Figure 7.1 Stair-stepping detail.

etation may germinate and appear vigorous, the compacted subsoil restricts root growth and infiltration of water, which will eventually result in the concentration of salts in the topsoil, which will in turn limit successful plant growth. In addition, the slower-growing deeper-rooted plants will not establish on the site (Brown et al. 1986).

The soils may require deeper conditioning prior to seeding, which can be accomplished with equipment that is able to plow to the necessary depth. Once these cultural operations are complete, final seedbed preparation can begin. The optimal amounts of fertilizer, pH, and organic additives should be determined by the soil tests.

Blending soils to create a soil medium with a greater absorption capacity might also be part of the plan. Soils of different particle size can be combined to increase the permeability or moisture-retaining capabilities of a soil. Layering is another means that can be used in conjunction with a barrier as part of the revegetation plan. *Layering* involves placing layers of different soil materials over a barrier to encourage root growth and aeration and to also allow moisture to drain away from the impermeable cap materials.

Seeding should be performed as quickly as possible after final grading. Generally, the most efficient seeding method for large areas is *hydroseeding*. In this method, seed, fertilizer, mulch, and lime are applied in a single operation.

Figure 7.2 Photograph of tracking on a slope.

For level areas, *seed drills* are often used, but several passes are required to apply all the constituents. All cultural seeding operations should be performed at right angles to the slope (parallel to the contours). If the plan includes shrubbery, trees, or seedlings, these must be planted by hand (Darmer 1979).

Mulches are generally recommended for all revegetation projects. The choices of mulch materials are broad and have a very wide range of characteristics. The complexity of materials aside, the role of mulch is important and should not be overlooked. The mulch is used as insulation against abrupt soil temperature changes and as protection from runoff and precipitation to reduce erosion and evaporation. Mulch also encourages infiltration and holds seed in place.

Using sod

Although sod material can be expensive and require additional installation efforts, it is appropriate for certain locations and applications. The range of plant materials available in sod has grown appreciably and now includes wildflowers as well as special-order mixtures in addition to the familiar fine turf sods. When using sod, it is important to confirm that the material conforms to the requirements set forth by the governing certification agency, which is usually the state department of agriculture. Sods are generally categorized by the quality of the root development as compared to the top growth.

The advantages of sod besides the immediate visual impact are the resistance of sodded areas to erosion and the quick cover of areas in the summer when cool season grasses will not grow. The preparation of the site to be sodded is similar to the seedbed process already described. The surface is graded to reflect the final grade and cultivated to a depth of at least 7.5 cm (3 in). The soil amendments and fertilizers may be placed before the cultivation. The sod is placed by hand so that the edges abut. Open joints and gaps should be filled with sod cut and shaped to fill the openings. After placing the sod, the area is rolled or lightly tamped to "seat" the sod onto the prepared surface. The keys to establishing a healthy sod on the site are the preparation and the followup. The followup must include regular water and care until the site is established.

Using sod in swales or slopes requires some extra precautions. The pieces should be installed from the bottom up toward the top in horizontal strips with the long edges of the sod running parallel to the contours. Individual pieces should be staggered to offset the vertical joints. On steep slopes greater than 5:1, it may be appropriate to anchor the sod in place using wood stakes driven flush with the surface of the sod.

Enhancing slope stabilization with trees

Although the grasses and forbs used to stabilize a slope immediately after a disturbance may become established and even thrive on a slope, the long-term success of the stabilization can be improved by incorporating trees into the stabilization plan. The deeper root penetration serves to bind the slope soils together and to provide additional cover and slope protection from precipitation. The use of nitrogen-fixing "nurse trees" should be included as part of the planting mix (Table 7.3). The nurse trees should represent approximately 25 percent of the tree and shrub component (Vogel 1987). The use of selected trees and shrubs also gives the designer some control over the slope as it matures. The evidence from the field of the volunteers which will establish over time suggests that trees and shrubs are a natural and perhaps even necessary element of the long-term stabilization of the disturbed slope (Figs. 7.3 and 7.4).

TABLE 7.3 Nitrogen-Fixing Trees and Shrubs

Indigo bush (*Amorpha fruticosa*)
Lespedeza (*Lespedeza* spp.)
Bristly locust (*Robinia fertilis*)
Rose acacia (*Robinia hispida*)
Autumn olive (*Eleagnus umbellate*)
European black alder (*Alnus glutinosa*)
Black locust (*Robinia pseudoacacia*)

Figure 7.3 Photograph of trees planted on a slope as part of a stabilization plan.

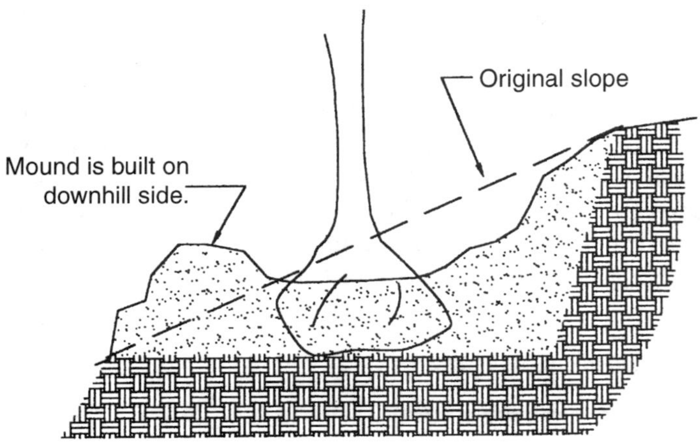

Figure 7.4 Tree planting on a slope detail.

Table 7.4 lists some trees and shrubs, along with their characteristics, that are often used in restoration projects.

Although a tree has very little value in protecting a "new" slope from erosion, it does have a role to play over time. The canopy provides protection from the sun and rain to the soils and shade to the understory. The leaves from deciduous trees contribute to the ground litter, which again serve to

TABLE 7.4 Characteristics of Selected Trees and Shrubs for Use in Restoration

Plant type and characteristics
Eastern Region of the United States
Washington Hawthorne (*Crataegnus phaenophrum*), tree. Deciduous. Medium growth rate to 9 m (30 ft). Dense twiggy upright growth. 1.5–3.0 m (5–9 ft) spacing on well drained to moderately well drained soils.
Scotch pine (*Pinus sylvestris*), tree. Evergreen. Rapid growth rate to 15 m (50 ft). Dry to somewhat poorly drained soils. Very rugged tree. Lower pH limit is 4.0. Throughout most of eastern United States.
Common juniper (*Juniperus communis*), tree. Evergreen. Slow growth. Uses spacing of 1.5–2.0 m (4–6 ft). Prefers limestone soils. Dry to moderately well drained soil.
Black locust (*Robinia pseudoacacia*), tree. Deciduous. Rapid growth to 15 m (50 ft). Can be direct seeded. Widely adapted to different soils. Good leaf litter. Used in planting mixes. Nitrogen fixing. Lower pH limit is 4.0.
White pine (*Pinus strobes*), tree. Evergreen. Rapid growth. Prefers rich, moist, especially heavy soils. Used for screens. Lower pH limit is 4.0.
Hackberry (*Celtis occidentalis*), tree. Deciduous. Moderate rate of growth to 8 m (25 ft). Tolerates acid soils to pH 5.0. Tolerates poorly drained soils to excessively drained soils.
Thornless honeylocust (*Gleditsia triacanthos enermis*), tree. Deciduous. Moderate growth rate to 11 m (35 ft). Moderately drained soils. Lower pH limit is 6.5.
Red oak (*Quercus rubra*), tree. Deciduous. Moderate rate of growth in acid soils.
Gray dogwood (*Cornus racemosa*), tree. Deciduous. Rapidly growing shrub. Prefers sunny location but will tolerate shades. Will grow in wide range of soils. Forms colonies. Lower pH limit is 5.0.
Red chickberry (*Aronia arbutfolia*), shrub. Deciduous. Moderate growth rate. Tolerates dry to somewhat poorly drained soils.
Tartarian honeysuckle (*Lonicera tatarica*), shrub. Deciduous. Rapidly growing in well-drained sunny location. Lower pH limit is 5.0.
Arrowwood (*Viburnum dentatum*), shrub. Deciduous. Rapid growth. Prefers well-drained to moist soils. Sunny location. Lower pH limit is 4.0.
Red osier dogwood (*Cornus stolonifera*), shrub. Deciduous. Rapid growth in well-drained soils. Sunny location. Forms thickets. Lower ph limit is 4.5.
Forsythia (*Forsythia intermedia*), shrub. Deciduous. Rapid growth in well-drained soils. Sunny location. Tolerates stony rough slopes. Vigorous growth.
Japanese juniper (*Juniperus procumbens*), shrub. Evergreen. Rapid growth. Sandy and loamy moderately moist soil. Prefers sun. Hardy, low-spreading shrub.
Sargent juniper (*Juniperus chinensis sargentis*), shrub. Evergreen. Moderate rate of growth. Prefers moist, slightly acid sandy soils. Tolerates droughty banks. Low-creeping shrub.
Indigo bush (*Amorpha fruticosa*), shrub. Deciduous. Adapted to wide conditions. Fairly slow growth rate. Can be seeded. Lower pH limit is 4.0.
Autumn olive (*Elaeagnus umbellate*), shrub. Deciduous. Competes well with established herbaceous layer. Used as a nurse plant. Lower pH limit is 4.0. Growth to 6 m (20 ft).
Western Region of the United States (Less Than 80 in of Rain)
Arizona cypress (*Cupressus Arizonica Greene*), tree. Evergreen. May be established from direct seeding. Persistent once established, but as a seedling, only poor tolerance for drought. Prefers gravelly northern or cut slopes. Tolerates high temperatures. Precipitation range from 16–20 in. From southern Texas to Arizona.

TABLE 7.4 Characteristics of Selected Trees and Shrubs for Use in Restoration (Continued)

Big tooth maple (*Acer grandidentatum Nutt.*), tree. Deciduous. May be established by seeding, but slow rate of growth for seedlings. Has a moderate rate of spread once established. Prefers well-drained soils of porous sandy to gravelly loams. Will grow on steep slopes. Tolerates moderately acidic to slightly basic soils. From Utah and western Wyoming to southeastern Arizona and New Mexico.

Bur oak (*Quercus macrocarpa*), tree. May be established by seed or by container stock on difficult sites. Adapted to a wide range of soils. Will tolerate soils from a pH of 4.0 to moderately basic. Drought resistant but intolerant of floods. Precipitation range from 15–40 in. From the Dakotas and northeast Wyoming to Midwest and South to Texas.

Green ashe (*Fraxinus pennsylvanica*), tree. Deciduous. Although it is slow to establish and may need protection from competition initially, it is tolerant of moderately basic to strongly acid (ph 4.0) soils. Prefers alluvial soils. Tolerant of periodic flooding and drought. Precipitation range 15–45 in. From central Montana and Wyoming to Dakotas, throughout eastern United States.

New Mexico locust (*Robinia neomexicana A. Grey*), tree. Deciduous. Fair rate of success from seed. Seedlings drought resistant. Good spread particularly on harsh sites. Thicket forming. Prefers moist soils on canyon bottoms, bottom of north slopes. From west Texas to Arizona through Utah and Colorado.

Pinyon pine (*Pinus edulis*), tree. Evergreen. Best results with nursery stock. Very long lived. Good rates of natural spread. Can become a pest once established. Adapted to calcerous caliche soils. Good on harsh eroded sites. Tolerant of drought and heat. Precipitation range 12–18 in.

Ponderosa pine (*Pinus ponderosa*), tree. Evergreen. Best results with nursery stock. Good drought and fire tolerance. Not tolerant of shade or saline and sodic soils. Precipitation range from 15–25 in but can survive on as little as 7 in. From western Dakotas to Montana to Arizona and Texas.

Quaking aspen (*Popolus tremoloides*), tree. From container or nursery stock. Short lived but forms thickets or colonies with extensive shallow root system. Prefers deep sandy to silty loam soils ranging from moderately basic to moderately acidic. Widely adapted to western United States. Precipitation from 15–30 in.

Antelope bitterbrush (*Purshia tridentate*), shrub. Evergreen. Persistent on a range of soil pH. Good for stabilization. Useful as browse. Resists drought and moderate salt. Intolerant of high water table, flooding. New Mexico to California, north to British Columbia. Precipitation 10–25 in.

Apache-plume (*Fallugia paradoxa*), shrub. Semievergreen. Weak competitive ability, but establishes quickly on disturbed slopes. Prefers full sun. Moderately basic, well-drained soil. Tolerates drought and salt. West Texas through Arizona.

Big sagebrush (*Artemsia tridentate*), shrub. Long lived, persistent competitor. Prefers well-drained, deep, fertile soils. Will tolerate a range of pH conditions. Tolerant of drought and salt. Intolerant of high water table. From Arizona and New Mexico to Nebraska. Precipitation range from 7.5–17 in.

Chokecherry (*Prunus virginiana*), shrub. Fair to poor germination when seeded, but spreads quickly from roots to form thickets. Does well with grasses and forbs. Prefers well-drained, moderately acidic to moderately basic, silty to sand soils. Intolerant of clayey or poorly drained soils. From 12–30 in of precipitation. Widely distributed in cooler areas of northern and western United States.

Curlleaf mountain mahogany (*Cercocarpus ledifolius*), shrub. May be difficult to establish from seed, but persistent once established. Prefers basic, well-drained clayey soils. Tolerant of drought and salt. New Mexico and Arizona to Montana to 10,000 ft. Precipitation 6–20 in.

TABLE 7.4 Characteristics of Selected Trees and Shrubs for Use in Restoration (Continued)

Desert bitterbrush (*Pursha glandosa*), shrub. Establishes well from seed. Tolerant of drought. Persistent and spreading. Prefers well-drained sandy to clayey soils. Southwestern Utah to southeastern Nevada and California.

Douglas rabbitbrush (*Chrysothamnus viscidiflorus*), shrub. Persistent and excellent spreader. May compete with grasses and forbs. Prefers basic, well-drained clayey to coarse-textured soils. Broad range of adaptability. Weak growth in acid soils. Precipitation range from 6–20 in. From New Mexico and Arizona to Montana.

Fringed sagebrush (*Artemisia frigida*), shrub. Fair competitor, but may be slow to establish from seed. Prefers well-drained neutral to slightly basic soils. Tolerant of drought. Fair salt tolerance. Precipitation range 8–20 in.

Gambel oak (*Quercus gambelii*), shrub. Persistent once established, but spreads slowly. Prefers sandy and gravelly loams on slopes. From west Texas to Arizona to Utah and Colorado, southern Wyoming. Precipitation range 16–20 in.

Golden current (*Ribes aureum*), shrub. Persistent spreader. Good compatibility with grasses and forbs. Adapted to well-drained alkaline soils on shallow slopes. From Utah to eastern California. Precipitation range 8–14 in.

Gray molly summer cypress (*Kochia Americana* var.), shrub. Prefers alkaline or saline clay soils. Excellent drought and salt tolerance. Precipitation range 6–10 in. From New Mexico and Arizona to Montana and Idaho.

Green ephedra (*Ephreda viridis Coville*), shrub. May be established by prepared direct seeding. Poor rate of spread. Prefers well-drained alkaline soils. Adapted to dry shallow soils on slopes. Tolerant of salt. Excellent drought resistance. Intolerant of high water table or floods. Precipitation range from 8–14 in. From Utah and northern Arizona to eastern California.

Longleaf snowberry (*Symphoricarpos longiflorus Gray*), shrub. May be difficult to establish by seed, but good persistence and compatibility. Prefers well-drained to dry soils. Will tolerate acid and basic conditions. Fair drought tolerance. Intolerant of salt. Oregon south to Texas and California.

Saskatoon serviceberry (*Amelanchier alnifolia*), shrub. Persistent with fair spread after only fair establishment by either seed or cutting. Prefers medium-textured, well-drained soils. Fairly resistant to drought, but intolerant of flooding or salt. Precipitation from 14–20 in. Western Texas and New Mexico to Montana to West Coast.

Shadscale (*Atriplex confertifolia*), shrub. Establish from cuttings. Good rate of spread after established. Prefers alkaline soils. Adapted to a range of soil textures. Excellent tolerance of salt and drought. From New Mexico to Canada, west to eastern California, Oregon and Washington. Precipitation range from 4–8 in.

Siberian pea shrub (*Carangana arborescans*), shrub. May be established by seeding. Persistent with fair rate of spread once established. Fair compatibility with grasses and forbs. Will tolerate soil pH across broad range (pH 4.0–12.0). Prefers well-drained soils. Adapted to shallow, infertile, and rocky soils. From northern Great Plains to central Utah and Colorado. Introduced from Siberia and Manchuria.

SOURCE: Adapted from Willis G. Vogel, *A Manual for Training Reclamation Inspectors in the Fundamentals of Soils and Revegetation,* prepared for the Office of Surface Mining and Enforcement by the U.S. Department of Agriculture (Berea, Ky., 1987), p. 67; D. Brown, C. L. Hallman, J. Skogerbee, K. Eskern, and R. Price, *Reclamation and Vegetative Restoration of Problem Soils and Disturbed Lands* (Park Ridge, N.J.: Noyes Data Corp., 1986); and P. Rogoshewski, H. Bryson, and R. Wagner, *Remediation Action Technology for Waste Disposal Sites* (Park Ridge, N.J.: Noyes Data Corp., 1983).

stabilize the surface and contribute to the evolving soil structure (Vogel 1987, p. 67).

The expense of tree planting on very large sites might be offset by using plantings of seedlings. The individual seedling and installation costs are relatively low when compared to larger trees, but some investigation into mortality rates for various species and long-term costs is still justified. The solution may be to plant a combination of sizes and species. The mix can take advantage of the lower cost of younger trees and the greater vigor and success rate of larger trees. The final plan is of course a reflection of budget, site characteristics, and available plant species.

The competition between herbaceous plants and trees and shrubs may be a matter of concern. *Allelopathy* (that is, the chemical competition between plants) or even the shading of young trees and shrubs by a layer of vigorous herbaceous plants may require intervention to prevent damage from competitive stress on the slower-growing woody components of the planting plan. The use of larger trees and/or shrubs to overcome the influence of the layer of herbaceous plants is cost prohibitive. A more workable approach would be to plant the site in alternating strips. The herbaceous strip would be fertilized, but the alternating strip containing shrubs and/or trees would be fertilized for the woody plants and planted with a layer of unfertilized herbaceous plants (Vogel 1987, p. 37).

Trees and shrubs will do better in a soil that is about 30 in thick as opposed to a thinner soil, which is appropriate for the grasses and legumes. Compacted soils will impede root growth, so that even routine root ball planting methods may be inadequate in disturbed soils. The heavily compacted soils, which do not allow root penetration anyway, are smoothed during the excavation of the planting pit, which further reduces the already limited pore space available for the transmission of air and water. The "teacup" effect occurs when the backfilled soils in the planting pit become saturated with water that cannot drain away through the compacted surrounding soil. Ultimately, a plant caught in this circumstance will die.

Contemporary recommendations for planting trees discourage using the "pot" method even in native soils. The new understanding of plant growth recommends that the planting pit be five times the width of the root ball, but only deep enough to situate the root ball at the proper depth. The slope stabilization area should be fertilized with approximately 500 lb of 10-10-10 fertilizer per acre, which should be worked into the slope prior to the application of mulch. Organic matter can be added to the soil. The root should sit on undisturbed soil to reduce settling. The root area should be carefully backfilled to eliminate large voids without compacting the soil in the pit too much. Water can be used to settle the ground naturally (Fig. 7.5).

The slope should then be mulched with woodchips to a depth of 4 to 6 in on the slope immediately upon completion of grading. Woodchips should be approximately 2 in^2 in size, and the mulch should be applied uniformly over the planted area.

Figure 7.5 Photograph of a watering system for a newly planted tree.

Trees should be of a species that has adapted to growing on slopes. Seedlings should have had two full growing seasons in nursery beds prior to planting. Seedlings should be set vertically and roots spread carefully in a natural position in the planting hole. All trees should be thoroughly watered the day they are planted. All excess excavated material should be used to make curbs for water retention.

Streams

In most urban and suburban areas, stream buffers are routinely eliminated or severely minimized as part of the development process. The developed landscape quickly concentrates accumulating runoff and conveys it to streams. Most of the runoff that reaches stream buffers is in the form of a concentrated flow rather than a sheet flow that might occur in an undeveloped drainage condition. In this concentrated form, the runoff crosses the area that would have been the stream buffer in a pipe or channel. The concentrated runoff also conveys directly to the stream the particulates and pollutants that would have been filtered and trapped by the buffer.

Protecting existing stream buffers and banks is far preferable to having to restore them and trying to mimic what nature had already established (Table 7.5). Stream buffers should be designed to provide at least a minimum width indicated by the specific site and stream conditions (Fig. 7.6). Effective stream buffers in developed or urbanized areas may range from as narrow as 20 ft to more than 200 ft, depending on the topography, the amount of impermeable area, and the degree to which runoff is concentrated. Most local ordinances and standards are amalgamations of experience and liberal borrowing from other standards. Communities that have stream buffer standards require a minimum total width of at least 100 ft, or they require the buffer to include the 100-year floodplain.

Stream buffers are usually designed to incorporate three distinct zones or functions. The zone nearest the stream usually extends a minimum of 25 ft from the stream bank, and improvements are significantly restricted to such

TABLE 7.5 Benefits of Urban Stream Buffers

1.	Reduce small drainage problems and complaints
2.	Allow for lateral movement of the stream
3.	Provide flood control
4.	Protect from stream bank erosion*
5.	Increase property values*
6.	Enhance pollutant removal
7.	Provide food and habitat for wildlife*
8.	Protect associated wetlands
9.	Prevent disturbances to steep slopes*
10.	Mitigate stream warming*
11.	Preserve important terrestrial habitat*
12.	Supply corridors for conservation*
13.	Provide essential habitat for amphibians
14.	Reduce barriers to fish migration
15.	Discourage excessive storm drain enclosures and/or channel hardening
16.	Provide space for storm water ponds
17.	Allow for future restoration

*Benefit is amplified by or requires forest cover.

SOURCE: Center for Watershed Protection, *Better Site Design: A Handbook for Changing Development Rules in Your Community,* prepared for the Site Planning Roundtable, Ellicott City, MD, 1998, p. 130. Used with permission from The Center for Watershed Protection.

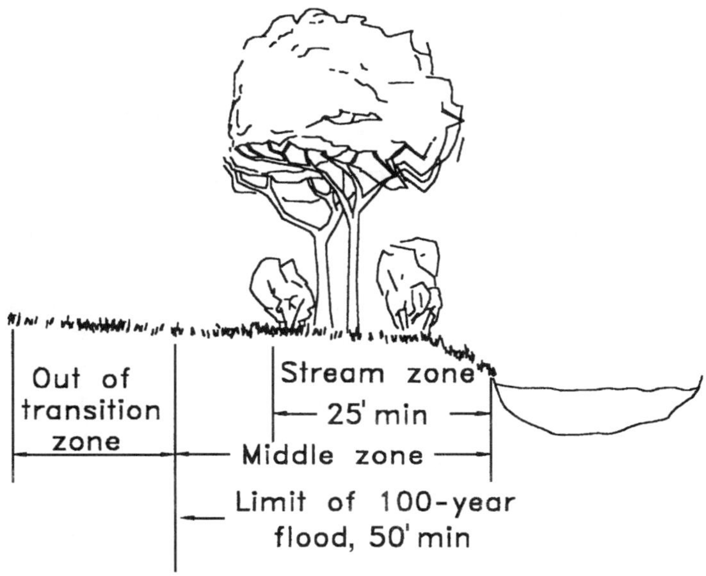

Figure 7.6 Stream buffer design detail.

minimal encroachments as unpaved footpaths and swales. Utilities and paved crossings are kept to a minimum, and the stream zone remains in mature and dense vegetative cover. Beyond the stream zone, the middle zone is often used for complementary purposes such as bike paths, or perhaps storm water appurtenances. The middle zone contains the 100-year floodplain and often contains seasonal wetlands and other habitat features. The middle zone is usually a minimum of 50 ft, but its width in practice is a function of the floodplain width, the presence of critical habitat or wetlands, and the topography. The outermost zone is an area of initial transition from the more developed landscape to the stream. It may be a lawn with shrubs and trees in which a variety of activities may be conducted. Gardens and recreation activities are completely compatible with the transition zone.

The goal of the stream buffer is to re-create or to maintain to the extent possible the predeveloped conditions or overland sheet flow, infiltration, and the process of filtration and deposition provided by vegetation. In general, native species of vegetation should be encouraged and invasive exotics removed. The shape of the stream buffer may be fairly irregular if it is properly designed to account for the floodplain, important habitat, and topographic features. The development planning can accommodate the variations in the buffer by varying lot width and depth or by modifying the site layout. Stream crossings should always be kept to a necessary minimum.

Streams are formed by water and gravity and influenced by geology and climate. Less obvious are the biological elements of a stream that contribute to its function. Climate dictates the amount of water available through precipitation and evaporation, and it influences the landscape in terms of vegetation, the richness and texture of the soil, and so on. Geology affects the character of soil and the slope of land. The amount of water and the distribution of water ultimately affect the character of a stream. Climate also dictates the biological character of the watershed, which has important influences on the stream. The amount of rain is important, but the distribution of rain over time and geography is also important. The richness and texture of soils (which are functions of climate and geology) and the interactions of soil with plants and animals determine the rate of infiltration, as well as the amount of sediment that is washed into a stream, the amount of water supplied to the stream in summer or dry months, and to some extent, the temperature of surface water.

In humid climates the action of precipitation and running water coupled with the biological activity weather rock into soil very quickly. The presence of decaying vegetation contributes an important organic element, which in turn creates a rich complex soil texture that results in limited overland flow, high infiltration rates, and ultimately the slow passage of water through the soil to streams. In arid climates there is less precipitation and less vegetation, and the infiltration rates are slower. In such climates, there is also more runoff, and the landscape reveals the erosive power of high-velocity concentrated flows.

The velocity of flowing water is a function of the slope of the channel, the resistance offered by the stream bed and banks, and the depth of flow. The steepness of the stream gradient gives flowing water the velocity and the power to erode channels. Velocity is always faster in the deepest parts of the channel and away from the sides. A deep river with the same gradient or slope as a shallow stream will have a much greater velocity.

Gravity and friction also play important roles. *Friction* is the resistance to flow or the presence of obstructions that tend to impede the downward flow of water. Friction in streams is not easily modeled or well understood. The resistance to flow is influenced by the stream's roughness. Roughness is affected by the size of materials in the channel that make up the bed and banks, the amount and type of vegetation on the bed and banks, and the amount of curvature in stream. The shearing action of water running over a stream bed results from the interaction of gravity and friction and causes the erosion of the bed and banks. As velocity increases, shear and erosive actions tend to increase.

When there is an obstruction in the stream, the obstruction causes an increase in turbulence and directs flow toward the stream bottom or sides. As a result of turbulence, shear stresses increase on the stream bank or bottom and the bank erodes or the bottom is scoured.

In the course of forming channels, streams will flood, change channels, scour pools, fill pools, erode banks, meander, and so on, as a result of changes in flow and sediment carried by the stream (Figs. 7.7 and 7.8). A stream is always trying to reach *dynamic equilibrium,* which means the amount of water and sediment that enter a stream are equal to the amount of water and sediment that leave it. To reach and maintain equilibrium, the stream will adjust its channel to reflect changing conditions in the watershed. A stream may erode the stream bed to create a deeper channel or erode the banks to make a wider

Figure 7.7 Photograph of step pools.

Figure 7.8 Photograph of natural stream meanders.

stream or deposit sediment to create islands, and so on. These changes are always in response to the stream's tendency toward dynamic equilibrium—to balance the energy of the water and the sediment load.

A purely hydraulic system could operate without a gradient because accumulated surplus water can generate its own surface slope and is capable of flow on a horizontal surface. The *energy grade line* is a graphic representation of the potential energy (*head*) possessed by the river along its longitudinal profile. The loss in head with distance reflects the amount of energy consumed by resisting elements in the system. Head loss is greater over riffles than through pools.

The transport of bed load requires a gradient, and it is in response to this requirement that a stream channel system adjusts its gradient and achieves an average steady state of operation year in and year out. In nature the balance between a stream's load and its capacity exists only as an average condition. In terms of the moment, a stream is rarely in equilibrium, but it is in a constant dynamic change toward equilibrium. The stream never stays near equilibrium for long because its conditions are subject to constant change—floods, droughts, and manmade alterations to the stream or its channel. A stream in a state of dynamic equilibrium is called a *graded stream.* Alternate deepening by scouring and shallowing by deposition are responses to changes in the stream's ability to transport its load. This process is known as *degradation.* In circumstances in which there is capacity for more bed load, rivers will scour and deepen. Where there is too much bed load, the stream gradient increases, velocity increases, and channels widen. This process is known as *aggradation.*

Sinuosity

The line connecting the deepest parts of a channel is known as the *thalweg.* Water does not flow in a straight line. Even in channels that appear to be straight, the thalweg will be shifting from side to side, which will ultimately create an alternating series of bars. Straight stretches often contain alternate bars formed by material deposited on the channel bed along the sides alternating down the channel. Deep pools are formed opposite of the alternate bars, and the shallow riffles are found midway between. Meanders serve to lengthen the channel, dissipating the stream's energy over longer distances than would a straight channel, which results in a more stable stream. The distance along the centerline of a channel (*channel length*) divided by the distance between meanders is used as a measure of stream sinuosity. One effect of meandering is to increase resistance and with it energy dissipation at the pools, making the grade line more uniform; thus meandering serves to help the stream reach a condition of near equilibrium. It should be noted that *meandering* is synonymous with *bank erosion.*

Sinuosity refers to the classification of streams by their pattern. There are three types of stream patterns: sinuous, braided, and meandering. However, it should be noted that these are relative patterns, and there is no bright-line distinction between them. Streams with a channel versus valley length of less than 1.5 are considered to be *sinuous,* those with 1.5 or greater are considered to be *meandering,* and those greater than 2.1 are said to be *tortuous.* A straight stream would have a sinuosity of 1. *Braided streams* are those that do not have single main channel. Braiding occurs in circumstances in which material has been deposited in a channel, perhaps during a high-flow condition, where it forms an island. Under lower-flow conditions, the accumulated material becomes stabilized by vegetation with which it resists displacement in later floods.

The process of stream meandering is complex and difficult to accurately model. Stream behavior is a function of gradient, volume, the quantity and character of sediment, and channel roughness and composition. As volume changes, the stream's capacity to do work changes as well so that an accurate model must consider many different conditions. In spite of the difficulties, some practical methodologies have been developed. In general, a stream meander radius of from 2.7 to 2.8 the bankful width is recommended.

Stream assessment

The most commonly used method of stream assessment for restoration purposes in the United States today is the *Rosgen method* developed by David Rosgen (Fig. 7.9). The Rosgen method is performed on four prescribed levels and is fairly sophisticated. Use of the method requires specific training or study. The Rosgen method is fairly comprehensive, yet it is sufficiently flexible that it can be modified to fit a particular stream as needed.

The first level of assessment is the *geomorphic characterization,* which uses aerial photography and topographic mapping to identify the stream as one of 11 valley types characterized by the gradient of the stream and the topography of the valley. Streams are identified as one of 8 general types based on the geometry of their channel and the surrounding floodplain.

Figure 7.9 Photograph of a stream restored using the Rosgen stream assessment method as a guide.

The *morphological description* is the second Rosgen level, and it involves a more detailed assessment using actual measurements of the stream. At this level, the stream is characterized into 1 of 94 categories based on the degree of the stream's entrenchment, the ratio of its width to depth, its surface gradient, its bed materials, and its sinuosity. Stream bed materials include organic and inorganic components and may range from large boulders to fine organic sediments. Organic material is important to the living members of the stream ecosystem since it provides the basis for the food chain. The ratio of width to depth is considered to be a critical indicator of stream stability.

The third level involves using the findings and experiences of the first two levels to summarize the existing conditions and to evaluate the stream's stability potential based on existing riparian vegetation, patterns of in-stream deposition and meander, and the quality of in-stream habitat. From these first three levels of evaluation preliminary conclusions can be drawn. These conclusions inform the fourth level of inquiry that utilizes measurements of stream flow, stability, and sediment over time.

It should be noted that not all stream restoration projects require an approach as sophisticated as the Rosgen method. Some very serious stream degradation situations may require immediate attention that should not wait upon a Rosgen full assessment. In the opposite scenario, some fairly simple restoration efforts may be completely adequate for only nominally degraded streams.

Riparian zones

Healthy functioning streams are not only hydraulic or hydrologic systems but also biological systems (Figs. 7.10 and 7.11). Therefore, a key element of stream quality is the health of the contributing watershed. Some ways in which vegetation and other biotic elements of riparian systems contribute to stream quality are the following:

1. Roots of trees, sedges, shrubs, and so on bind the soils of banks to increase the stability of stream banks and to resist erosion.
2. Overhanging vegetation shades streams, which keeps water along the stream's edges cooler.
3. Biotic debris decays and provides nutrients to water.
4. Biota contribute to the health of riparian and upland soils to increase infiltration while decreasing erosion and sedimentation.
5. Animals build dams, wallows, and so on.

Stream bank stabilization

The stabilization of a stream bank requires careful consideration of the stream under various conditions. Stabilization elements must be designed with consideration of the various flow conditions that should be anticipated in the stream. For example, wingbars or revetments installed to protect the

Figure 7.10 Photograph of a well-developed riparian zone.

Figure 7.11 Photograph of a stream section with a dysfunctional riparian zone.

stream bank under normal flow conditions may increase erosion under flood conditions. Likewise an underdesigned stabilization feature may not be able to resist the velocities and energy of a flood and may be damaged or destroyed. The variability of streams makes prescriptive design impractical.

In the past many stabilization projects have consisted of merely armoring the stream with various materials specified to be heavy enough to resist flood conditions. In heavily developed watersheds with high-velocity frequent floods, the designer may have little choice. Unfortunately, a stream stabilized in such a manner offers little in the way of stream functions except to convey the runoff away. Armoring eliminates many of the valuable functions of the stream corridor by limiting the biotic and hydrologic interaction between the stream and the riparian zone.

Under circumstances where armor is not necessary, there are numerous methods and strategies that have been successfully applied. Many of these methods have evolved from the experiences of community activists and outdoor sports organizations interested in stream quality. In fact, in most communities in the United States today, there are watershed interest groups. Many of these groups have acquired a great deal of expertise in watershed and stream corridor protection. The focus of these groups is usually the protection or restoration of stream function.

Restoring overhang and bank conditions that favor improved habitat and riparian function begins with an inventory of habitat areas, debris locations, and stream transects to identify vegetation types, vegetation overhang, measure of shaded area, condition of stream banks (angle and height), undercut banks, channel width and water width, depth and velocity, gradient, substrate composition, and pool-to-riffle ratio (Hunter 1991) (Figs. 7.12 through 7.14).

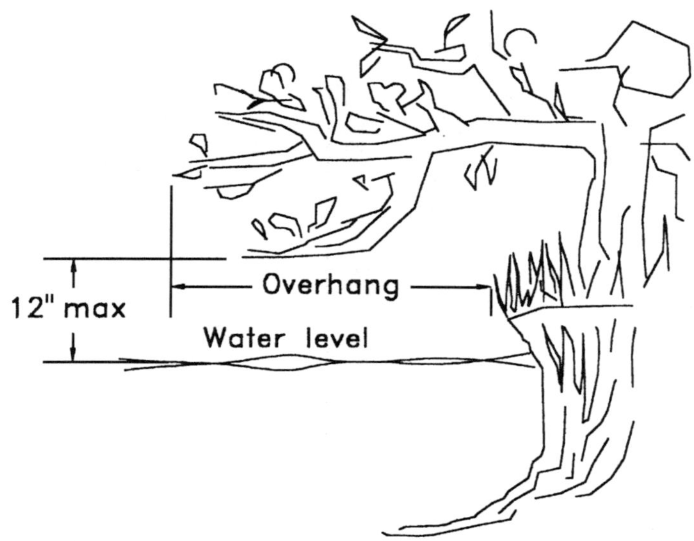

Figure 7.12 Measuring stream overhang detail.

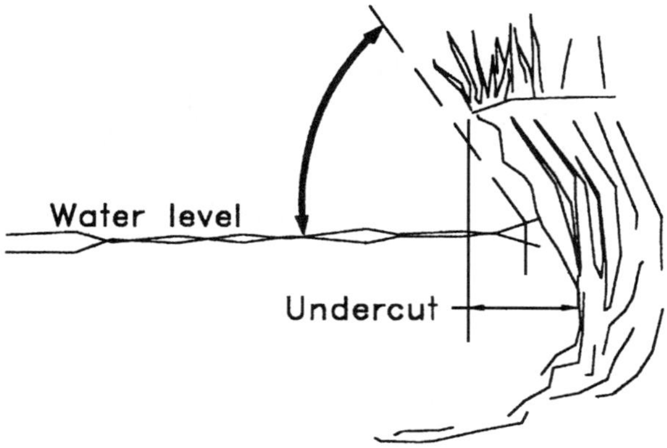

Figure 7.13 Measuring bank angle detail.

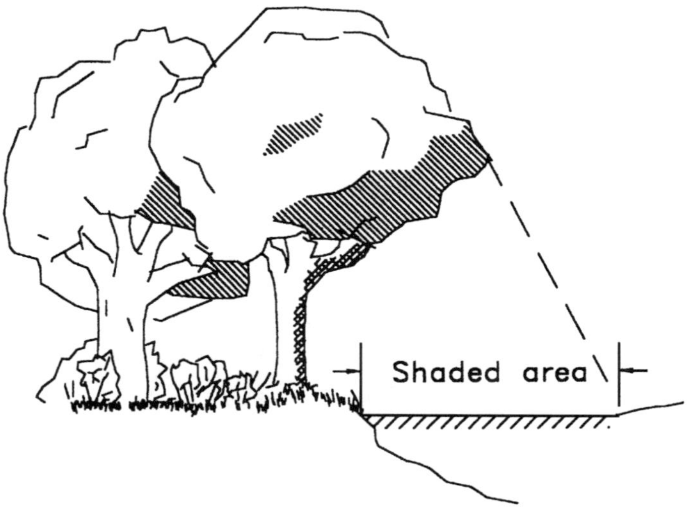

Figure 7.14 Measuring shade detail.

The protection and cooling effects of overhanging vegetation and bank angle cannot be overlooked when assessing stream banks (Figs. 7.15 and 7.16). Overhanging vegetation protects aquatic life from observation from above and shades the water from the sun. To be effective as cover, overhanging vegetation should be no more than a foot above the surface of the water. The angle of the banks also contributes to the protection and cooling effects. Overhanging banks are particularly important to habitat. A vertical bank or banks with an angle greater than 90° offer no protection from predators or the sun. Very often degraded streams have suffered as the bank erodes and the overhanging bank is lost. In urban streams, the stream bank may over time "retreat" from the stream, leaving the stream more and more exposed.

As part of the stream assessment, it is important to determine the stream width as opposed to the water width on the day of the assessment. The stream assessor

Figure 7.15 Photograph of an urban stream in which the bank is retreating from the stream.

Figure 7.16 Photograph of an impacted stream.

usually will measure the stream or channel width as the distance across the channel from the edge of terrestrial vegetation to the opposite edge. In an impacted stream, the channel width may be significantly different from the water surface width. Variations between channel width and surface water width may also be attributed to seasonal or annual variations in flow that must be accounted for.

Bank restoration for habitat purposes involves restoring the relationship of bank angle and height. Many innovative ways of re-creating this relationship have been developed using local available materials or manufactured materials. Many excellent projects have been completed using timbers or logs (Fig. 7.17). In general, if and when the logs decompose, the bank will have already reestablished itself and will no longer need the structural support. Fiber fascine has been developed for this purpose (Fig. 7.18). The *fascine* is a bundle of live cuttings wired or lashed together and secured, usually at the toe of a bank at or near the water edge. The fiber fascine allows a maximum of flexibility, strength, and permeability much like an established mass of roots might. The materials are lightweight and easily handled and provide an effective medium for encouraging root growth.

The fascine is used in conjunction with other stabilization methods and is an excellent bank protection method for stabilizing the toe of slopes especially where there is outflow from the bank or where water levels fluctuate. The fascine allows water to pass through it and provides protection and stabilization even before the cuttings begin to establish themselves. Fascines are not used in places where surface water or drainage will run over them. If cutting materials are locally available, the fascine is fairly inexpensive. Also, they are easy to install but do require knowledgeable installers. The flexible, sausage-like character of the fascine allows it to be fitted to the conditions found in the field. Fascines are constructed from plants that root easily such as willows. Cuttings are bundles with all the butt ends placed in the same direction and are wired together every foot to foot and a half. The cuttings are usually about 2 to 3 ft long.

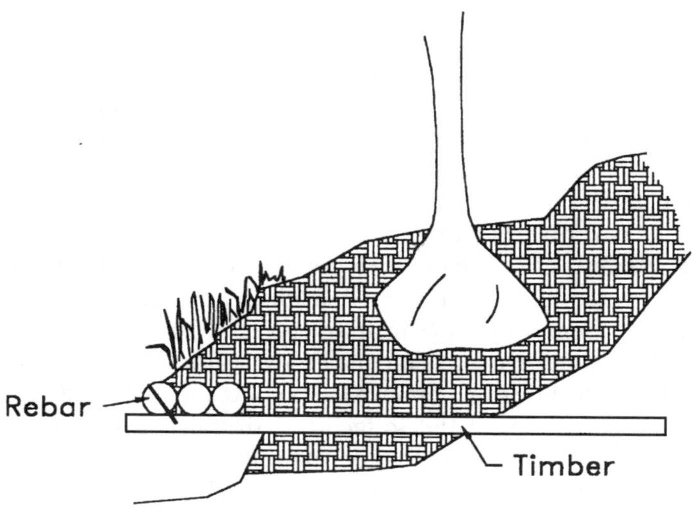

Figure 7.17 Timber bank restoration detail.

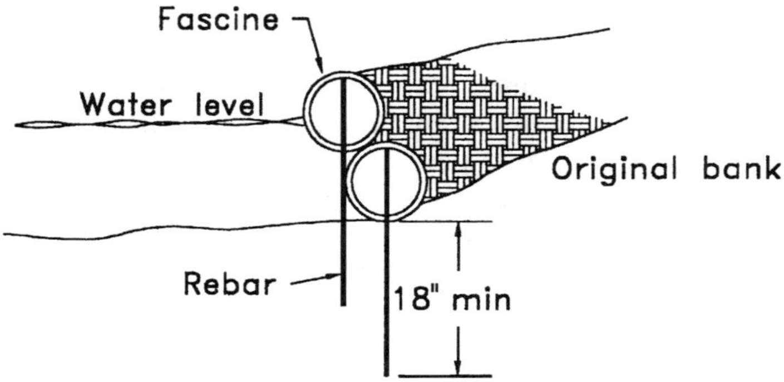

Figure 7.18 Fiber fascine bank restoration detail.

Establishing stream bank vegetation

Stream bank vegetation provides mechanical stabilization of the bank by its roots and acts to absorb some of the energy of flooding. It is sometimes difficult to reestablish vegetation on eroded or damaged stream banks if there is substantial traffic (pedestrian or agricultural) or if there is a great deal of shade from overhanging trees. Of course the newly planted bank is subject to damage by floods and periods of significant rain until it is established. Generally speaking, the strategies for reestablishing vegetation are limited to planting cuttings or seedlings of woody plants or direct seeding. Within the general strategies are a variety of methods.

Live stakes. *Live stakes* are, as the name suggests, living woody plant cuttings that will tolerate cutting and still be capable of quickly establishing a new root system. They are usually fairly sturdy and will withstand being lightly driven into the stream bank. The live stake is substantial enough to withstand light flooding and traffic and will develop into a fairly robust shrub or tree in short order. Even live stakes, however, will not resist much traffic or active erosion. It is a fairly common strategy because it is inexpensive if cuttings are locally available, it takes little time or skill to install, it can be done quickly, and, if done properly, it results in a permanent solution. Live stakes often are not sufficient in themselves to stabilize a stream bank or to reestablish effective vegetative cover so they are often used in conjunction with other methods. Live stakes are most effective on fairly moderate banks with a slope of 4 to 1 or flatter. They are installed in stabilized original bank soil, not on fill.

Live stakes of from 1/2 to 1 1/2 in diameter and from 2 ft to 2 1/2 ft long are installed during the dormant season and at low water (Fig. 7.19). Stakes are alive with bark intact and branches removed. The butt end is usually cleanly cut at a 45° angle. The top end is cut flat to facilitate driving. Live stakes should be cut with at least two bud scars near the top to promise growth and development. All cuttings should be fresh and moist and not stored for more than a day prior to installation. It is preferable if they are installed the day of cutting.

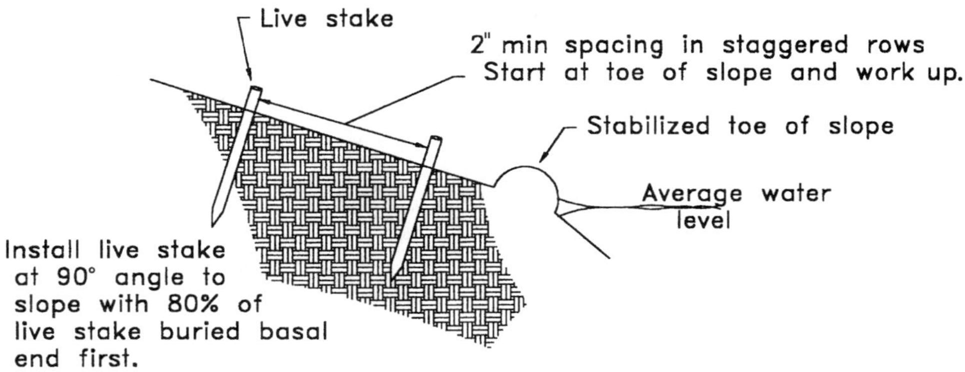

Figure 7.19 Live stake installation detail.

Branch packing. *Branch packing* is often used to stabilize small washed-out sections of bank, and it involves filling a washed-out or excavated area with alternating layers of soil and live branches (Figs. 7.20 and 7.21). Branch packing is used for relatively small areas rarely larger that 10 to 15 ft long, 5 or 6 ft wide, and more than 4 ft deep. It requires quite a bit of material and labor. The method has been used underwater. Branch packing is done during the dormant season and when water levels are low. Branches may vary in size ranging up to 3 in in diameter. Branch length varies with the size of the washout and the point of installation, but all stakes should be long enough to extend from the stabilized face of the new slope back into the original bank soil, as shown in Fig. 7.20.

A more robust version of branch packing is the construction of *cribwalls* using logs (Fig. 7.22). This construction is used in areas larger than branch packing might address or in places where strong currents reduce the effectiveness of other methods. The cribwall involves building a rectangular structure of logs and backfill similar to branch packing except that the materials tend to be larger and heavier.

Construction involves preparing a foundation 2 to 3 ft below the existing stream bed. The first logs are placed parallel to the direction of flow with the second series perpendicular to the first and so on with each successive layer installed at right angles to the preceding one until the crib is about 60 percent more or less of the finished height of the bank. Each layer of logs should slightly overlap similar to the construction of a log cabin. Logs should be at least 6 in in diameter and should be secured in place using a rebar driven through a hole drilled at the corners. Since the cribwall is used in areas where strong currents or high flows are expected, the leading and trailing edges should be protected with riprap.

As the structure rises, it is filled with a soil and gravel mixture. Live cuttings should be installed at the top of each log facing the stream. The size of the cribwall structure may limit the opportunity to bring the live stake in touch with the original bank material. Instead, the live stake should be securely driven into the fill material at a slightly downward angle.

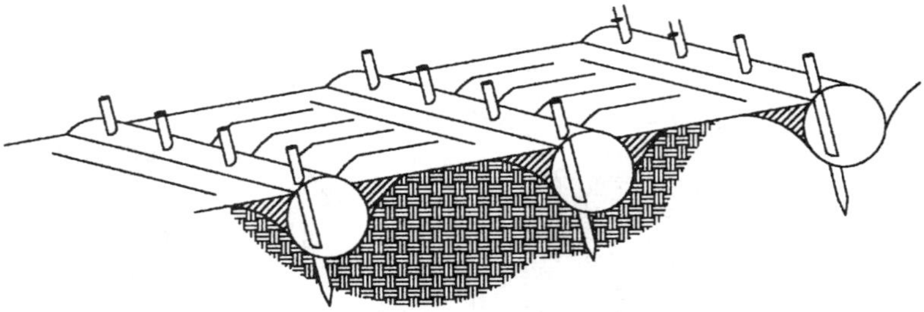

Figure 7.20 Branch packing detail.

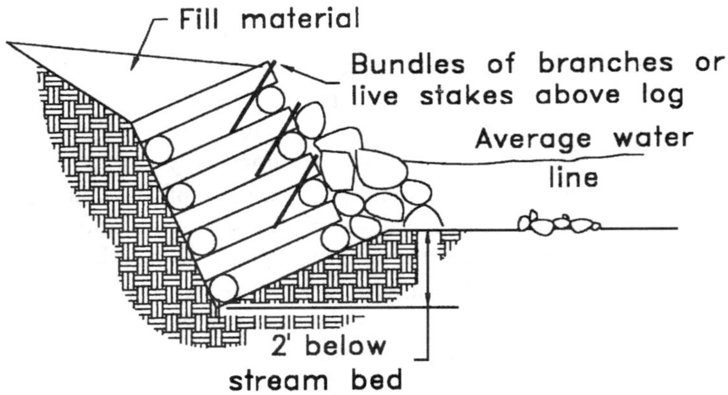

Figure 7.21 Branch packing detail alternate.

Nonvegetative bank stabilization

There are circumstances in which a vegetative bank is not a practical option either for natural or anthropogenic reasons. Such cases require "hardening" and the use of different strategies. Stream "hardening" in this case does not refer to the capture and channelization of the stream into concrete troughs for the most part.

Gabions. Gabions are effective in a number of stream bank stabilization applications, and they are sometimes used as armor or walls on slopes. They are more elaborate than the stabilization methods just described, and they require the use of heavier construction equipment and materials than some of the live stabilization methods (Fig. 7.23). *Gabions* are rock-filled wire baskets that are wired together. The great weight of the stone structure creates a gravity mass that is designed to resist expected flows. Once installed, they create

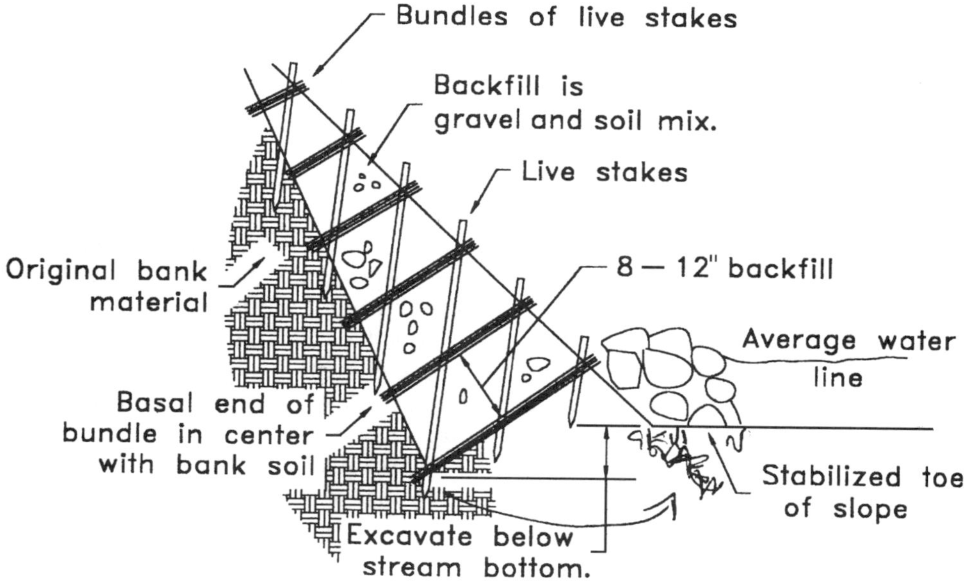

Figure 7.22 Cribwall detail.

Figure 7.23 Photograph of a gabion-stabilized stream.

a permeable, somewhat flexible, and stable structure. They may be used with live plantings or live stakes to offset their structural, unnatural appearance.

Gabion walls are keyed into place below the stream bed and so are installed during periods of low water. If an installation is using live cuttings, the gabion must be installed during the dormant season as well. There are many combinations and dimensions for gabions that allow the designer to select the best combination. Gabions are usually described as *baskets* or *mattresses.* Baskets tend to be fairly boxlike—that is, they come in lengths of 3 ft and heights ranging from 1 to 3 ft. Mattresses are fairly flat but long, usually less then a foot high, and 2 or 3 ft wide, and they come in lengths of up to 12 ft.

The installation of a gabion involves keying in the first layers of the structure. The bed is excavated and compacted to prevent settling and slumping. The gabion is set into place empty, and then it is wired together and filled with stone in 1-ft layers. If the filled basket requires a slight adjustment, it is usually winched into place. Baskets usually are filled with stone of about 6 to 8 in, and mattresses use a smaller ballast of 3 to 4 in of stone. In locations where it is expected that water will seep through the gabion or where erosion might occur, it is recommended that geotextile fabric be placed between the gabion and the backfill.

Deflectors. *Stream flow deflectors* are devices used to divert flow away from an eroding bank or to deepen a channel (Fig. 7.24). They are fairly simple to construct. Many have been installed by local conservation groups to improve habitat and stabilize streams. They can protect banks from erosion by directing the stream's energy away from an area. Deflectors can have negative impacts on streams if they are not carefully designed and installed. For example, diverting or concentrating too much flow may increase erosion downstream. Many deflectors are built from logs that must be replaced periodically. Deflectors are ineffective or even harmful on narrow or very fast streams or on banks being stabilized at deep pools.

The objective of most deflectors is to direct the current away from a bank and into the center of the channel. The deflector must be substantial enough to resist the energy of the stream at various levels of flow. The most common design used involves constructing a triangle of 30°, 60°, and 90° angles with the short leg facing downstream and the long leg anchored to the bank. Deflectors can be built of logs or stone. In either case the materials must be heavy enough to stand up to anticipated flows. Most deflectors have a fairly low profile that allows high flows to pass over them. Logs are anchored in place and to each other using ½- to ⅝-in steel rods or rebars driven at least 3 ft into the stream bed. Log deflectors are usually filled with stone after construction.

Regardless of the type of construction, the deflector should be keyed into the stream bed and bank. The stream bed should be excavated to at least 1 to 2 ft, and the deflector should be keyed into the bank at least 6 ft. Gabions may also be used.

Figure 7.24 Photograph of a stream deflector.

Wetlands

Wetlands are defined as those areas that are inundated or saturated by water often enough and long enough to support vegetation that is typically adapted for life in wet soil conditions (Fig. 7.25). Wetlands play an important role in the hydrologic cycle, and they are very productive environments. Nutrients collect in wetlands, and accordingly they generally display a great deal of biodiversity in plants and animals. Wetlands are identified by the presence of hydric soils and hydrophytic vegetation, along with the hydrology necessary to support the vegetation. *Hydric soils* are identified by color and include *gleyed soils* typical of wetlands (dark clays) as well as rich organic soils. Hydric soils develop under sufficiently wet conditions to support the growth and regeneration of hydrophytic plants. A hydric soil evolves under conditions of saturation, in areas that are flooded or ponded long enough during the growing season to develop anaerobic conditions in the upper part.

Wetland hydrology is usually identified as areas ranging from saturated soils (within 18 in of the surface) or submerged up to 2 ft. The most important source of water for most wetlands is groundwater. *Hydrophytic vegetation* is vegetation that can live in water or on a substrate that is submerged or anaerobic at least part of the time. Wetlands can be classified as *tidal* or *non-tidal, forested, scrub-shrub,* or *emergent.* Coastal wetlands make up only

Figure 7.25 Photograph of a natural wetland.

about 5 percent of the wetlands in the United States. The remaining 95 percent are inland types of wetlands. It has been estimated that 43 percent of the threatened and endangered species in the United States rely directly or indirectly on wetlands for survival. In addition, 80 percent of the breeding bird population requires for their survival *bottomland hardwoods,* which are wooded swamps found primarily in the southeastern United States. Some 22 states have lost at least 50 percent of their original wetlands. From mid-1970 to mid-1980, wetlands were lost at a rate of 290,000 acres per year.

A study of the Charles River in Massachusetts compared conditions with and without 8422 acres of wetlands within the Charles River basin and predicted the annual flood damage without wetlands would exceed $17 million. In this case the Army Corps of Engineers elected to preserve the wetlands rather than engineer and construct a large flood control facility. Wetlands are important to flood control because they support abundant plant life, which produces a root mass, which traps and retains water-borne sediment. The root-and-sediment agglomeration absorbs some of the energy of waves breaking onshore, and it also slows the velocity of runoff water.

The Congaree Bottomland Hardwood Swamp in South Carolina is a floodplain that acts as a water treatment facility to remove toxins, sediment, and excess nutrients, from water entering the groundwater. The least-cost substitute water treatment plant that would be needed to replace this swamp would cost $5 million. In the Chesapeake Bay area, a riparian forest within an agricultural watershed has been shown to remove 80 percent of the phosphorus and 89 percent of the nitrogen from agricultural runoff before the water entered the bay. It is estimated that 71 percent of the value is derived from species that use the wetland in some manner during their life cycle. A study in 1977 in Michigan estimated that each acre of wetland generates the equivalent of $489.69 in economic value associated with recreation, fishing, hunting, and so on. Waterfowl hunters spend more than an estimated $300 million annually on this sport. Bird watchers and bird photographers spend an estimated $10 million each year. Over one-half of the wetlands in the continental United States were lost between the 1700s and the mid-1970s. Approximately 100 million acres of wetlands remain. The results of wetland destruction are the obvious loss of the animals and plants but also the loss of soil through unchecked erosion, the loss of water storage capacity, and the loss of water-purifying infiltration.

Through Section 404 of the Clean Water Act, the United States is attempting to salvage and maintain our remaining wetlands. The act accordingly requires permits for the discharge of dredged or fill material into "waters of the United States" and "wetlands." The 404 program is administered jointly by the Army Corps of Engineers, which issues the necessary permits, and the EPA, which retains oversight and veto power. The program allows for two types of permits: individual and general. *Individual permits* are provided for special circumstances not covered by the array of general permits and are subject to fairly intense review. *General permits,* or nationwide permits (NWPs), are provided for many of the activities that are incidental to development and that are perceived to represent collectively relatively minor impacts. NWPs are available for maintenance (NWP 3), utility crossings (NWP 12), minor road crossings (NWP 14), stream and wetland restoration activites (NWP 27), reshaping existing drainage ditches (NWP 41), recreation activities (NWP 42), and storm water management facilities (NWP 43).

Wetlands are mapped by the U.S. Department of the Interior using high-level infrared photography. Soils maps can also be used to locate wetlands because the maps identify hydric soils. Wetlands are also identified in the field. A delineator determines the number and general location of transects through the site to be delineated. He or she makes shallow excavations determine if hydric soil conditions are present. The delineator then surveys the vegetation around each sample location to identify the predominant types of trees, shrubs, and forbs. Delineators keep copious field notes, and they usually establish photographic records of conditions. Based on the field conditions, delineators interpolate between nonwetland and wetland areas to find current limits. While hydrologic conditions may vary from year to year, soils and mature vegetation do not vary as much and can often be used as reliable indicators of wetland conditions.

Constructed wetlands

Since much of the hydrology associated with wetlands is groundwater in nature, it is very difficult to create artificial wetlands where there are not any.

However, it is not impossible to do so. Key to the long-term success of an artificial wetland is the successful design of the hydrologic element.

Choosing a supportive location is both difficult and critical. The right location has to have the hydrology or can be adapted to have the hydrology. Also the location cannot have some existing wetland or other feature of environmental value. Likely areas might be adjacent to existing wetlands or areas where there is significant other earthwork undertaken so that the new hydrology can be designed and built in (such as highway cloverleafs). Selected sites should not be subjected to significant grading due to the loss of soil structure associated with those activities. Experience with supplementing existing soil with hydric soils has been successful in helping to establish vegetation and micro flora and fauna in an artificial wetland.

Wetlands are natural sinks of surface waters and nutrients. They act as natural filters and water treatment systems and so may offer important capabilities to designers. When organic matter from livestock waste decomposes in water, oxygen is usually depleted, potentially suffocating the natural aquatic life. The biological oxygen demand (BOD) of untreated animal waste is about 100 times that of treated wastewater from a sewage treatment plant. Today wetlands are being built especially to collect and treat wastewater and runoff. Since 1988 there have been experiments in the United States in using artificial wetlands to collect and treat runoff from feedlots and barnyards.

Figure 7.26 Photograph of an artificial wetland.

One experiment in Texas uses a series of specially designed wetland cells and a 10-acre-foot, 100-day retention time artificial wetland to treat all of the wastes from a 400 to 450 cow and dairy operation. Though the wetlands are characterized by vegetation, it is not the plants in most cases that clean the wastewater. Rather, it is the bacteria and microorganisms the wetland supports that break down the solids. The plants then use the released nitrogen from the decomposing solids as a primary nutrient.

In an artificial wetland system, solids are collected in a holding tank, and the system is flushed twice each day into a primary wetland cell. The water moves via gravity to a second cell in about 10 to 12 days. Clean water is eventually collected into a pond, which is then used to irrigate fields and to supply "grey water" to the dairy operation. The system can also accommodate a 10-year storm without any impact on operations. The cost of construction in the Texas experiment was less than $10,000 compared with a minimum of $25,000 for a wastewater treatment lagoon system.

Constructed wetlands have been notorious for having problems and failures. This is not to say that human-made wetlands cannot work, but it should be recognized that there is a high degree of failure or unsatisfactory performance. Restoration of natural wetlands has been more successful in practice. In places where wetlands have been filled or drained, the original hydrology and hydric soils may still be present or they can be restored. In many cases even desirable native plant seeds and materials still remain viable as a sort of seed stock in the soil. While careful site evaluation using historic topographic maps can reveal former elevations and drainage patterns to restoration designers, successful restoration usually requires more than simply removing the fill or filling the drainage ditches. In other cases though, simply stopping the drainage leaving a former wetland is all that is necessary to begin the restoration process.

To assure success, the restoration project requires careful planning and establishing realistic performance goals. The performance goals provide a basis for clear measurement of project progress and success. Without clear quantitative measures, it may be difficult to determine whether the effort and cost have been worthwhile. The professional should clearly identify benchmarks of progress and success to demonstrate the value of their work to current and future clients. A successful restoration project will ultimately assume a natural appearance, and it will be assumed by many that the restoration was simply a matter of "nature healing itself" when in fact it was the result of careful project design and implementation by a restoration professional. Without a clearly articulated plan and performance measures, the role of the designer is ultimately lost.

To determine the measures of success, it is first necessary to understand what the functions of the wetland will be. In most cases the wetland will serve several purposes in the landscape, among them wildlife habitat, storm water collection and treatment, flood buffer, stabilization of riparian or littoral zones, water treatment, and recreation. Each purpose adds its own concerns and requirements to the restoration project.

Restoration planning

Wetlands are naturally diverse systems, which enables them to serve various purposes simultaneously. In fact, the diversity is necessary for a wetland to function at even its simplest roles in water purification. The purification process involves an array of biotic and abiotic interactions, each supporting the other, each drawing upon the resources of the other and providing resources to the next. A high functioning wetland is an extremely complex system probably beyond our current ability to design from scratch. Most restoration designs understand and rely on the system's capacity to add complexity to itself as it matures. While we may not be able to design or build such a system, we are able to measure it.

The supporting hydrology is critical to the wetland's success. Wetlands draw water from several sources including the groundwater, tidal water, runoff, and precipitation. The key to wetland hydrology is that whatever the source of water, variations in supply must not be so great as to deprive the wetland of water for too long a period. Understanding the water budget of a restored wetland is a critical part of planning. A careful assessment of the amount, timing, and character of the available hydrology should be the first order of business in a restoration evaluation. Surface waters should be carefully assessed in terms of pollutant and sediment loading. The source of surface water should also be understood. Storm water runoff is generally not considered to be an acceptable sole source for a wetland though exceptions to this guideline exist. Such a project may benefit from lining the project area to reduce infiltration. A rule of thumb is that for every acre-foot of water in the wetland, there should be about 7.5 acres of contributing watershed for projects east of the Mississippi. West of the Mississippi evaporation rates are much higher and precipitation rates generally lower so that more contributing watershed is necessary. Wetlands using only surface water as a source are particularly subject to damage during droughts and other seasonal effects.

A natural spring, a pond, or groundwater is usually considered to be the best source of supply for a wetland restoration. Springs and seeps are often subject to disturbance during construction and should be protected if it is intended that they be used to supply the wetland. Seasonal high and low water tables should be carefully evaluated to understand the nature of the available hydrology. Wetland restoration is further complicated by salt content in many western soils. Riparian or littoral zone wetlands enjoy a ready source of water from both the surface water and the interflow of groundwater.

Too much water may be as much of a problem as not enough. A project located low in a watershed may have to include control structures to manage high-flow situations and to retain water on the site. The expected water loss must also be considered. Annual and monthly precipitation data are collected from the U.S. Weather Bureau or agricultural stations and compared to evapotranspiration. This comparison is available from the Natural Resources Conservation Service or the National Climatological Center. The precipitation to evapotranspiration comparison is expressed as a P/E ratio. This is simply a measure of the amount of water that would evaporate from an open container; the *P/E* ratio is sometimes called the *pan evaporation ratio.* In addition to the *P/E* ratio, designers should factor in any expected loss from infiltration through the soils. Wetlands typically have soils with a very high organic content, and they retain a substantial amount of water—as much as 50 percent of wetland soils can be water by volume. This allows the wetland some moisture buffer in periods of low water availability.

Grasses play an important role in most natural wetlands. These plants vary considerably in the water depth they prefer and the duration of flooding they can tolerate. The best way to obtain information on which grasses are best suited for an application is to observe other functioning wetlands in the surrounding area. Table 7.6 is a partial listing of plants types according to their preferred depth of water.

TABLE 7.6 Wetland Plants According to Preferred Water Depth

Scientific name	Common name	Distribution
	Seasonal Flooding	
Bidens spp.	Beggarticks	Alaska to Quebec to southernmost states
Echinochloa crusgalli	Barnyard grass	
Hymenocallis spp.	Spider lily	
Lysimachia spp.	Loosestrife	
Hordeum jubatum	Foxtail barley	
Polygonum lapathifolium	Pale smartweed	
Iris fulva	Red iris	
Setaria spp.	Foxtail grass	
Spartina pectinata	Prairie cordgrass	Saskatchewan to Newfoundland to Texas and North Carolina
Panicum virgatum	Switchgrass	
Calamagrstis inexpansa	Reedgrass	
Distchlis spicata	Saltgrass	
Alopecurus arundinaceus	Foxtail	
Scolocarpus foetidus	Skunk cabbage	
Hibiscus moscheutos	Swamp rose mallow	California to Massachusetts, Texas to Florida
	Seasonal Flooding to Permanent Flooding to 6 in	
Leersia oryzoides	Rice cutgrass	
Juncus effuses	Soft rush	
Carex spp.	Sedge	
Eriophorum polystachion	Cotton grass	
Cyperus spp.	Sedge	
Iris virginicus	Blue iris	
Iris pseudacourus	Yellow iris	
Dulichium arundinaceum	Three-way sedge	
Beckmannia syzigachne	Sloughgrass	

TABLE 7.6 Wetland Plants According to Preferred Water Depth (Continued)

Panicum agrostoides	Panic grass	
Scirpus cyperinus	Woolgrass	
Habanaria spp.	Swamp orchids	
Cypripedium spp.	Lady's slipper	
Hydrocotyle umbellate	Water pennywort	
Calth leptosepaia	Marsh marigold	
Phalaris arundinacea	Reed canarygrass	Alaska to Newfoundland, California, New Mexico, North Carolina
Polygonum coccineum	Swamp smartweed	British Columbia to Quebec to California to South Carolina
Polygonum pensylvanicum	Pennsylvania smartweed	
	Flooded from 6–20 in	
Polygonum amphibium	Water smartweed	Alaska to Quebec, California to New Jersey
Cladium jamaicensis	Sawgrass	California to Virginia, southern states
Acorus calamus	*Sweetflag*	
Calia palustris	Water arum	
Zizania aquatica	Wild rice	Manitoba to Nova Scotia, Texas to Florida, Washington to Alberta, difficult to establish from seeds
Alisma spp.	Water plantain	Southern Canada to southern USA
Glyceria pauciflora	Western mannagrass	Alaska to South Dakota to California and New Mexico
Typha latifolia	Wide-leaved cattail	Alaska to Newfoundland, to the southernmost states
Typha angustifolia	Narrow-leaved cattail	Washington to Nova Scotia, to southernmost states, most common in northeastern states
Typha domingensis	Southern cattail	California to Delaware near coasts
Typha glauca	Blue cattail	Washington to Maine, common in central New York and along Delaware and Chesapeake Bays
Scirpus fluviatilis	River bulrush	
Saggittari latifolia	Broadleaf arrowhead	British Columbia to Quebec, to southernmost states
Pontederia cordata	Pickerelweed	Minnesota to Nova Scotia, Texas to Florida
Glyceria spp.	Mannagrass	
Nasturtium officinale	Watercress	
Peltandra cordata	Arrow arum	
Vaccinium macrocarpon	Cranberry	
Juncus balticus	Saltmarsh fimbristylis	Coastal salt marsh, New York to Florida
	Flooded from 20–75 in	
Potamogeton pectinatus	Sago pondweed	

TABLE 7.6 Wetland Plants According to Preferred Water Depth (Continued)

Ranunculus flabellaris	Yellow water buttercup	
Ranunculus aquatilis	White water buttercup	
Phragmites phragmites	Phragmities	Nova Scotia to southernmost states
Deschampsia cespitosa	Tufted hairgrass	Coastal marshes Alaska to California, east to south Dakota and North Carolina
Scirpus validus	Softstem bulrush	Alaska to Newfoundland to southernmost states
Myriophylum	Milfoil	
Elodea	Water weed	
Zizaniopsis miliacea	Giant cutgrass	Illinois to Maryland, Texas and Florida
Nuphar luteum	Spatterdock	Alaska to Newfoundland, California to Florida
	Floating	
Lemna spp.	Duckweed	Alaska to Quebec, California to Florida
Azolla spp.	Water fern	
Spirodela spp.	Giant duckweed	

Wetland protection

Although regulatory restrictions exist to protect wetlands, much of the damage to wetlands occurs because of changes in the contributing upland area. There are site design considerations that will reduce the impacts of development on downstream wetlands. Storm water should be managed to retain as much water on a site as possible by minimizing the practices that result in runoff and using features to increase infiltration as discussed in Chap. 6. If storm water is to be discharged into a wetland, it is best to use many small outlets or gabion outlets to disperse the concentrated discharge and reduce its velocity. Buffers should be planted between the discharge point and the wetland to filter and provide some “polish” to the discharge. The construction of an artificial wetland to act as a filter might be considered.

Erosion Damage

Damage caused by erosion is often difficult to repair because the impacted area is usually subject to repeated episodes of erosive flows of water. Simply replacing eroded materials generally is not sufficient to stabilize the eroded area. In most cases it is necessary to excavate the eroded channel in the fashion shown in Fig. 7.27, and to tamp into place new material. For areas with velocities that exceed the ability of the soil to resist shearing it is necessary to provide at least temporary geotextile protection. Some places may require permanent protection. Similar steps are required for eroded slopes. It may also be necessary in some cases to divert the runoff from the impacted area during construction and even for a short time afterward.

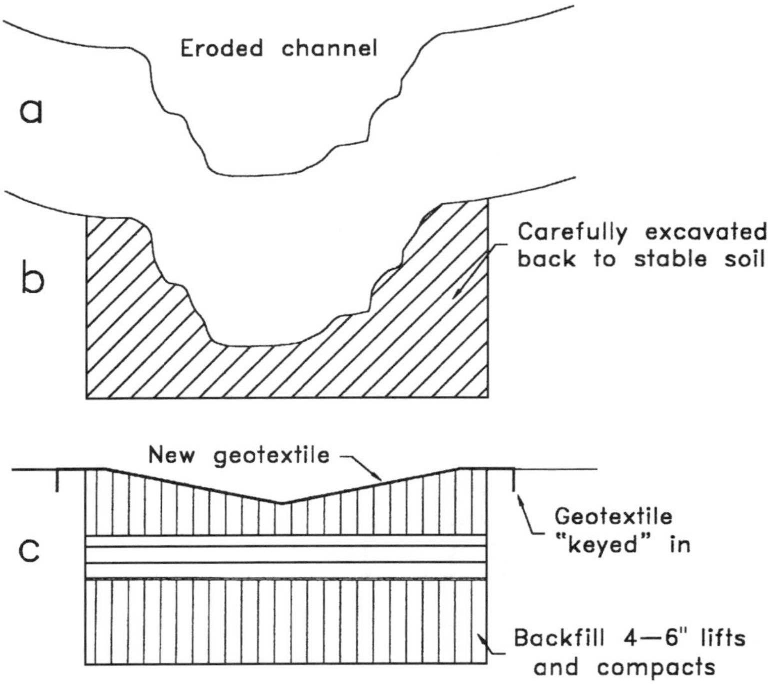

Figure 7.27 Eroded channel repair detail.

Brownfield Redevelopment

The risk associated with owning contaminated property was exacerbated by the Comprehensive Environmental Response Compensation and Liability Act (CERCLA), often referred to as the Superfund, which made landowners liable for contamination found on the property whether the landowner had caused the contamination or not. Purchasers became appropriately wary of buying a property that might be contaminated, often electing to purchase an undeveloped site instead.

The practice of reusing previously developed sites has become more common in recent years because of a more favorable public policy environment. Often these sites are referred to as *brownfields,* and they are generally thought of as abandoned sites or properties that are significantly underused and undervalued because of environmental contamination or the general perception of contamination (Fig. 7.28). The definition of *brownfield* is ambiguous and can include many sites with little or no significant environmental issue. In the past, property owners vigorously sought to keep their property from being referred to as a "brownfield," but with the advent of financial incentives, it is now sometimes considered to be a favorable designation.

Site design and planning practices sometimes must address residual contamination as part of the site development. The state of brownfield practice is to determine a management and design strategy that will minimize the risks associated with the contamination at an acceptable cost to benefit. On such

Figure 7.28 Photograph of a brownfield.

sites environmental issues must be accounted for on the construction site as well as on the built site. Most states have *voluntary cleanup programs* (VCPs) or brownfield programs of one sort or another. Since this is an emerging and developing area of practice, the elements and success of these programs vary significantly. The shift in public policy is to encourage development in communities where these places exist and to bring private money to the cleanup process. The interest from the private development community is simply whether there is a good real estate deal.

General elements of state programs

A complete analysis of the various state programs is beyond the scope of this book; however, a general discussion will provide a sense of brownfield activities today. The most common elements of state programs are some form of protection from liability for past acts or contamination, predictability in cleanup standards and process, risk-based standards for assessment and design, and financial incentives.

Liability protection

State programs may provide buyers with protection from the liability of the state's own superfund program. Since state laws do not offer protection from federal liability, many states have entered into *memorandums of understanding*

(MOU) with the federal agencies involved. Through the MOU, the federal government essentially acknowledges the state program and agrees to honor its arrangements with developers.

Liability protection from the state program is managed in several different ways. Some states provide a document that acknowledges the state's agreement to not pursue any further action on the site. Letters of cleanup completion may serve the same purpose. Under some programs the state agrees to assume liability for contamination, in essence standing between the property owner and the risk of undiscovered acts of the past. These protections may be restricted to new purchasers and may not be available to existing landowners under some programs.

Cleanup standards

Among the concerns of property owners and buyers is the uncertainty of cleanup standards. It is difficult to determine a comprehensive standard for site cleanup that is not so conservative as to be impractical in most cases. The most successful programs have provided developers with a menu of choices that range from very conservative standards such as the Safe Drinking Water Act's *maximum contaminant levels* (MCLs) to standards that are based on risks and specific site conditions. A range of choices allows the developer to prepare a strategy that balances the site issues and the resources. In general, the more stringent standards are more costly, but they result in greater risk reduction. An owner of a site with a very small amount of contaminated material might choose to use the most stringent standards available because the liability relief provided is worth the expense. Developers of properties with more contamination may elect to follow a less costly approach but will have to manage the risk of onsite contamination.

Risk and risk management

We routinely assess risk in our daily lives; we weigh the risks or potential costs of an action against some benefit. We proceed or not, usually based on whether the probable benefit is greater than the potential cost. Determining environmental risk is a scientific process of evaluating adverse effects of lifestyle choices, exposure to a substance, or some specific activity. There is always a degree of uncertainty in risk assessment because our efforts are limited by how much we know or the accuracy of what we know. It is important to recognize, however, that "uncertainty" in this case refers to a lack of precision in the numbers, and so risk is most often expressed as a range of risk rather than a specific number.

The risk assessment process for environmental contaminants is fairly straightforward, usually consisting of four steps beginning with a clear definition of the problem. Since all possible issues and risks cannot be simultaneously considered, narrowing the questions is critical to being able to measure the risk. Usually just one source of risk is considered at a time, although recent efforts have been made to calculate the risk from combined multiple sources or materials. The next step is to determine the amount of exposure a person would normally have to a substance through inhalation, ingestion, or absorbtion. The length of time of exposure, the pathway of exposure, and the way a chemical behaves in the environment are all evaluated. The actual toxicity of the substance is known from evidence gathered through animal studies, in vitro studies, comparison studies, and epidemiological studies. Studies using animals are not the same as studies on humans, and for environmental risk-assessment purposes, extrapolation of such studies is not an acceptable practice.

Brownfield site designers are likely to become involved with risk as part of the project review or public hearing process. Understanding the dynamics of risk communication is an important skill in such circumstances. Foremost is the need to understand and expect public outrage and anger. People will become angry when confronted by risks imposed upon them by others. The language and concepts of environmental risks are unfamiliar, outside the control of the public in many cases. Confronted with uncertainty, an emotional response is common. People are concerned about the likelihood and the effects of exposure. They will want to know the legal standards and what the health risks are.

Risk must be communicated carefully. Information should serve to help people understand and evaluate risk. Risk communication includes explaining what is not known and providing people with as much information as is available to answer their questions and concerns. There is a tendency to want to communicate risk in familiar terms to help people understand. However, caution should always be used when using comparisons. Comparisons of involuntary and voluntary risks, for example, are to be avoided. Lifestyle risks such as smoking or drinking are voluntary and are not valid comparisons with risks that have been imposed on someone. Presumably a person assumes a voluntary risk because of some perceived benefit that is greater than the risk. Involuntary risks are rarely balanced with a benefit.

As for any other presentation risk, brownfield site development communication should be carefully planned and prepared with a clear understanding of the target audience. Information should be prepared in an easily understandable and concise format. Concentration analogies such as those presented in Table 7.7 are one way to help people understand quantitative measures such as parts per million or parts per billion. It is important that risks be straightforwardly addressed and not downplayed. The presentation should acknowledge uncertainty where it exists. Above all it is important to be accurate, complete, and honest. The presentation should be sensitive to voluntary and involuntary risks and avoid inappropriate comparisons. All questions should be answered, if not during the presentation then with a followup contact. Misinformation is sure to evolve in the absence of accurate information.

General strategies

The general strategies for brownfield projects are usually limited to doing nothing, the use of administrative or institutional controls, use of engineering

TABLE 7.7 Concentration Analogies

One Part per Million
One automobile in bumper-to-bumper traffic from Cleveland to San Francisco
One pancake in a stack 4 mi high
1 in in 16 mi
1 oz in 32 tons
1 cent in $10,000.00
One Part per Billion
One 4-in hamburger in a chain of hamburgers circling the globe at the equator two and a half times
One silver dollar in a roll of silver dollars from Detroit to Salt Lake City
One kernel of corn in a 45-ft-high, 16-ft-diameter silo
One sheet on a roll of toilet paper from New York to London
1 s of time in 32 years
One Part Per Trillion
1 ft^2 in the state of Indiana
One drop of detergent in enough dishwater to fill a string of railroad cars 10 mi long
1 in^2 in 250 mi^2

controls, onsite remedial action, or offsite disposal or treatment. These are listed in a general order of increasing cost and liability relief. While doing nothing may be desirable, it is not often an option on a site that is actually contaminated in some way. *Administrative or institutional controls* include steps that restrict the use of the property in some way. Among other measures, these include restrictions on what activities may be conducted on the property and/or restricting access to all or parts of the site. Owners might have to carry specific insurance or provide some kind of performance bond.

Engineering or *technical controls* refer to strategies such as caps and active or passive remedial actions. Engineering strategies are usually directed toward isolating or containing the contamination or treating it in some way over a period of time that continues after construction. Engineering solutions differ from remedial actions in that the contaminated material is treated to acceptable levels before construction begins.

There is a broad range of technologies and strategies for *onsite treatment.* For example, *pump-and-treat methods* are common for addressing groundwater concerns, and the improvement in success for *bioremediation* promises to increase the use of that approach. Methods such as *vitrification* that increase the impermeability of soil—either by adding cement to it or by heating and melting—are less commonly used primarily because of their expense. Methods such as *phytoremediation* and *natural attenuation* are being used more often and will become more common as our experience with them grows. The least common and most expensive method on a large scale is removing material from the site whether for disposal or treatment. In the case of disposal, the landowner will maintain liability for material he or she places in a landfill.

Design concerns

For the most part brownfield redevelopment employs all of the techniques and materials normally used in site design. Depending on the site, some additional extraordinary steps or elements might be required. For most brownfield sites, site design issues are limited to dealing with capping, installing utilities in contaminated materials, using vegetation, allowing for drainage, controlling the risk of exposure to workers and users, and postconstruction remediation issues. Soil conditions can be demanding (Fig. 7.29). Many soils on brownfield sites consist of artificial land and unconsolidated fills, and many soils are expansive.

Development on a capped site

Development on a capped site poses some challenges for the designer. Constructing a cap on a contaminated site isolates and contains contaminated material from receptors and from natural transport mechanisms. Most

Figure 7.29 Photograph of brownfield soil conditions.

development, however, involves the installation of underground utilities and infrastructure that by definition penetrate the zone of contamination. Plans requiring the installation of utilities in contaminated materials should be carefully considered. The best practice may be to isolate the utility trench or foundation excavation from the contaminated material by lining the trench with the cap material or other suitable alternative (Figs. 7.30 and 7.31). Caps

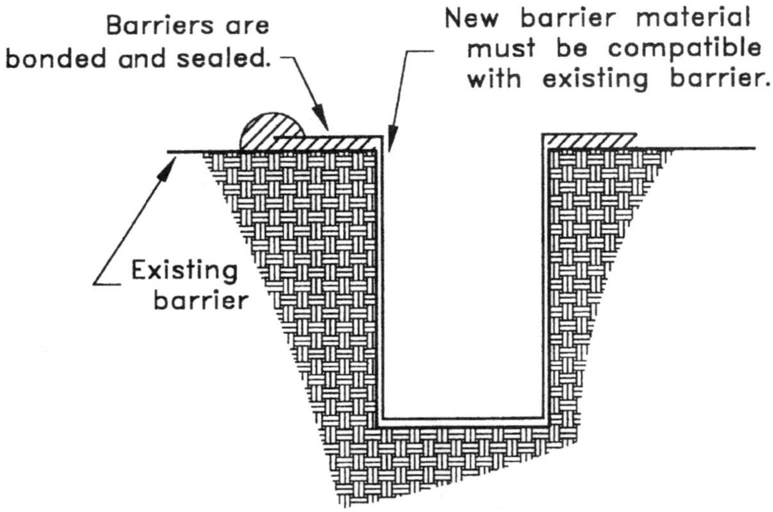

Figure 7.30 Lined utility trench detail.

Figure 7.31 Photograph of lined foundation excavation.

are constructed of a variety of materials but today most are made of impermeable geotextiles that lend themselves to lining trenches and excavations.

Capping may restrict plant root penetration, which might require the designer to treat the site as if it were in effect, a rooftop. Rooftop strategies actually adapt well to capped sites for the most part. Caps are commonly designed and intended to be at least as permanent as the contamination itself. The practical fact of the matter is, however, that caps are under assault from natural processes from the moment they are installed, and therefore they require maintenance. The life cycle cost of a cap that is to be used as a parking lot, for example, will be higher than the cost for a typical parking lot. Potholes and cracks that might occur in the normal wear and tear of a parking lot may not be acceptable on a parking lot over a cap. Sites with clay caps rely on the expansive character of specific clays, and these caps will require continuous moisture to be added to maintain the integrity of the cap. Paving or introducing impermeable elements to a clay capped site may interfere with the cap's performance.

Managing storm water may be a challenge on any type of cap. On a capped site, infiltration methods may not be allowed. In such cases it may be necessary to use less desirable management methods such as a lined detention basin. Pipes conveying storm water will have to be installed using tight leakproof joints to avoid infiltration and exfiltration.

Chapter

8

Site Layout

Sustainable site design must include broad considerations of the environmental role of the site. The site exists as part of a larger landscape and ecosystem. Sustainable design must recognize and retain as many of the functional elements of a site as is possible.

Landscape Ecology

The study of landscape ecology has made important contributions to our understanding of how development impacts the environment. Still a fairly young science, *landscape ecology* is concerned with the fragmentation of habitat, biological diversity, the design and management of the land resources, and sustainable development. Landscape ecology studies the cause-and-effect relationships between elements in the landscape. Having emerged as a distinct science only as recently as the 1960s or 1970s, it has done well to promote a recognition that the reductionist approach to science is inadequate to wholly describe the complexities of landscape ecosystems. Landscape ecologists also work closely with and promulgate the concept of the *total human ecosystem*—that is, the understanding that humans, along with all of our activities and cultural complexities, are an integral part of the landscape. To attempt to describe or study the landscape without considering the influence of human activity upon it would be a pointless exercise. For these reasons the products of study in landscape ecology are substantially more than just flowcharts of energy and materials.

A reading of the underlying theory of landscape ecology reveals it to be a broadly interdisciplinary science. Ecologists, physicists, biologists, and geographers are as concerned with values and ethics as they are with the systems-thinking approach to the new science. There appears to be an expectation that a new language, a new means of describing the ecosystem, will emerge as our understanding improves. For example, there is recognition in landscape ecology that order as human beings commonly understand it to exist in nature probably does not exist at all. Instead, we human beings look for order and select those things that best support our understanding of it. In fact, order is probably more complex and is better studied and expressed through fuzzy set theory or fuzzy logic. Using the fuzzy approach, qualitative elements of the landscape can begin to be described mathematically while quantitative values can begin to be expressed in words.

Landscape ecology offers important information for those engaged in site planning and design. For example, understanding some of the principles of landscape integrity will allow the planner to incorporate them into designs. In general, the landscape can be described in terms of four general elements of vegetative mass and form: patches, edges, connecting corridors, and mosaics (Dramstad 1996). The *mosaic* of a landscape refers to its overall pattern of patches and connectivity. For the most part, the landscape mosaic is beyond the scope of a given land development project. It is important, however, for the designer to understand how a site fits into the larger landscape mosaic around it before the planning process begins. Once understood, the site planner is better able to incorporate the principles of patches, edges, and connections into the plan.

Patches are concentrations of habitat type, most commonly visualized, for example, as a woodland patch in the midst of farmland or an urban area in the middle of a national forest. A patch may be small such as an undeveloped city lot or large like a park. The origin of patches could include human activity such as farming or urbanization or natural activity such as resource distribution as from a desert spring or emerging wetland, the natural succession or homeorhesis (Naveh 1993) or some intervening disturbance. Not long ago patches were somehow degraded or less than optimal conditions, often described as "islands," disconnected areas isolated from the rest of the ecosystem. It has been found, however, that patches may in fact be beneficial conditions depending on their character, size, and location. A patch contaminated with industrial wastes is a degraded site with little redeeming value, but a relatively small patch of woodland adjacent to a farm field or even in a suburban or urban area may very productive.

What is known is that diversity declines and extirpation increases as patch sizes decrease and isolation or the distance between patches increases. The quality of the habitat is also a function of size; the smaller the patch size, the less diverse it is simply because the number of possible habitat types is reduced. There is no simple minimum patch size recommendation. There are some positive aspects to some small patches—for example, the preservation of small specialized or unusual habitats like wetlands can be beneficial—but careful assessment is necessary to be sure the small size is sustainable.

There is interaction between habitat types. *Ecotones* are the *edges* or boundaries between different habitat types. In most cases edges between habitat types are blurred—there is no bright line between the meadow and deep woods, or between upland and open water. Instead, there are zones of transition: ecotones. These boundary areas are often the most productive parts of a landscape. As transition areas, they contain aspects of both areas that they connect. The *littoral zones* that are the transition from upland to open water are usually more productive than either the upland or the open water. Both terrestrial and aquatic animals use the littoral zone to feed and to mate. Likewise the transition from deep woods to meadow is more productive than either of the habitats connected by it. In fact, edges are buffers where changes in water elevation can be absorbed or areas of increased competition where the forest attempts to invade the meadow or both. As buffers, they act as filters as well. Natural ecotones, or edges, tend to be curvilinear and soft transitions full of the complexity provided by the areas they border. By contrast, the boundaries of human activity tend to be straight and abrupt, providing little buffer or transition.

With the encroachment of development and farmland, habitat has routinely been cut up into patches and isolated. Even patches located near one another can be isolated by barriers such as major highways or fences. Thus the value of *connecting corridors* between patches increases with the fragmentation that is associated with human activity. Although some highways or rail lines or other linear developments act to prevent the movement of wildlife, some other linear developments like electric transmission corridors or pipelines or floodplain areas may actually serve as pathways from one patch to another. The key is to recognize these patches and to provide connections between them. To be effective, the planner must understand the behaviors and habits of the animals or plants that are expected to use the corridor. For example, some species require a visual connection so connections must be visible from one to another.

Site Layout

The most obvious aspect of the site design is how the proposed project lays upon the land—that is, how the buildings and facilities will be organized. The way in which that organization will occur is determined first by the land itself and then, to varying degrees, by the values of the developer, the local ordinances, the community standards, and the nature of the project as these are all perceived and balanced by the designer. In synthesizing these diverse parameters, the designer will be able to visualize the actual layout of landscape features on the site. To this mix the designer will add the important design practices and standards that guide him or her as a professional.

Thus the designer's analysis and sensitivity to the site inform the entire design process. An awareness of a site might include knowing its history, its place in a larger landscape ecosystem, its real estate value, and its local political or economic importance. Site design is among other things a synthesis of all these concerns in the context of the design objectives or program. For example, redevelopment may be a preferred environmental strategy in some projects since it involves the reuse of already disturbed land, and it is therefore in a sense, recycling the site. Honoring particularly important historical aspects of a site in some way should also be considered if appropriate. Defining the role of the site in the larger landscape fabric—the mosaic—is critical. In developing a new site, the environmental functions it performs should be preserved, and its impact on the environment should be minimized. Redevelopment projects should attempt to restore function wherever possible.

Site development practices have started to change with the recognition of the negative impacts of urban sprawl and the increasing awareness of smart growth alternatives. There is, of course, no simple, one-size-fits-all solution to the problems of urban sprawl or the challenges of smart growth, and the range of solutions available to designers can be expected to grow with the sophistication and needs of the marketplace. Community standards are a critical component of the design process. Much has been said and written about the homogeneity of the modern built landscape. Every place looks pretty much like every other place, or so it is said, but there are communities that have established an identity if not clear standards. In these communities even the mundane is made to meet community expectations, as shown in Figs. 8.1 and 8.2.

Residence and Residential Community Design

The primary objective in residence and residential community site planning is to provide a site that is a desirable place to live for the intended users. Different parameters will be applied and different elements will be selected depending on who the end user is to be. Developments targeted toward young

Figure 8.1 Photograph of commercial site.

Figure 8.2 Photograph of commercial site.

families with children, for example, will include some features that would be out of place in a project designed for empty-nesters. House size, lot size, common spaces, and recreation facilities would all be quite different, but there are some commonalities found in quality residential development.

In his book *Save Our Lands, Save Our Towns,* Thomas Hylton lists 10 attributes of quality development and communities, which are summarized in Table 8.1. In essence, Hylton and others have found that quality for the most part means the careful adaptation of traditional, even archetypal forms of a community's design with important accommodations for modern life.

The most desirable communities are most often those that allow a maximum of pedestrian access to the necessaries of life—schools, work, shops, and the like—but also provide for easy transit in and out of the neighborhood. Other characteristics of such communities are the presence of human-scale streets and buildings, lots of well-developed trees to soften and temper the streetscape, and diversity in the social and architectural makeup of the community. While security and safety are often cited as important, achieving these simply through the hardening of buildings and sites is not desirable.

Local streets should be designed in such a way that a coherent pattern of circulation can be recognized. In addition, development should be laid out in a fashion that is sensitive to the land and that does not require substantial alteration to it or diminish its character. Houses or residential units should be arranged to provide variation and visual interest. After the style and

TABLE 8.1 Characteristics of a Quality Community

1.	A sense of place
2.	Human scale
3.	Self-contained neighborhoods
4.	Diversity
5.	Transit-friendly design
6.	Trees
7.	Alleys and parking lots to the rear
8.	Humane architecture
9.	Outdoor rooms
10.	Maintenance and safety

SOURCE: Adapted from Thomas Hylton, *Save Our Lands, Save Our Towns: A Plan for Pennsylvania,* photography by Blair Seitz, Pennsylvania's Cultural & Natural Heritage Series (Harrisburg, Pa.: Rb books, 1995).

affordability of houses, the lot layouts and character are the most important elements of the typical residential development project. The number of lots is of critical concern to the developer where the lot size and character are of importance to buyers. In a competitive residential market, developers may compete on the basis of price, the quality or character of their units or amenities. Valuable lot amenities may include the presence of trees, lot shapes and sizes, views, and accesses to water.

At the early stages of the site analysis, it is important to begin to identify home sites. Generally, this is done using topographic mapping of the site and walking the entire site to identify valuable locations or site features. The identification of home sites and the related issues will drive the planning and design of the site. Home sites are found by determining where it would be nice to live; it is fundamentally a simple process. A good location is a combination of its surroundings, access, and amenities, as well as more subjective attributes. These less tangible attributes might include the extent to which the location of a lot conveys a sense of neighborhood, or security (Table 8.2). In the development of sites for more affordable homes, lot sizes tend to be smaller and the linear feet of road per unit lower. Hillside lots will tend to be irregular in shape, reflecting the physical aspects of the site. The lot configurations on hillsides are designed to encompass a desirable living space in a more difficult development condition.

Lot configurations may range from tight clusters that leave large undeveloped portions of the site as open space, to individual lots with dwelling units separated by varying degrees of distance. In either case, the sites and layout are designed to minimize impact and maximize site value.

The layout of a residential development is part of an overall community design that also includes recreation facilities, schools, shops, offices, and religious institutions. The larger community thus forms a context into which any new development project must fit, and fit well. Site planners and developers need to consider how a proposed project will relate to existing or future fea-

TABLE 8.2 Homeowner Preference for Proximity to Open-Space Features

Open-space feature	Mean score
Adjacent to wet pond	4.44
Adjacent to natural area	4.27
On a cul-de-sac	3.83
Adjacent to golf course	3.67
Adjacent to public park	3.10
Adjacent to dry pond	2.05

tures within the lifestyle choices of the future inhabitants as well. Chapter 4 discusses the ratio of population to different recreational and open-space facilities, and that type of objective information is invaluable to site designers.

Emerging practices

In the 1980s the Massachusetts Department of Environmental Management (DEM) created the Connecticut Valley Action Program, which was overseen and administered by the Center for Rural Massachusetts. The purpose of the action program was to find or develop an approach to planning in the Connecticut Valley that would preserve the scenic, historic, and environmental qualities of the area and still allow for development. The Connecticut River Valley has been inhabited for thousands of years because of its location on the river and its rich fertile soils and pleasant climate. Europeans began to settle in the valley beginning in 1634, and as their numbers grew, they first established and later incorporated the many beautiful small towns for which this area is still known. In the prosperous years that followed World War II, however, the rapidly developing urban regions nearby began to encroach on the valley, and its existing ordinances and planning regulations did not provide the protection that residents had thought they would.

From 1951 to 1972, the land converted from farmland to development tripled, and projections indicated that this trend would continue. The river valley area is made up of 19 towns in three counties. No coordinated effort existed to direct development or the construction of infrastructure to which future development would be attracted. In addition, the very features and attributes that were considered the most desirable (rural character, views, and access to open space and river front) were the first features that were being compromised. By forming the Connecticut Valley Action Program, the DEM set the stage for the establishment of a regional approach to planning and conservation.

Among the tools developed by the DEM for use in the valley was the Agricultural Preservation Restriction (APR) program. Through the APR program, landowners could sell their development rights to the state. In addition, the DEM, in cooperation with other state agencies, increased the support of agricultural activities by providing training in integrated pest management and intensive farming practices to make the agricultural use of the land more effective, more profitable, and more sustainable. The action program has been most widely recognized, however, for the publication of *Dealing with Change in the Connecticut River Valley: A Design Manual for Conservation and Development* (Yaro et al. 1990), which describes plausible and effective rural design parameters and policies that have a very broad application.

The Center for Rural Massachusetts, under a grant from the DEM, developed the manual in the belief that the only way to avoid a hodge-podge of low-density suburban developments and "islands" of preservation was to formulate a comprehensive approach to development and preservation. From the outset the center recognized that the solutions must be practical. Complete faith in land preservation efforts would ultimately be unsatisfactory simply due to the limits of preservation resources. Reliance on unregulated market forces was surely not an acceptable alternative. Thus the comprehensive plan would have to encompass both development and preservation in a practical fashion. To accomplish this goal, the center devised design guidelines that were based on preserving the character of the region and those elements most prized by its residents while providing for viable residential and commercial development.

Much of the center's plan, as it is being implemented, involves turning classic planning tools upside down. For example, it is common for zoning regulations to require a commercial shopping center to be set back 100 or even 200 ft from a public right-of-way. This arrangement forces the shopping center into the familiar "strip" with a sea of parking and macadam in front of it. The center's recommendations require instead a maximum setback of 25 ft from the public right-of-way. The result is that all of the parking is now to be behind or alongside the building. This new arrangement also encourages people to walk from store to store and brings the store into a more "human" scale akin to the old downtown area. This design is more in keeping with the character of the existing towns and streetscapes of the valley. Merchants have accepted the new design because it gives them two places to advertise (front and back entrances).

In its landscape requirements for commercial development, the center eliminated the use of the classic juniper and bark mulch plantings in favor of native plants and wildflowers. Residential developers are encouraged through density bonuses to form clusters and preserve open space. The density bonuses allow the developer of land to purchase back some of the development rights purchased by the state from local farmers. By obtaining these development rights, a developer of a residential project might be able to build 15 houses in a cluster on 12 acres of ground; without those development rights, the builder might be restricted to 10 or 11 units. The density bonus encourages the builder to cluster the 15 houses onto a portion of the site, say, 4 or 5 acres, leaving the rest of the site in open space.

The Connecticut River Valley project group did not invent these ideas, but they promoted them actively in the community and gained broad support for them among the residents and landowners. The project group encouraged the implementation of the ideas through public meetings and the use of their book. *Dealing with Change* identifies eight different landscape types and illustrates possible development scenarios, from development under traditional zoning regulations to development under the revised rural landscape planning guidelines. The book includes sample ordinances for the municipalities to consider and adopt. The success of the effort by the DEM has been linked to several factors: (a) The public had a vested interest in preserving the character of the area in which they lived and owned land. (b) Because the Center for Rural Massachusetts dealt with the municipalities and the residents in developing the design guidelines, the values and objectives of the project were the values and objectives of the residents. (c) The plan was practical; it allowed for, and even encouraged, development that was consistent with the guidelines. Performance requirements for new development were published and straightforward. (d) The guidelines and ordinances were communicated in a readable, friendly style, using graphics and photos that were easily understood by any interested person, rather than in the "legalese" and "technobabble" normally used in documents of this type. The guidelines suggested by the center have been adopted by most of the municipal governments in the Connecticut River Valley, and the region has become a model of effective rural planning.

Lot layout alternatives

The rapid post-World War II expansion of suburbs and edge cities has had several undesirable impacts. The cost of housing has continued to rise in real terms over the last 30 years so that affordability has become a major issue in some communities. Rapid development has also been criticized for consuming agricultural land, reducing environmental quality, and abandoning urban centers to deal with poverty and other social ills. These complaints generally are grouped under the umbrella of "urban sprawl." Resolving most of these issues will require a shift in public policy and political will that are outside the scope of the site design professionals' immediate influence. However, through their site planning work, designers can influence communities by addressing concerns for environmental quality, development density, and housing affordability. Design professionals should expect the push for development to continue in response to the growth and resettlement of our population. They should also expect the concerns over the environment, sustainable development, and the character of a community to continue and grow.

The key to a successful residential design regardless of the cost of the housing lies in how effectively the design creates a sense of place and relates to the end user. These objectives are achieved in many ways. Some sites have certain natural features that will automatically connect the site to its users and that may be worked into the design. Other projects must rely on the combination of the housing type and landscape architecture to create the feeling on the site that attracts and holds residents. The familiar grid layout of the post-World War II–era residential development is not the preferred approach in most communities today. Contemporary site layouts rely on more curvilinear street designs and a greater mix of building styles and types than were prevalent in those early suburban projects. The focus of contemporary residential site design is on the efficient use of the site—that is, balancing the number of dwelling units and the development costs with the interests of the community and the environment. This has been described as a *quality-of life approach.* Numerous projects have demonstrated that high-quality development can be affordable and efficient.

It has been shown that if all the costs are considered, large-lot developments consume more resources than they contribute to a community. Consequently, the trend in development is now toward the smaller-lot, higher-density developments that are generally consistent with the goals of the various smart growth initiatives around the United States. The use of a variety of densities and housing types has proven to be a successful community development formula, and it has replaced the familiar uniform density and housing types of the past in some communities.

Research conducted by the National Association of Home Builders, Randall Arendt, Andres Duany and others have all found that smaller lots and cluster developments have an equal or greater initial and resale value than traditional residential development (Arendt 1991 and NAHB 1986). *Small-lot development* usually refers to projects with 6 to 12 units per acre. It has been common to encounter resistance to higher densities in communities with the more traditional density of 1 to 4 units per acre, but the experience of numerous communities has been that many buyers find the higher-densities attractive. People like living in the higher density communities as long as the amenities and privacy are there. The primary target market includes young professionals, couples without children, or empty-nesters, but there are successful small-lot communities for every demographic target.

Small-lot development tends to work best on relatively flat sites. Small-lot clusters are an ideal way of reaching a gross density and preserving important open-space features of a site. In general, site layout is critical to project success, and it should be focused on lifestyle amenities, emphasizing outdoor living and indoor privacy. Small-lot developments, especially those with more unusual configurations, work best in more sophisticated real estate markets (Kreager 1992).

The familiar grid layout as shown in Fig. 8.3 is an efficient way to subdivide property, but it can be monotonous, especially for residential areas. Grid layouts are familiar forms of development to most people and provide a certain level of comfort for many people. The key advantage of the grid layout is the relative ease it provides for finding one's way and the maximization of lots it allows per linear foot of street. However, the straight streets of the familiar grid layout often invite higher vehicle speeds than are desirable, especially where wide cartways are used. In contrast, curvilinear streets are far more interesting visually and may help to manage vehicle speeds. However, this design is somewhat less efficient with regard to lot count. Also, the degree to which streets curve and the way they are laid out make finding one's way through some communities confusing and difficult. This may be of concern in

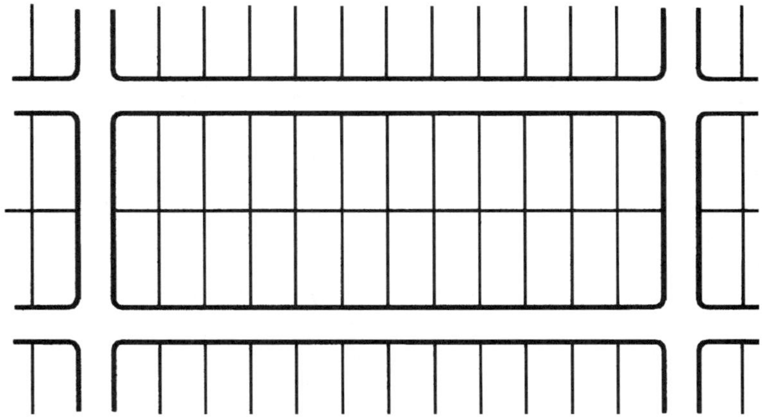

Figure 8.3 A traditional grid layout detail.

particular in communities designed for older residents. There are alternatives to the traditional grid layout, however, as shown in Figs. 8.4 and 8.5, which depict several versions of alternative grid designs.

As higher house-lot densities have become more common, a variety of different lot configurations have evolved to accommodate the smaller lot sizes and traditional or familiar housing types. Some projects use a combination of these lot arrangement strategies while others design around a single lot and housing type. For the most part, these small-lot single-family home strategies are of one of five types: deep narrow lots, wide shallow lots, alley lots, Z lots, and clustered lots.

Deep narrow lots. The deep, narrow lot configuration allows for a familiar lot and house pattern with the garage and front of the house facing the street (Fig. 8.6). Lots typically range from 3000 to 4800 ft^2, about 6 to 8.5 lots per acre. The typical 40-ft-wide lot allows for a total of about 10 ft of side yard, which leaves 20 ft for the garage and 25 ft for the house. Garages are often designed close to the front of the lot, often in front of the house façade to maximize the amount of yard space behind the house. This tends to create an unattractive street view of all garage doors. Also, the deep, narrow lot provides for only minimal backyard privacy, especially in housing with two or more floors. This may be offset if special attention is paid to the location of windows in adjacent units and if visual landscape barriers are used, but it is difficult to anticipate the location of windows and site lines in projects where different housing models are possible.

Wide shallow lots. An alternative to the deep narrow lot is the wide shallow configuration that allows for a standard-width house and garage and conveys the feeling of a traditional neighborhood (Fig. 8.7). The wide shallow lot cre-

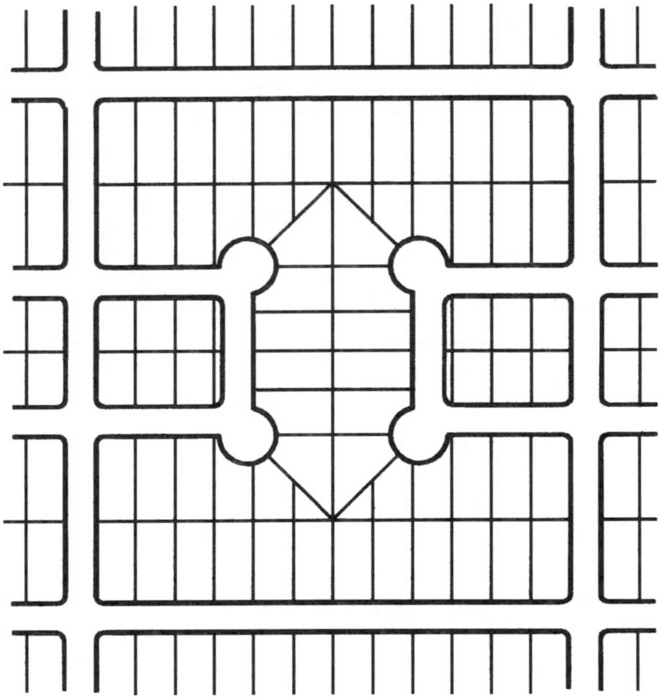

Figure 8.4 An alternative grid layout detail.

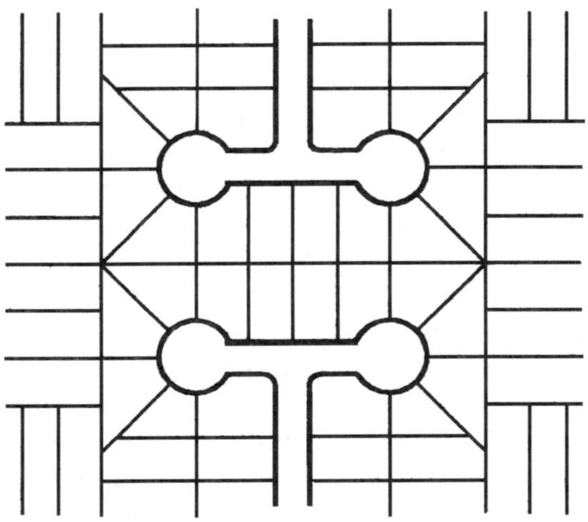

Figure 8.5 An alternative grid layout detail.

ates a feeling of a larger lot and space between units by presenting its longest dimensions along the street frontage. These lots generally yield about 6 to 7 units per acre with lot sizes from 3500 ft^2. In general, wide shallow lots are not as desirable as deep narrow lots because the wider lots are more expensive and

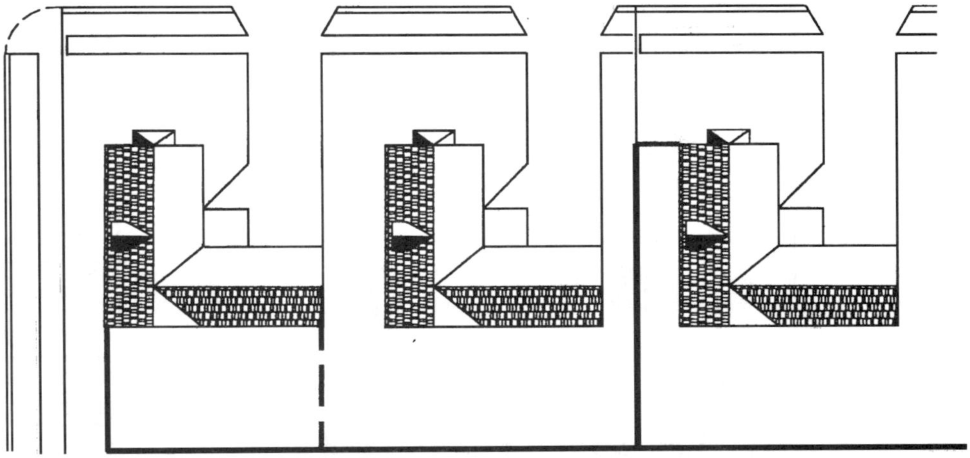

Figure 8.6 Deep narrow lots detail.

there is less useful yard space. Development costs may be higher since there are fewer units per linear foot of road and utilities. An increase in lot width of 20 ft will result in an increase of almost 50 percent higher utility costs per unit over the deep narrow plan. The backyard of the wide shallow lot offers little privacy, especially if two-story homes are constructed. However, the use of fencing and appropriate landscaping can increase privacy.

Alley lots. Another small-lot layout alternative returns to the use of alleys behind the house (Fig. 8.8). Alleys were common in cities many years ago. Garages were located in the back of properties, and access was over a common alley. The alley design allows for lots of 3300 to 4500 ft^2 yielding 4 to 8 units per acre. Many older, desirable neighborhoods built in this configuration exist in cities throughout the United States. By locating the garage in the rear, the streetscape is all house fronts—no driveways and no garage doors. The alley is usually 16 to 18 ft wide. The paved alley increases development costs somewhat, but many of the traditional neighborhoods using the alley layout have narrower streets and lots that offset the additional cost of the alley. Some municipalities resist the alley arrangement because of increased maintenance, but in other cities, the alleys are not public rights-of-way but are held in common by the neighbors through a common access easement and maintenance covenant. Projects with alleys provide ideal utility corridors.

Z lots. The term *Z lot* is used to refer to a layout in which the house is placed on or very near to one property line and is called a *zero-lot line.* In some configurations the lot on lines may jog around the building to create a more interesting space. Such lots are said to resemble a Z—hence its name. The Z lot is often slanted relative to the street to increase the appearance of lot width. Houses are designed to increase light and maximize privacy with the use of strategically located windows and entranceways. Some Z-lot developments

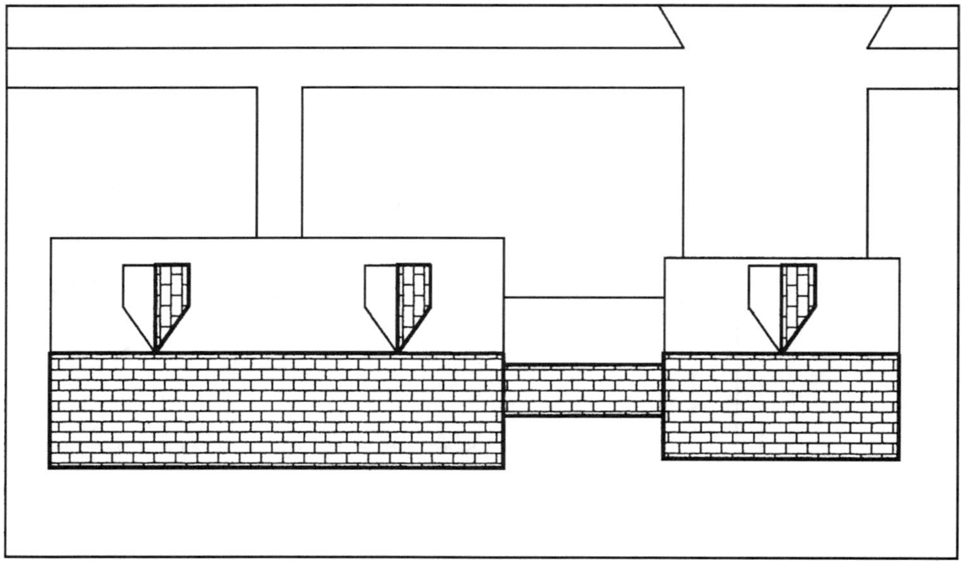

Figure 8.7 Wide shallow lots detail.

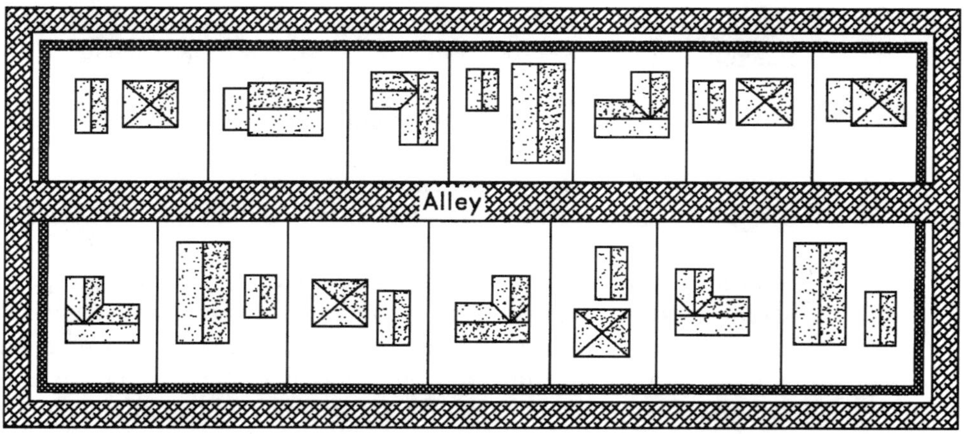

Figure 8.8 Alley lots detail.

provide special maintenance easements or even contractual arrangements like condominium agreements to provide access to buildings for maintenance. Easements along lot lines may be difficult for Z-lot configurations.

Clustered lots. Cluster designs have become more common in recent years because they work with, rather than against, the planning goals of communities (Table 8.3). In general, the principle behind the cluster design is to allow the same number of units on a tract as would be there normally but to group the units into clusters of greater density (Fig. 8.10). A density bonus is sometimes

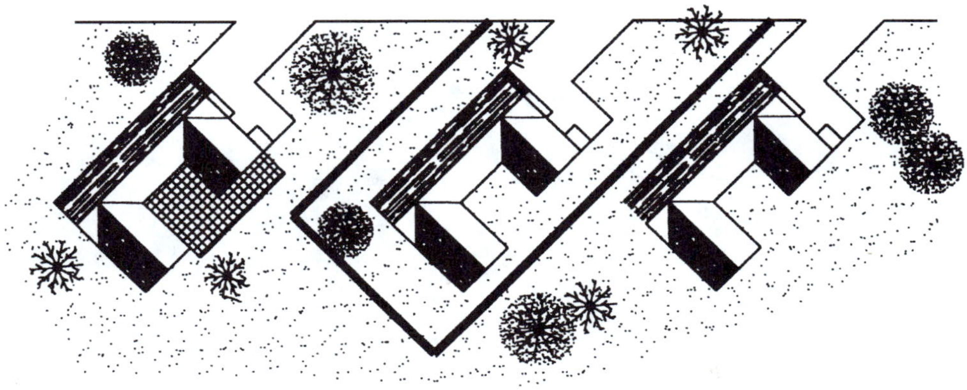

Figure 8.9 Z lots detail.

TABLE 8.3 Cluster Design Attributes

Allows the same number of units in a smaller space, which creates more open space
Creates a low-profile visual impact on an existing community
Allows for open-space buffers between areas supporting incompatible uses
Preserves important natural functions of landscape
Contributes to the rural character of an area
Supports rather than contradicts the character of the site
Establishes a benchmark for future projects

allowed to encourage the preservation of open space. Cluster development can reduce the visual impact of new development on a community as well as reduce the amount of negative environmental effects. It allows developers to utilize the land and preserves valuable natural areas, agricultural land, riparian zones, and so on. Cluster developments are usually welcome because they minimize the impact of the development and are sensitive to rural character, the nature of the site, and the community. Effective and successful cluster developments may also serve to establish a quality threshold for other future projects.

Easements and rights-of-way

Allowances for easements and rights-of-way in higher-density developments may require more planning and thought than they would in less dense projects. With smaller front and side yards, easements may take a significant portion of the street side of individual lots. Utility easements may restrict the planting of large trees or fences. Some utilities prefer easements outside of the cartway to reduce the cost of maintenance and repair. In other cases, the proximity of one utility to another may require extraordinary construction methods and increase development costs. Easements along the back property are possible for some utilities, but access is required, which may have a negative

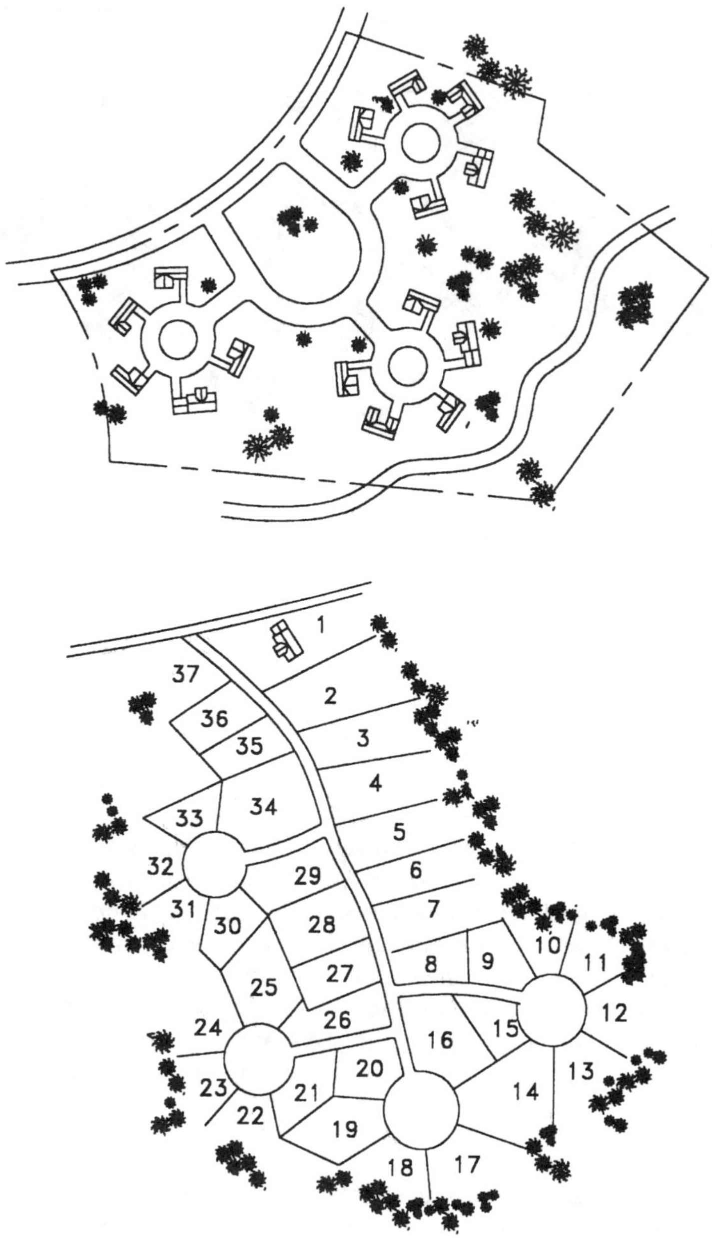

Figure 8.10 Clustered lots detail.

impact on the use and enjoyment of the lot. Many small-lot projects are designed to allow utilities to be installed within the public right-of-way on the street, usually between the curb and sidewalk. Still other projects provide a utility corridor easement across front lawns and restrict the amount and type of landscaping that can be used.

Affordable housing site design

The cost of new development is a concern for many communities and many have found that key people in the community can no longer afford to live there. Zoning and land development ordinances are prescriptions for development. Development costs are a function of many factors, but among them are the local development standards, which ultimately are passed on to new-home buyers. All the while that communities have been trying to find ways to increase the number of affordable homes, they have been learning that the developments produced by their local standards are not only inconsistent with the character of their communities, they are also contributing to unwelcome sprawl.

At the same time homes in many older communities continue to be considered valuable and command high market prices. A visit to some of these older communities too often reveals that many of the features that contribute to their continuing appeal and market value would not be allowed under current ordinances and practices. Many of the standards for community development that were in place up to World War II were revised shortly afterward. Street widths, lot sizes, setbacks, and many other aspects of postwar community development were enlarged and modeled on the grid type of street and lot layout. Wider streets and larger lots reached their peak in the 1980s and 1990s. The growing awareness of the negative environmental impacts as well as the increased cost of the initial development and life cycle costs of unnecessary pavement and oversized lots have encouraged a shift toward a more affordable and lower-impact design that does not sacrifice public safety or environmental function. Increasing the number of affordable housing units remains a priority in many communities. Affordability can be improved dramatically by specific changes in local development standards and practices (see Tables 8.4 and 8.5).

Designing for Security

In recent years there has been an increasing awareness of the role the design of public spaces plays in crime prevention and general security. While it is important to note that there are many social and economic influences with more impact, it has been demonstrated that design may play an important role in heightening the security of a community, particularly as part of efforts to improve distressed communities.

Crime prevention through environmental design

The Crime Prevention Through Environmental Design (CPTED) effort is most successfully conducted in conjunction with other community efforts such as community policing and neighborhood awareness programs. Design is only one element among several. Rob White of the University of Melbourne observes that there are two schools of thought regarding CPTED. One approach studies how places can be designed and built to be safer simply to create better-quality places. The other he describes as "situational prevention," which is remedial work addressed to specific trouble or hot spots (White

TABLE 8.4 Standards for Affordable Residential Development Design

Standard	Single-family detached	Single-family attached or townhouse
Lot size	4500–5000 ft^2	3500–5000 ft^2
Lot width (min)	No minimum, 50 ft	No minimum to 16 ft
Lot coverage (max)	40–50%	50–75%
Setbacks, front	10–20 ft	5–20 ft
Back	5–15 ft	5–10 ft
Side, each/total	5/10 ft	0/5 ft
Right-of-way width	35–0 ft	30–50 ft
Cartway width*	18–28 ft	22–32 ft

*9-ft-minimum travel lane on low-volume local street with 8 ft for each parking lane.
SOURCE: Adapted from Welford Sanders, Judith Getzels, David Mosena, and JoAnn Butler, *Affordable Single Family Housing,* Planning Advisory Service Report No. 385 (Washington, D.C.: American Planning Association, 1984); Welford Sanders and David Mosena, *Changing Development Standards for Affordable Housing,* Planning Advisory Service Report No. 371 (Washington, D.C.), The Joint Venture for Affordable Housing, American Planning Association, 1982); and Stephen S. Fehr, "Reducing Land Use Barriers to Affordable Housing," *Planning Series No. 10* (Harrisburg, PA: Planning Services Division of the Bureau of Community Planning, Pennsylvania Department of Community Affairs, 1991).

1998). It is necessary to recognize that while environmental conditions may encourage or discourage crime, design alone is not an answer. To discourage crime, we must create environments that make it hard for criminals to do their work and encourage other acceptable or desirable activities. Design is part of a larger strategy that must include management as well as social and community development.

There is no single formula for the design of defensible space. Therefore, each planning effort requires a thorough understanding of the physical environment and social environment of a neighborhood. What is the layout physically? But also, who is coming and going? Who belongs and who doesn't? What are the dynamics of the problem? Is it traffic? Automobile or pedestrian? Night or day? What are the neighborhood routines? Thefts are higher near schools, rapes are higher near hospitals, and so on because of predictable routines that in turn create opportunity.

The design solutions to problems communities face range from improving security and safety elements to helping to increase neighborhood identity and pride (see Table 8.6). Adam Graycar has observed that crime is not an equal opportunity endeavor—where you live, how you live, and who you are have a great deal to do with your chances of becoming the victim of a crime. Not all crime is considered equal either. Predatory crimes such as homicide or assault are more serious and less common than crimes such as drug-related activities, violence, and theft. Graycar observes that for a crime to occur, there are three necessary elements: a likely offender, a target, and the absence of a capable guardian. The "capable guardian" refers to all social, political, and design strategies used to prevent crime. Situational CPTED focuses on providing physical evidence of the capable guardian by creating spaces that reduce

TABLE 8.5 Elements of Better Residential Site Design

Narrower, shorter streets
Smaller lots with less restrictive setbacks and lot width requirements
Increased allowable lot coverage
Increased use of effective stream buffers
Increased infiltration of storm water
Grass-lined swales used instead of pipes and paved gutters

TABLE 8.6 Site Design Strategies for Crime Prevention

Provide effective lighting.
Design to assure good lines of sight along streets and paths, near buildings.
Consider crime prevention when selecting plant materials.
Use traffic calming measures and circulation planning to reduce joy-riding.
Look for and anticipate escape routes.
Encourage people to observe streets and public spaces.
Use vandal-proof materials, and assure quick repair and replacement of damaged materials.
Restrict traffic on residential streets (one-way streets, traffic calming devices).
Increase the evidence of formal and informal surveillance.
Restrict vehicle movement.

opportunity and increase the risks and effort required to pursue a criminal activity (Figs. 8.11 and 8.12).

A community that is aware of what is going on within it, in which activities in public spaces are readily observed, is less likely to have a crime problem primarily because the community itself, through its interaction and behavior, represents a capable guardian. While much of the CPTED effort is geared toward physically modifying space, this effort should be a product of community desire and interest. The key to CPTED is the involvement of the community. In some ways, designers working with communities become facilitators of the community's goals. In most instances, the budget for implementing design solutions is limited so it is important to have the greatest impact with the resources available. To determine the scope of the problems and to develop a design strategy, the CPTED process usually begins with an assessment of the neighborhood. In many cases determining the boundaries of the neighborhood and the study area is very difficult, but it is important to have a finite area for consideration.

Working with neighbors and local businesses, the CPTED team identifies the attributes and the problems of the neighborhood. The team looks to find the positive elements—that is, the points of stability such as schools, churches, or long-standing businesses—and then locate these places on a map. Next, the team identifies the problem areas and locates them on the map. The CPTED team may elect to map abandoned buildings, vacant lots, high crime areas, homeownership, parking areas, areas that are poorly lit, traffic patterns or anything else that contributes to the character and concerns of the neighborhood. From these maps and the juxtaposition of positive and negative elements, the CPTED team can work with neighbors to identify and prioritize steps toward improving community life.

Figure 8.11 Fences in a traditional neighborhood.

Figure 8.12 Bollards and plantings used in a city neighborhood separate public from private space and signal community activity and surveillance.

Territory, access, and surveillance

In general, there are three aspects of defensible-space design: territory, access, and surveillance. *Territory* refers to private and public spaces. Territory is established by establishing tangible distinctions between spaces. Distinctions can be made using textural changes in pavements of walls, elevation changes (a step up or down), barriers such as walls or fences, visual barriers such as low fences or shrubs, or psychological barriers such as consistent neighborhood organization or themes. When evaluating space, ask, "What could I get away with here?"

Access refers to providing and restricting access; in short, control. Blocking off streets is sometimes helpful, but it is usually not the preferred method. Through streets are preferable because they provide necessary access for pedestrians and vehicles. In addition, blocked streets are considered more threatening, and residents may not want to project that image of their neighborhood. Other street designs are usually better received by residents: intersection narrowing, S curves in the streets, dual-use streets, and traffic calming measures such as one-way streets, turn restrictions, or bollards. Physical access might also be restricted. This method is known as *target hardening,* and it involves installing fences and gates or other restrictions. Target-hardening measures are sometimes necessary as preliminary or temporary design elements used to gain control.

There are subtle ways of communicating boundaries and creating a sense of territory or neighborhood. In some situations, the use of low fences or walls, or signs or certain colors is enough to signal to people that this area is set apart, that they are passing through one area and entering another. Such symbolic barriers provide subtle identity to an area to both residents and visitors and make a sense of ownership more palpable. Space that does not indicate use or is not controlled within a neighborhood is an attractive nuisance and perhaps an invitation to unwanted behavior.

Surveillance refers to seeing and being seen. The suggestion of surveillance can be made simply by opening more windows and doors onto the street so that people are seen as both the observers and the observed. Points of congregation such as playgrounds and porches encourage residents to see and be seen, increasing the degree of visible surveillance in a neighborhood. While lighting is important, the sense that there are eyes on the street is more likely than lighting to be a deterrent to unwanted activity.

With the increased terrorist threat faced today, designers should expect to address security issues beyond crime in the site design plan, especially where public facilities are concerned. This may require consultation with a security expert or a design professional with specific security experience. The federal government has developed site security guidelines, but we should expect that this is an emerging area of practice and the standards are still evolving. The General Services Administration developed a set of security standards in the early 1990s, and every federal facility has been assessed and upgraded to minimum standards. Much of the focus of the standards deals with architectural issues, interior security, and technology, but site planning also plays an important role. Most state and local facilities and many private institutions that might also be targets for terrorists have little or no relevant security. In many cases the short-term answer has been target hardening or creating a fortresslike feeling in and around public buildings. Public buildings and public spaces adjoining them are more than the sum of activities that take place within them. They represent the values and the character of the people, and so they must remain accessible, attractive, and inviting. Security and safety are important concerns, but it is widely agreed that target hardening is not the best first line of defense.

Security planning and design are completely consistent with the CPTED principles: Good surveillance reduces opportunities to be unobserved and increases the risks of being caught, effective design limits the opportunities for access and escape and protects the building and people. Where crime may be directed to one or a few individuals per incident, terrorism is directed toward the greatest possible number of people per incident. Where criminals look for a means of escape, we have learned that terrorists may have no thought of escape. The assessment of the site, therefore, looks for a different type of vulnerability. The challenge is then to the designer to protect the site from intruders in a fashion that is more than simply hardening the facility. New facilities that might be the target of such attacks should incorporate security into the most basic design considerations. Redevelopment or retrofitting project planners should be aware of the vulnerability of the site and make appropriate recommendations. It is likely that site designers will work in conjunction with security experts, but they should develop an awareness and expertise of their own as well. To not consider these issues in one's design may be seen later, after an incident perhaps, to have constituted a breach in the standard of care expected from a design professional. Many of these concerns are not parts of building codes or design standards yet—they may not even be on the client's list of concerns—but they require the attention of the designer nonetheless.

New facilities should incorporate a setback from the street that allows observation of all approaching vehicles and pedestrians. Vehicles and pedestrians are directed into specific patterns of approach through the site design. The setback, however, presents an esthetic concern. An unadorned open space may facilitate surveillance, but it clearly speaks of a bunker attitude in terms of design. To improve both the appearance and function of the plaza created by setbacks, designers might consider incorporating changes in elevation to make access with a vehicle more difficult and the site more pleasing. Other low barriers in the plaza would make a direct path by a vehicle impossible. To protect the building further, the building could be raised above street grade and the plaza used as a transition over the change in grade. The plaza should be designed to function as a public space and should be filled with activities.

The key site design concern is access by pedestrians and vehicles. The points of access for pedestrians should be limited to provide a maximum amount of surveillance and control. Walkways should be set away from the building, and plant materials and landscape features should not obstruct a clear field of vision around the building. Approaches to entrances should be open but access controlled by vehicles by hardened bollard systems or other methods such as changes in elevation or direction. Separate entrance facilities might be considered. The separate facility isolates everyone entering the facility for a security check and could serve as a barrier to vehicles attempting to get to the entrance. The most common method of keeping vehicles away has been to rely on large planters or other heavy items, but large planters or tree masses may create blind spots or hiding places. Landscaping should be kept below 24 in in the security surveillance area.

Vehicle access should be on roads that are curvilinear to require vehicles to drive slowly. Parking should be kept well away from the building, and separate controlled parking may be advisable for key personnel. It is likely that new public facilities will not be built with public parking beneath them. Strict setbacks from the building should be observed for all vehicles. Loading and unloading areas should be large enough for needed queuing but not allow for any parking. Loading docks should be designed in accordance with the facility management's preferences.

Of course, all of this must be accomplished while meeting accessibility requirements and facilitating the smooth operation of the site. The site exterior should be well lighted to avoid having dark places near the building. Lighting should be coordinated with exterior closed-circuit television systems to keep obtrusive lighting to a minimum. The combined effects of these measures are to create a clear perimeter around the building with an obvious buffer to make terrorist or criminal acts more difficult. Design professionals should be cautious, however: Security is expensive. One should be careful with site development cost estimates if security costs are to be included.

Lighting

Lighting serves to improve security and way finding, but it also provides important visibility to commercial sites and can be used to create special effects and feelings in the nighttime landscape (Table 8.7). With the development of specialty lighting products and effects, lighting has become as creative as any aspect of site design, and it is a specialty of many designers. The design of site lighting is just as often performed by companies selling lighting equipment as by an ancillary service; however, finding the right combination of products, lighting types, and distribution can be a complex undertaking. The purpose for the lighting is the critical consideration; for example, lighting for security or surveillance will call for a different strategy than lighting for a more intimate space (see Table 8.8). Lighting is selected on the basis of the type of light, the distance from a light source to an object, the light of surrounding areas, and the nature of the activity being illuminated. Many organizations have specific lighting standards or preferences that will influence design. The nature of the lighting industry is such that new products and capabilities are being introduced all the time, and like so many other aspects of site planning, lighting requirements are often a matter of local ordinances.

The distribution and brightness of light are the fundamental elements of lighting design. *Distribution of light* refers to how much light is cast over an area. Lighting to accent certain areas or to create a mood or feeling requires

TABLE 8.7 Performance Characteristics of Different Sources of Light

Type of light	Lumens per watt	Life, h	Color	Notes
Fluorescent	70	6,000	Good color purity, white	Affected by cold weather
High-pressure sodium	130	16,000	Poor color purity, yellow to orange	Washes out colors in landscape
Incandescent	10–18	750–1,000	Very good color purity, yellow	
Low-pressure sodium	190	11,000	Poor color purity, pink to orange	Washes out colors in landscape (gray)
Metal halide	90	14,000	Cool white good color purity	
Mercury vapor	55	24,000	Cool white, good color purity	Strong in blue-green spectrum
Tungsten iodine	18–20	2,000		

TABLE 8.8 Recommended Levels of Illumination

Area and activity	Lux, lx	Footcandles, fc
Building exteriors		
Entries, active use	50	5.0
Entries, infrequent use	10	1.0
Vital locations or structures	50	5.0
Building surroundings	10	1.0
Buildings and monuments		
Bright surroundings	150–500	15.0–50.0
Dark surroundings	50–200	5.0–20.0
Bikeways		
Along roadside	2–10	0.2–1.0
Away from road	5	0.05
Bulletin boards, kiosks	500–1000	50.0–100.0
Major roads	10–20	1.0–2.0
Collector roads	6–13	0.6–1.2
Local roads	4–10	0.4–0.9
Walkways, open air	5–10	0.5–1.0
Walkways, enclosed	6–40	0.6–4.0
Park or garden walkways	20–40	2.0–4.0
Steps in park or garden	10	1.0
Stairways	200–600	20.0–60.0
Gardens	50	5.0

TABLE 8.8 Recommended Levels of Illumination (*Continued*)

Garden features	200	20.0
Loading areas	200	20.0
Parking areas	10–20	1.0–2.0
Outdoor athletic areas		
Badminton	200	20.0
Baseball, infield	110–300	11.0–30.0
Baseball, outfield	100–200	10.0–20.0
Basketball	100	10.0
Football	100–1000	10.0–100
Field hockey	100–200	10.0–20.0
Skating	100	10.0
Softball, infield	100–500	10.0–50.0
Softball, outfield	70–200	7.0–20.0
Tennis	200–500	20.0–50.0
Volleyball	100–200	10.0–20.0

a lighter and more elegant use of light. For such applications, the angle and position of the light are determined for their visual or esthetic effects as opposed to their roles in way finding or security. The use of uplighting, moonlighting, and backlighting to create a feel very different from the daylight landscape has become more common. Uplighting is most effectively used to feature objects that can be viewed from a limited point of view. The light source is located low and is pointed toward the object and away from the viewer. Uplighting is commonly used against walls or fences or in gardens that will be viewed from only one side. This orientation lights the object without any glare to the viewer. Uplighting is an unusual effect because the eye is not used to seeing things lighted from below in nature. This method is effective at creating dramatic textures and contrasts in the night landscape.

There are several methods for computing the *brightness* of different lighting choices. The *point illumination method* measures the illumination at a given point whereas the *average illumination method* measures a more general distribution of light. The *point distribution method* is described as follows (see also Fig. 8.13):

$$E = \frac{I \cos \theta}{d^2}$$

where E = illumination on a horizontal surface, fc
I = lamp intensity, lm
θ = angle between the fixture and a point on the ground, degrees
d = distance from the luminaire to the point

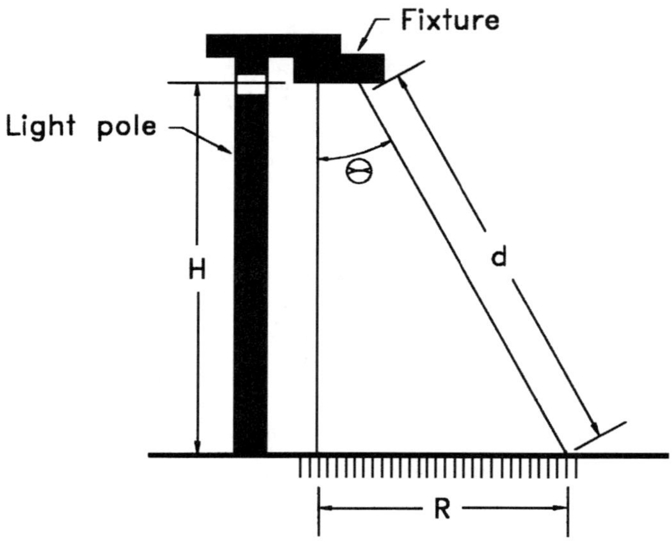

Figure 8.13 Point distribution calculation dimensions.

The point distribution calculation is useful to determine the constancy of light within its distribution, but it is a fairly intensive method. The *average illumination method* is a more general method:

$$F = \frac{luM}{LW}$$

where F = average illumination, fc
l = lamp intensity, lm
u = coefficient of utilization
M = maintenance factor
L = horizontal distance between fixtures, ft
W = width of the area illuminated, ft

To solve for L:

$$L = \frac{luM}{FW}$$

Designers should consider the extent to which the *efficiency,* sometimes referred to as the *maintenance factor,* of the lamp changes over its life. Some lamps may vary as much as 75 percent over their operating life. The maintenance factor includes variation as the light source ages as well as the effects of dust or dirt on lamp covers. Maintenance factors vary, but 50 percent is a common rule of thumb. Performance expectations for illumination and the coefficients of utilization are provided by the manufacturer as part of the photometric information for the luminaire.

Moonlighting is accomplished using combinations of lighting carefully located high up in trees and other low-wattage ground-level lighting used to illuminate branches and leaves from below. Moonlighting can create very dramatic effects, and it is especially good for transitions between lighted areas. The filtered light may provide adequate light for walking on marked paths and can provide for particularly beautiful effects. Designers should be aware, however, that these effects can be difficult to achieve if there are areas more brightly lit nearby.

Backlighting is sometimes used to feature a tree or shrub or other element with an unusual or visually pleasing silhouette. To minimize the risk of glare to the viewer, the height and angle of the lighting must be adjusted carefully. *Silhouette lighting* can also be effective by uplighting a wall or surface behind an object. *Indirect,* or *bounce, lighting* is achieved by directing light to a surface that reflects light into or onto a desired area. The development of *extruded fiber optic lighting* and other products has introduced the possibility of drama and beauty into the night landscape.

Commercial Site Design

Site location

The layout of commercial sites is driven by the nature of the enterprise in addition to the local ordinances and community practices and expectations. A key issue for the developer and tenants is always location in the community, and site selection is extremely important. The ideal commercial site seems to have a somewhat elusive but immutable character. Every community has sites that are successful despite a seemingly poor location and other sites that never succeed regardless of the tenant or business that locates there. A site analysis that studies only the development potential or the visibility and traffic past a site often cannot identify the underlying cause of success or failure by itself.

Some of the factors that contribute to the success of a commercial site are related to demographics: Is the site located near enough people with enough disposable income? Success is also correlated with the type of business or mix of tenants: Is the business mix able to draw people either as a destination or on an impulse? Does the mix of tenants work well so that together they draw more business than any one tenant would draw alone?

Other aspects of site success are well within the scope of the site professional's work. During the site analysis, the site location should be explored. Commercial development with retail shops will usually require a minimum amount of existing traffic at the proposed location. Very large retail projects may rely on becoming destinations themselves and be less concerned about existing traffic. In either case, existing intersections are prized locations for most commercial projects.

Access to the project site is critical. *Accessibility* in this sense refers to the ability of the customer or client to be able to get into the shop or business. *Visual access,* that is, a view into the site, is also important. While there does not appear to be a fixed standard, many retail operations require a minimum number of parking places to be located within a given distance from the door. One of the most lamentable aspects of retail development is the visual impact the projects have on the community. Even small corner stores can bring a significant change in character to a neighborhood if they are not designed carefully. Ultimately the most significant negative impacts are associated with automobile traffic and parking, but the intrusion of bright lights and noise can also be problematic.

There is a strong preference on the part of retail operations to be able to show the public the available parking and its proximity to the door. Retail operations often resist attempts to reduce the visual impact of parking (by putting it behind the building or screening it) in the belief that if customers do not see convenient parking, they will go somewhere else. Except in smaller projects, in-fill, high-end, or theme-related retail projects, it is difficult to overcome this preference. The designer must also accommodate delivery and distribution traffic on the site. In most projects these activities are located behind the building, further complicating relocating the parking.

The most important part of accessibility is visibility. Customers and clients generally need to see the development. The efforts to mitigate the visual impacts of developments are complicated by the need of commercial projects to be seen. In most cases it is necessary for the site professional to find a design that meets both needs. In many cases the expectation of the community is such that no screening is required or expected, but as time passes and the community gains experience with development, the expectation changes. Effective signage, combined with distinctive landscaping and lighting, can provide way-finding guides to customers without sacrificing visual or environmental quality (Figs. 8.14 through 8.16).

An alternative to the traditional strip layout may be used to reduce the amount of impermeable area dedicated to parking spaces. Developing commercial sites in a U shape rather than a strip may increase the number of parking spaces within a given distance from a merchant's door while reducing the coverage of the site. Store owners are concerned with the number of parking spaces within a given distance to the front door. In a strip center the parking lot is necessarily stretched out before the entire strip so that all stores have an appearance of adequate parking immediately in front of them. In a U-shape layout, a concentration of parking spaces would be in front of every store, so a parked car would be closer to more stores arranged in a U shape than in a strip shape. (The stores at the ends of the strip or the ends of the U will always have fewer nearby parking spaces than a store in the middle of the strip.) The U shape is therefore a more efficient arrangement of space, and it requires less coverage on the site.

Building location

The location of a building on the site is a critical element of site planning in terms of the building function. The site planner should locate the building in

Figure 8.14 Photograph of screening at a retail site.

Figure 8.15 Photograph of screening at a commercial site.

Figure 8.16 Photograph of screening at a retail site.

such a way that its impact on the site is minimized, while its functions and design are maximized. Selecting a location should be a combination of managing the solar influences of the site and balancing the earthwork so as to achieve a balance between the building's utility and its esthetics.

Locating proposed structures offers the designer the first opportunity to focus the design of the site in a sustainable direction. The building location fixes the limits and extent of the site disturbance. Clustering buildings reduces the size of the disturbed area and allows the designer to minimize road length and paving. Recognizing that the disturbance of the site impacts the entire landscape well beyond the limits of the property line, the building location decision allows the designer to look for ways to maintain or reestablish links to other parts of the landscape ecosystem. Care should be taken to protect stream corridors, wetlands, and other important landscape features.

The manner in which a building is situated on a site can have important implications in the energy costs of heating and cooling. For northern areas buildings should be located on the portions of the site that receive the most light during the hours of greatest sunshine—about 9:00 A.M. to 3:00 P.M., particularly in the winter months. The building should be located in the northern most part of this area, but adequate distance from neighboring properties should be maintained to allow for possible shading from nearby future development. Open space should be located on the southern side of the building.

Various studies have reported that open space with a southern exposure is preferred over open space with a northern exposure.

While orientation on the site is important, building shape may be more important, although site designers may not have influence over the shape of the building. Square buildings are inefficient shapes for heating and cooling, though they tend to be more efficient than long narrow buildings on a north-south axis. The best combination of shape and orientation is an elongated building on an east-west axis. In northern latitudes in the winter buildings orientated on an east-west axis receive almost three times as much solar radiation on the south side of the building than on the east or west. This situation is reversed in the summer where the roof, east, and west sides of the building receive more solar radiation than the southern side.

Chapter

9

Vegetation in the Site Plan

Plants are an integral part of most site plans. They contribute to the esthetics of a site, to its economic value, and to its ability to function. This chapter is directed toward the functional contributions and durability of plants in the landscape.

Planting Design

Planting plans are shaped by the underlying form of the site plan, but this is not to suggest they are mere eyewash or window dressing. Plants contribute a great deal to the quality of our experience and to the character of a place. The choice and arrangement of plants can be used to frame views, to accent or to hide other site features, to direct pedestrian traffic, to create outdoor spaces, to invite, to repel, to provide comfort, to encourage motion or pause, or to modify scale or the environment. Plantings may be formal or informal, simple or sophisticated, according to the objectives of the site.

Effective planting design is a synthesis of texture, color, line, form, and balance. *Lines* are formed in the landscape as edges that can be created using plants, paving, reoccurring patterns, or grading. *Landscape form* refers to the mass and shape of a group of plantings considered as a whole. *Texture* refers to the appearance of the form as gradations ranging from coarse to fine. *Colors* play a variety of roles in the landscape. The warmer colors such as reds, oranges, and yellows tend to appear closer to the viewer while the cooler colors such as blues and greens appear to recede. In addition, colors tend to evoke different emotional responses from people. Designers should be familiar with the use and effects of color in the landscape.

These elements are used in combinations to evoke a certain response to impart a desired character to the project site. Designers employ repeating patterns, lights and shadows, symmetries and asymmetries, and various nonliving materials to achieve a desired effect. In nature there are few straight lines so to us, plants appear to be inherently human. Planting plans using a symmetrical *balanced form* with a central axis are said to be *formal designs* (see Figs. 9.1 through 9.6). They are highly organized and speak to stability and structure, perhaps even of authority. Formal designs have been preferred in the past, which is why many older homes and neighborhoods have fairly formal lawns and gardens. Formal designs are today still preferred in many important symbolic civic landscapes because they are able to convey a sense of importance to the space or place.

For most other current applications, less formal asymmetrical forms are more common (see Figs. 9.7 through 9.11). Lines in the asymmetrical planting plan are still used to define space and provide way-finding information to pedestrians, but the asymmetrical design appears more natural and less austere. The asymmetrical, informal arrangement of plants tends to have softer edges, less definition. It is important to note, however, that while a design is asymmetrical, it is still balanced. Large masses may be offset with a number of smaller groups of plants or a longer line.

Plantings are particularly good at directing attention and activity (Figs. 9.12 and 9.13). Used in conjunction with other materials, they can identify a low-key transition from one area to the next. The well-designed transition acts as a subtle signal to the observer that there is a change occurring. In this way plants can be used to "describe" areas as private or public, accessible or out of bounds. If needed, plants can also be used to crease masses or lines to make a dramatic impression that provides important signals to visitors.

Figure 9.1 These street trees reinforce and also soften the formality of the straight boulevard.

Figure 9.2 The reflecting pool and the Washington Monument are a formal arrangement of line and form that speaks to stability and authority.

Native plants

Among the trends in landscape work and design is the awareness of the value of native species of plants and the damage caused by exotics. Using native plants contributes to some degree to biodiversity, reduces or even eliminates the need for pesticides and fertilizers, reduces maintenance costs, and may increase or improve wildlife habitat. Designs incorporating native species tend to be more natural in context—that is, they tend to use nature as a model—and so they increase many of the landscape functions missing or minimized in other landscape designs. Once established, native landscapes tend to require less care, less water, and fewer additives since the plants have evolved to survive and even flourish in the extremes of a particular region or zone (see Figs. 9.14 through 9.16).

This being said, it should be noted that what is deemed a "native plant" or an "exotic pest" is anything but a precise science. As plants extend their range, they necessarily move into new areas. Likewise once plants are introduced to

Figure 9.3 This formal space between condominiums is softened by the regularly planted trees. The trees also provide cooling shade and help to bring the common space into a human scale.

Figure 9.4 This less formal arrangement in another part of the same community shown in Figure 9.3 is designed to act as a visual buffer of adjacent properties.

Figure 9.5 While this courtyard in an elderly housing building is highly organized and formal in appearance, it is composed of a number of smaller less formal spaces in which the residents can meet or spend time alone. The predictability of the formal landscape contributes to orientation and wayfinding for some older residents.

Figure 9.6 This brick approach to a courthouse provides a fairly formal, highly organized public space clearly associated with the authority of the court. The formality is accented by the strong lines within the pavement and the trees that lead the eye to the mass created by the building.

Figure 9.7 In contrast to Figure 9.6, this informal landscape in Niagara Falls, N.Y. contains strong lines, and variations of texture and color to provide visual interest, balance, and direction to the pedestrian. The apparent lack of organization adds to the interest and curiosity evoked by the design.

a new range, they begin to compete with those plants already resident. At what point does an invader become a resident? Many familiar plants are not native, but it would be difficult to imagine the landscape without them. They are as American as apple pie so to speak, introduced to a new environment and flourishing. Perhaps the extent of a species' impact is best measured not by the fact alone that it is not native but by its contribution (Fig. 9.17) or harm in its nonnative environment.

Exotic and invasive species

Awareness of native plants and the undesirable impacts of some nonnative, or exotic, plant species has increased dramatically over the last 10 years. There are thousands of exotic species present in the landscape today, but attention is usually focused on the invasive exotic species that if left unchecked displace other species (Table 9.1). The threat from invasive exotic plants is expected to increase, and the damage to native plants will grow accordingly. Some exotic species thrive so well that they displace, extirpate, and even drive to extinction the native plants. Many of these undesirable species have been introduced as landscape plants, and many continue to be sold in nurseries around the country.

It should be noted that not all introduced plants are by definition undesirable or harmful (Table 9.1). Apple trees, for example, are hardly a threat to

Figure 9.8 This strong edge provides clear guidance and direction to pedestrians at the National Zoo in Washington, D.C. Note the various textures and colors present in the plants.

native forests and have become part of our culture. When selecting plants for the landscape, designers should evaluate their choices with regard to their potential impact on the environment. Plants that spread and establish easily by self-seeding or spreading roots should be reconsidered. Groundcovers that establish and spread quickly should not be used where they may "eşcape" into adjacent open space. Such ground covers should be confined by paved areas (see Fig. 9.18).

Using Trees in the Landscape

Trees bring numerous welcome attributes to the site. Tree masses can make significant positive impacts on microclimates by providing cooling shade, filtering dust and particulates, and buffering undesirable sounds and sights. For

Figure 9.9 This residential garden demonstrates the dramatic effects of texture, light, and shadow, and materials. (*Photo by Brent Baccene.*)

example, up to 30 percent improvement in energy efficiency is possible from properly selected and located trees. Trees contribute to cooling by shading buildings and cooling surfaces and also by providing evaporative cooling surfaces associated with transpiration and evaporation. Properly located trees may also reduce reflected light from the building surfaces and windows. Sunscreens are most effective when located on the western and southwestern sides of buildings to reduce heat from the summer setting sun. Deciduous trees on south sides of buildings will admit winter sun but block summer sun. Medium to large trees located 15 to 30 ft from buildings are most effective. As a rule of thumb, the distance between the building and the tree should be about $^1/_4$ to $^1/_3$ the mature height of the tree. Smaller trees may be planted closer, but the summer breezes they generate may be less than they would be farther away from the building. Buildings can also be cooled using arbors and vines. Arbors are used throughout the world for cooling. Vines will reduce summer heat by absorbing much of the light. Deciduous vines lose leaves and allow winter heat gain (Figs. 9.19 and 9.20).

Planted windscreens also reduce the cooling of buildings in winter by redirecting or blocking winter winds. Evergreen trees located on the north and west side of buildings will screen winter wind. The effective distance of a windbreak from the building to be protected is about 30 times the vertical height of the screen, but the maximum protection is only within 5 or 6 times the height. A windscreen should be designed to be at least 60 percent dense all the way to

Figure 9.10 Photograph of plantings on an arbor.

the ground, especially on the windward side. Evergreen windscreens should be at least three rows thick while deciduous should be up to six rows thick.

Among the effects of land development, there is the inevitable mixture of land uses. Areas of transition from residential to commercial or industrial uses often require careful planning to offset the negative impacts of conflicting uses. The use of trees and other plantings to screen or buffer the unwanted impacts of these areas is a common practice (Table 9.2). To be effective, a planted buffer must be designed to accomplish the specific task or tasks required, and the selection of trees and plant types and characteristics is a key element of the design.

The design of the visual screen is probably the most common purpose for buffers along residential areas (Table 9.3). The function of the visual screen is commonly to block an unwanted view, which is usually accomplished with one or more simple rows of shrubs and trees. Unless carefully planned, such

Figure 9.11 This eclectic collection of materials, located at Baltimore's Inner Harbor, is arranged in an informal fashion but provides an inviting place for visitors to congregate, to sit and rest.

Figure 9.12 This collection of native plants clearly separates this private yard from the adjoining space.

Figure 9.13 The massing of street trees and shrubs at this street closure in Baltimore signals a change in the character of the street, from public thoroughfare to more private space.

Figure 9.14 Photograph of native vegetation in a Xeriscape.

Figure 9.15 Photograph of a pond constructed using native plants. Domaine Chandon Vineyard, Napa, California.

Figure 9.16 Native wildflowers were used to landscape around the protective structure in Petroglyphs Provincial Park, Ontario, Canada.

Figure 9.17 Photograph of a phtyoremediation project using a nonnative species of tree.

arrangements may not be effective at actually restricting the view; instead, they may simply serve to frame the unwanted view. In addition to screening the unwanted view, the well-designed buffer can affect other intrusive influences such as highway noises or fugitive dusts from adjacent commercial or industrial sites. Through the use of screens and buffers, it is often possible to eliminate a negative off-site influence and enhance the desirability of a difficult lot.

Tree masses have characteristics that have several significant impacts on their immediate environment (Figs. 9.21 through 9.23). The shade from trees will lower temperatures by as much as 10°F from surrounding areas. Shade also reduces evaporation from the area affected. The combination of these effects is a localized reduction in the relative humidity. Of particular interest to the design of buffers is the size and location of plants in order to take advantage of this localized influence.

Plants with compact, tight growth patterns will tend to be better screening plants (Fig. 9.24). These plants create a dense, "soft" collection of surfaces (leaves) that tend to absorb sound and provide surfaces for the deposition and filtering of dusts. A basic element of the design of buffers is the location of the buffer with regard to the source of the nuisance and the point of observation. Locating the screen is a site-specific consideration, but, generally speaking, the buffers are more effective if located closer to the source of dust or noise. In the case of using trees to meet energy needs by acting as a windbreak or providing shade, the buffer should be located closer to the house.

TABLE 9.1 Some Common Invasive Plants

Common name	Genus and species	Region
	Trees and Shrubs	
Australian pine	*Casuarina equisetifolia*	Southwest
Autumn olive	*Eleagnus umbellata*	Widely distributed
Bradford pear	*Pyrus calleryana*	Mid-Atlantic region
Burning bush	*Euonymus alatus*	Mid-Atlantic region
Brazilian peppertree	*Schinus terebinthifolius*	Southeast
Camphor tree	*Cinnamomum camphora*	Southeast
Chinaberry	*Melia azedarach*	Southeast
Chinese tallow	*Sapium sebiferum*	Southeast
Downy rose myrtle	*Rhodomyrtus tomentosa*	Southeast
Empress tree	*Paulwnia tomentosa*	Mid-Atlantic region
Honeysuckles	*Lonicera* spp.	Mid-Atlantic region
Japanese barberry	*Berberis thunbergii*	Mid-Atlantic region
Japanese Spirea	*Spiraea japonica*	Mid-Atlantic region
Mimosa	*Albizia julibrissin*	Mid-Atlantic region
Multiflora rose	*Rosa multiflorum*	Mid-Atlantic region
Norway maple	*Acer platanoides*	Mid-Atlantic region
Privet	*Ligustrum species*	Mid-Atlantic region
Tree of heaven	*Ailanthus altissma*	Mid-Atlantic region
Russian olive	*Eleagnus angustifolium*	Mid-Atlantic region
Sawtooth oak	*Quercus acutissima*	Mid-Atlantic region
Siberian elm	*Ulmus pumila*	Mid-Atlantic region
Winged euonymus	*Euonymus alatus*	Mid-Atlantic region
White mulberry	*Morus alba*	Mid-Atlantic region
	Vines and Groundcovers	
Air potato	*Dioscorea bulbifera*	Southeast
Bamboo (all)	*Phyllostachys, Bambusa, Pseudosasa*	Southwest
	Cenchrus ciliaris	
Chinese wisteria	*Wisteria sinensis*	Mid-Atlantic region
Climbing euonymus	*Euonymus fortunei*	Southeast
Creeping bugleweed	*Ajuga reptans*	Northeast, Mid-Atlantic region
Crown vetch	*Coronilla varia*	Mid-Atlantic region
English ivy	*Hedera helix*	Mid-Atlantic region
Fountain grass	*Pennisetum setaceum*	Widely distributed
Japanese honeysuckle	*Lonicera japonica*	Mid-Atlantic region

TABLE 9.1 Some Common Invasive Plants (*Continued*)

Common name	Genus and species	Region
	Vines and Groundcovers	
Japanese wisteria	*Wisteria floribunda*	Mid-Atlantic region
Japanese climbing fern	*Lygodium japonicum*	Southeast
Eurasian water milfoil	*Myriophyllum spicatum*	Southeast
Garlic mustard	*Alliaria petiolata*	Southwest
Giant salvinia	*Salvinia molesta*	Southeast
Giant sensitive plant	*Mimosa pigra*	Southeast
Hydrilla	*Hydrilla verticillata*	Southeast
Kudzu	*Pueraria lobata*	Southeast
Melaleuca	*Melaleuca quinquenervia*	Southeast
Mint (all)	*Mentha* spp.	Widely distributed
Old World climbing fern	*Lygodium microphyllum*	Southeast
Periwinkle	*Vinca minor*	Mid-Atlantic region
Purple loosestrife	*Lythrum salicaria*	Widely distributed
Skunk vine	*Paederia foetida*	Southeast
Torpedograss	*Panicum repens*	Southeast
Water fern	*Salvinia molesta*	Southwest
Water hyacinth	*Eichhornia crassipes*	Southeast
Water lettuce	*Pistia stratiotes*	Southeast
Wetland nightshade	*Solanum tampicense*	Southeast
Winged yam	*Dioscorea alata*	Southeast
Winter creeper	*Euonymus fortunei*	Mid-Atlantic region

Sound will attenuate over distance; therefore, the buffer is more effective closer to the source. This is also true of fugitive dusts or airborne particulate. The dimensions of the screen are also important. Width may be constrained by property limits, but ideally screens will not be limited to single properties and will extend as deeply as required to be effective. The height of the buffer is also important if it is to screen views, winds, sounds, sun, or dust. Sound dissipates at a predictable rate over distance. Dusts and particulate settle out of the air at a predictable rate. By understanding these measurable characteristics, a designer can use the materials and site characteristics effectively to the advantage of the project.

There are cases in which a poorly located row of trees has actually made a problem worse—directing a sound or diverting the prevailing wind to conduct

Figure 9.18 Photograph of kudzu, an Asian vine that was often used for groundcover and erosion control but that is considered a noxious exotic plant in the southeast states.

a nuisance where it is not wanted. It is known that several rows of trees are more effective than a single row and that several rows of combinations of different plants is more effective still. These increases in the density of the buffer can sometimes be accentuated even further through the use of graded berms to elevate the screen and provide a dense base for the screen.

In choosing plants, the design should specify a material that will mature relatively quickly and that will not become a maintenance problem. Of course, the plant materials chosen must be able to tolerate the nature of the nuisance. The buffer must be designed with the impact of the seasons in mind. A solid wall of evergreens is not the only solution to screening issues. Although deciduous trees will not offer any significant screening from views, sound, or dust in January, it may be in some cases that there will be no activity to be screened at that time. People tend to remain indoors more of the time, and windows and doors are shut much of the time. Distance can be used to some advantage by the designer to determine a blend of conifers and deciduous trees.

Consideration must also be given to the aspect of the screen with regard to the winter's sun warming the area that was shaded in the summer. Actual distances and plant heights are a function of a site's latitude, but generally speaking, a site is shaded on the south to southwestern side in the summer. In winter these exposures would provide valuable warming from the sun. The plan might also include the planting of successive plant types—that is, a com-

Figure 9.19 Photograph of Moscone Center.

bination of plants that would include fast growing plants that would ultimately be removed and replaced by slower-growing but more desirable species.

The presence of mature trees on sites is generally considered to be desirable. In residential projects people often pay a premium for a site with trees, especially mature trees. Designers may enhance the value or desirability of lots by saving existing trees. The decision to save or remove trees, however, should be approached by carefully evaluating the trees and the project.

Tree and Shrub Planting

Contemporary standards for planting trees are quite different from the old tree pit planting method. Research has led to the modification of techniques that take the site conditions into account. Three different categories of planting have been identified: street lawn, residential, and pit. These methods each

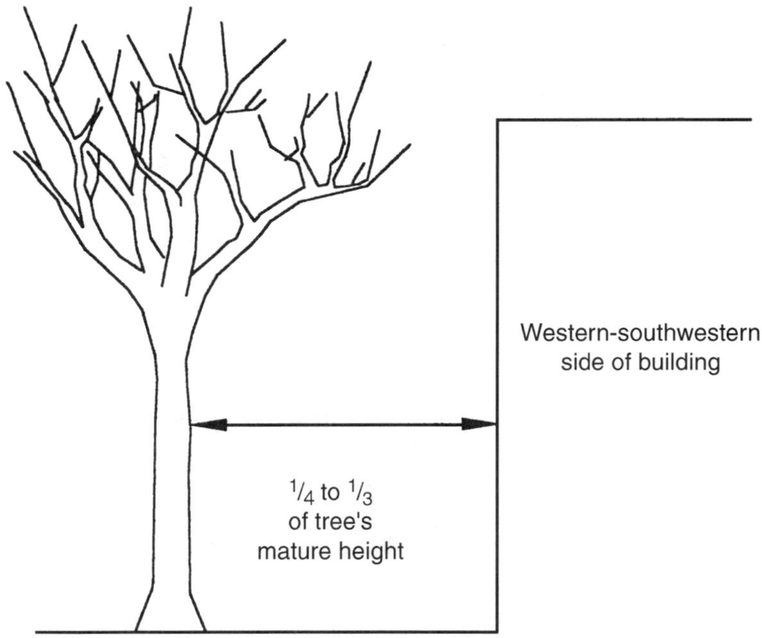

Figure 9.20 Locating trees to shade buildings.

TABLE 9.2 Applications for Planted Screens (Buffer Plantings)

1. A visual screen to block unwanted views, to mask glare, or to direct the viewer to a particular feature
2. A barrier to deflect or absorb sound
3. A filter to collect airborne dust and particulates
4. A source of shade and protection from the sun for purposes of comfort and/or energy efficiency
5. A windscreen

TABLE 9.3 Screening Design Considerations (Buffer Plantings)

1. The buffer should be close to the source of the unwanted noise or dust.
2. The depth of the buffer mass should be relative to the strength or magnitude of the nuisance.
3. Different types of plants should be combined because together they are more effective than single types of plants.
4. Grading should be used to enhance the effectiveness and visual interest of the buffer.
5. The height of the screen is as important as its width or depth.
6. The buffer should be visually pleasing.

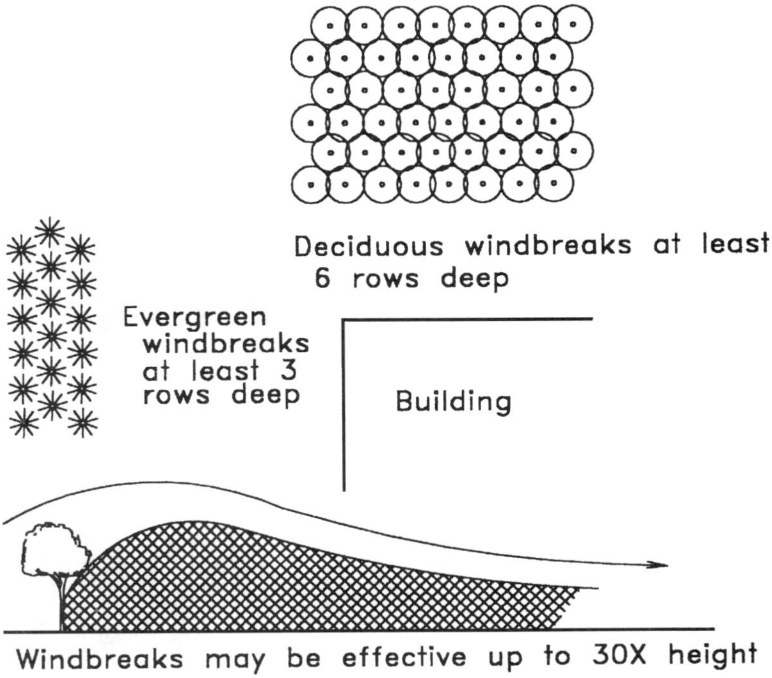

Figure 9.21 Windscreen design detail.

represent a condition that is far different from "estate planting" on which the old method was based.

The primary difference among these methods is the amount of soil space available to the tree. A great deal is known about the way in which trees grow and the requirements of growth. Most roots of trees are very small, ranging in size from a pencil thickness to a hair. These are the feeder roots that absorb and transmit nutrients and moisture to the plant. These roots grow *up* toward the surface to form mats in the first few inches of soil. These roots grow and die back in response to conditions near the surface. Periods of root growth occur in moist seasons, and dieback occurs in the hot dry summer and cold winter months.

The fundamental needs of tree health that should be considered for mature as well as for young newly planted trees include adequate room to grow. For new trees the location of planting should consider the tree's size in 5, 10, and 25 years. In the case of mature trees, the designer must consider the location of proposed improvements because they may restrict growth or the growth of the tree may become a nuisance or cause damage. Landscape plants grow and change over time. This aspect of living site elements should prompt some consideration of the impact of the plant over time in a given location. The actual selection of a specific specimen should be done with a critical eye. When selecting a tree, the designer should look for a straight trunk with well-balanced

Figure 9.22 Kudzu is an extremely invasive exotic. In this photograph the kudzu vine has covered trees, streetlights, and wires. (*Photograph used with permission of Karla Baccene Russ.*)

Figure 9.23 The Moscone Center in San Francisco uses trees and plantings on the surface of the building to reduce energy load.

Figure 9.24 Plantings as part of a sound screen strategy.

growth and symmetry throughout the tree. Trees with double leaders or deep Y's should be avoided. Bark should be intact and not swollen, cut, bruised, or cracked. A quick way to measure individual vitality is to compare the ball size, tree height, and caliper size.

Urban Trees

Although some plants do well in urban areas (Table 9.4), the average life span of city trees is less than 10 years. Some lessons can be learned from the causes of these tree losses. The single most common cause of city tree mortality is poor drainage. Tree pits along city streets or in some compacted urban soils are simply "pots" that have no drainage. Water collected in these pits does not drain away, and the tree is drowned. Tree pits designed for city environments or environments with poor drainage should include a means of draining excess water from the pit.

Concerns that may influence tree selection include exposure to pollution in urban environments. Even within a given city or community, environmental quality can vary widely and so tree selection must vary as well. The surfaces of buildings and pavement absorb and reflect heat that tends to cause droughty conditions in the urban tree pit.

A new tool in urban tree health is the use of structural soil, which will contribute to the health and longevity of urban trees (Fig. 9.25). The term *structural soil* refers to the use of a soil compacted to a degree that makes it an effective subbase for paving but will still allow for the penetration and growth of roots. Various mixes of structural soils have been suggested. Some people successfully combine a soil mix of about 25 percent silt of clay, 25 percent organic matter, and 50 percent sand with crushed stone ($^1/_2$ to $1^1/_2$ in) at a ratio of four parts stone to one part soil mix. Others use less stone or more sand. Structural soil contributes to the health of the plant by providing a volume for the roots to expand into, and it reduces sidewalk heave, which is to be expected as trees mature. In some cases polymer gels are added to the soil mix to absorb and hold water.

TABLE 9.4 Trees That Tolerate City Conditions

Hedge maple (*Acer campestre*)
Norway maple (*Acer platanoides*)
Columnar maple (*Acer platanoides* `Columnare')
Ruby horsechestnut (*Aesculus carnea brioti*)
Lavalle hawthorne (*Crataegnus lavelli*)
Washington hawthorne (*Crataegnus phaenopyrum*)
Russian olive (*Eleagnus angustifolia*)
Modesto ash (*Fraxinus velutinum glabra*)
Gingko (*Gingko biloba fastgiata*)
Thornless honey locust (*Gleditsia triacanthos enermis*)
Golden raintree (*Koelreuteria paniculata*)
Amur cork tree (*Phellodendron amurense*)
Red oak (*Quercus boreallis*)
Little leaf linden (*Tilia cordata*)

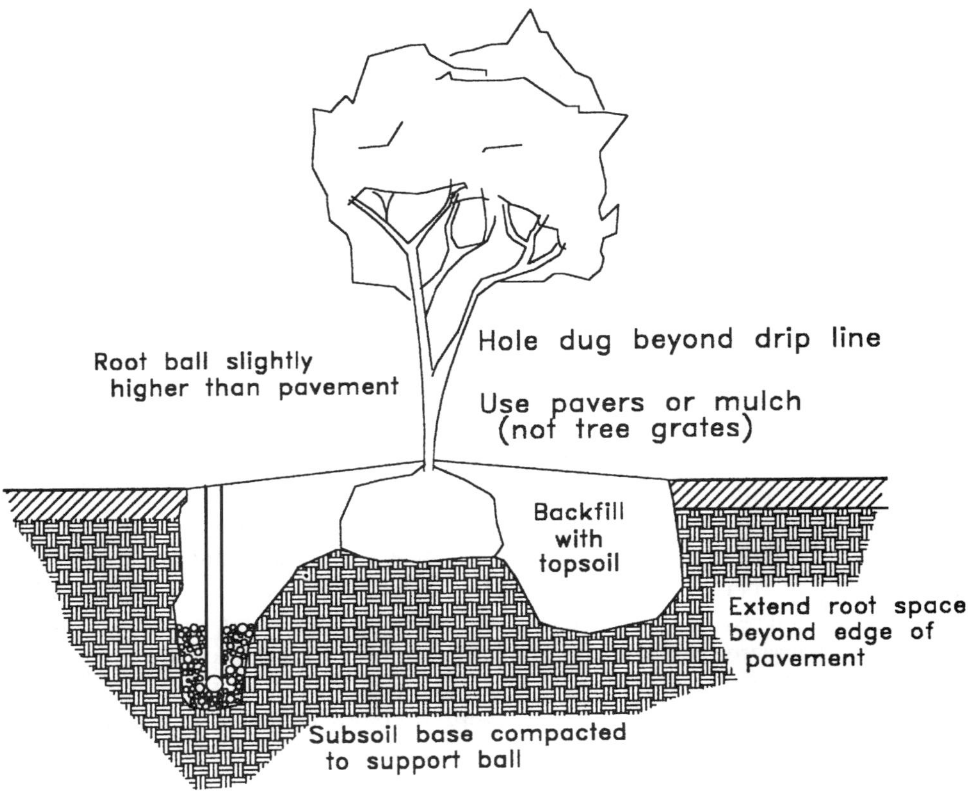

Figure 9.25 An urban tree pit.

The next most common cause of tree losses is mechanical damage from wire baskets, wire from staking, tree grates, or tree wrap. All of these devices are intended to support or protect the tree at some point in its move from the nursery to its ultimate location, but if they are installed improperly or left in place too long, they will become the cause of death. All wire or wrapping around a root ball should be cut away to allow the roots to grow beyond the root ball without restriction. Even biodegradable materials such as burlap remain in the soil years after the plant has been installed.

Tree staking is a practice that is debated. Staking a tree is a practice left over from the time when most planting was of bare root plants. A balled specimen should not require staking in most cases; however, if stakes are used for plantings on slopes or for security reasons (to avoid plant theft), they should be necessary for only 6 months or so. Tree wrap is used to "protect" the tree from "animals and vandals."

Tree grates are common in urban environments and are used to protect the tree from the damage of pedestrian traffic (Fig. 9.26). The rings are designed so that, as a tree grows, the ring can be cut back, allowing the trunk room to grow; however, in today's cities with shrinking maintenance budgets, cutting back the ring as needed is usually not done. One alternative to the tree ring is an installation of pavers over the tree root zone. The pavers allow some water to penetrate to the roots and can easily be removed as the tree grows.

The installation of trees in urban tree pits can be designed to increase the life expectancy of the trees and to reduce the cost of replacing dead trees. With this design, the city would create continuous tree pits or troughs that extended the length of the street. Each tree root zone would be connected to the oth-

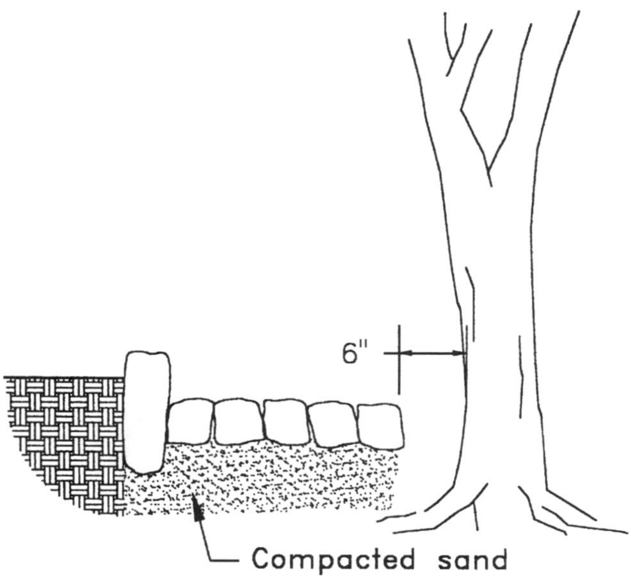

Figure 9.26 Pavers used over tree roots.

ers. The pavement over this tree pit could be of pavers. A study by the Cornell Urban Horticultural Institute found using pavers over these tree pits was a viable method both from a plant vitality standpoint and a long-term feasibility standpoint. Although some decrease in initial permeability was noted, the long-term effectiveness of the paver system was only nominally affected.

The study made several recommendations to be considered in the design and installation of these systems. A dimensionally small paver increases the number of joints and the permeability (Fig. 9.27). A joint thickness of $^1/_4$ inch filled with coarse sand contributes to infiltration. If a base course is used, it should be a mixture course of noncompacting sand over crushed stone. The pavers should be installed with the primary joint running parallel to the contours to intercept more runoff.

Selecting and Placing New Trees

It may be appropriate to remove the existing trees and replace them with new trees. Selecting a tree begins with selecting a site for the tree. A tree for a lawn

Figure 9.27 Photograph of pavers installed over the root zone of trees.

will be selected for different reasons than a tree for a city street. Although shallow-rooted trees are ideal for city conditions, they will tend to damage sidewalks and curbs. Also, the shallow roots exposed to the surface are often damaged by the pedestrian traffic in an urban environment. Tall trees may interfere with overhead wires. Some trees are grown for their shape or beauty, but they must be viewed from a distance to appreciate. Each site must be recognized for its characteristics and constraints when selecting the right tree. Fortunately there are so many species and varieties of trees that a fit is usually available for most combinations of site and purpose.

A tree should be located with an understanding of the cultural requirements of the tree and its intended impact or value to the site. There is no single source of plant information available or in general use, and nearly every book includes its list of recommended trees. Often these lists have a strong regional flavor, which can be valuable. When choosing what type of tree to use, it is best to consult a knowledgeable local professional. A local arborist, nursery staff, or landscape architect can assist the designer in selecting a tree that enhances the site and will tolerate the conditions on the site. General information is included in Tables 9.5 and 9.6 to illustrate some common trees that have specific tolerances or intolerances for some of the limiting factors found on development sites (Tables 9.5 and 9.6).

Preserving Trees

Experience has taught builders that homebuyers will pay a premium for a well-landscaped property. Polls reveal that people view a house lot with trees

TABLE 9.5 The Tolerance of Trees and Shrubs to Road Salt

Intolerant	Some tolerance	Tolerant
Sugar maple	Birch	Mulberry
Red maple	Hard maple	Hawthorne
Lombardy poplar	Beech	Red oak
Sycamore	Balsam fir	Tamarix
Larch	Douglas fir	Russian olive
Viburnum	Blue spruce	Black locust
European beech	Green ash	Oleander
Spirea	Pyracanthra	White acacia
Winged euonymus	Ponderosa pine	English oak
Black walnut	Arborvitae	Gray poplar
Little leaf linden	Eastern red cedar	Silver poplar
Barberry	Japanese honeysuckle	Osier willow
Rose	Boxelder maple	Bottlebrush

TABLE 9.6 The Tolerance of Some Common Trees to Fill

Most affected	Less affected	Least affected
Sugar maple	Birch	Elm
Beech	Hickory	Poplar
Dogwood	Hemlock	Willow
Oak	—	Plane tree
Tulip tree	—	Pine oak
Conifers	—	Locust

as more valuable than a house lot without trees, and they will pay a higher price for the lot with trees. In fact, people viewed mature trees as the most desirable aspect of the residential landscape. Trees can increase the value of a residential property from $3000 to $15,000, depending on the size, condition, number, and location of the trees. Even the value of existing homes can be increased by as much as 15 percent by the addition of trees and landscaping (Builder 1990). As home sites are developed, existing trees can add to the esthetic as well as to the economic value of the home.

Unfortunately, mature trees are often destroyed or damaged in the course of construction, or the effort to save a poor-quality tree is greater than the value of the tree. A careful evaluation of the site before construction begins is the first step in avoiding either of these mistakes. In many areas of the country projects are being developed with very tight restrictions on tree removal. These regulations may require that a very tight building envelope be maintained, with vegetation and earth outside the envelope undisturbed. On these projects, the minimum-disturbance restrictions are part of the sales appeal. Builders working on these sites are required to meet some strict operating guidelines.

Trees are damaged from cuts and fills because the balance between the roots and the soil is disturbed. The disturbance between the tree roots and the soil essentially interrupts the balance the tree had established with its supply of air and water. In some cases the disturbances may weaken the structural base of the tree as well. Tree roots grow and develop partially as a function of the air and water available in a given soil. The depth of a fill is important in determining its impact on selected trees. Soils on construction sites are generally left compacted and nearly impermeable from the trucks and equipment driving over them. Even without removing or adding soil to the base of a tree, the compaction from construction vehicles can damage trees. When a "blanket" of soil is added to the top of a grade, air and water are restricted from the root zone; generally the deeper the fill, the greater the restriction.

The depth of a fill is only one issue in determining its effect on a particular tree and the steps that must be taken overcome the negative impact. Other factors include the type and the health of the tree and the type of soil. Some species of trees are more tolerant than other species, and a healthy tree of any variety will withstand the stress of a fill better than a damaged or weak tree (Table 9.6).

The soil is a dynamic ecosystem in which there are complex interrelationships among the microorganisms, organic and inorganic matter, soil structure, moisture, and chemistry. The soil texture of the fill will be at best minimal simply because of the mechanical action of disturbances. *Soil structure* is the arrangement of soil aggregates—that is, distinct clumps of soil—and its composition is the result of the activity of the organic and mineral constituents and the beneficial effects of plant and microorganism life processes. The soil structure is a very important factor in how a plant is able to grow. Soils with a fine texture or particle size, such as clays, will tend to have a greater impact as fills because their fine particle size will fill available pore space through which air and water would travel to the tree roots. Even shallow fills of clay can severely damage a tree. Soils with a coarse texture, such as sandy or gravelly soils, cause the least amount of damage to trees because air and water move through the soil more readily. In most cases a shallow fill of several inches of gravelly soil, or soils of the same kind as the tree is growing in, will have no long-term effect on a tree. The tree will be able to compensate by extending its roots into the new layer. The upward extension of roots is more difficult in a deeper fill because of the loss of soil water and air and the absence of pore space.

Trees in fill

Constructing tree wells can save existing trees, but the trees chosen for the construction of tree wells should be those whose value and contribution to the landscape justify the cost and effort of the tree well (see Figs. 9.28 through 9.31). Old or damaged trees may not offer the longevity necessary to justify the additional expense; a young tree could be planted at less expense. The number, size, and quality of trees on a given site must be considered in making the decision to construct a fill protection. On a lot with many trees, the cost of saving one or two may not be attractive whereas the cost of saving a specimen on a lot without any other trees may be very attractive.

The site should be prepared before the grades are raised. All vegetation should be removed from the area affected and the soil worked. Fertilizer and soil additives should be added in accordance with specifications provided by manufacturers, nursery personnel, or a landscape architect. Once the soil is worked and the amendments have been introduced, care should be taken to not disturb the area with construction equipment or vehicles. The best method to protect the root zone is to isolate the area with a temporary fence. Tree wells must be designed to provide the tree with air and water as well as drainage away from the trunk. There are some fundamental principles common to all successful designs. To provide drainage away from the tree trunk and allow air to the area of the root zone, a series of 4- to 6-in perforated plastic pipes are laid radially from the root zone. The drain tiles should be installed with a positive slope *away* from the tree, and they should extend to or go just beyond the drip line of the tree.

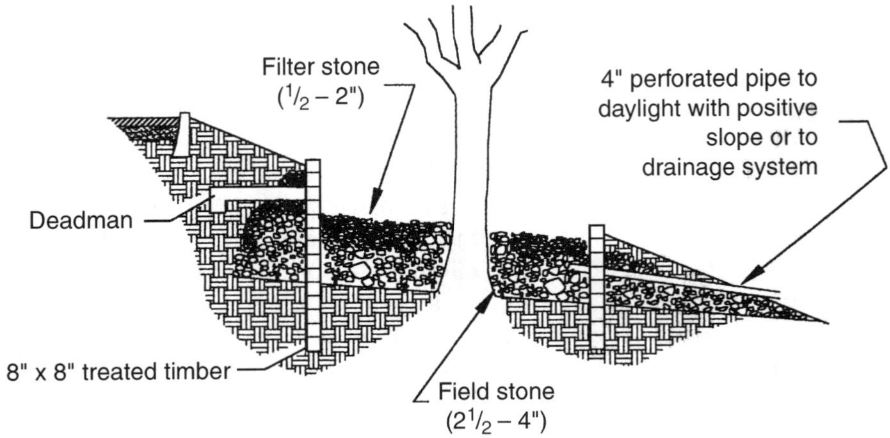

Figure 9.28 A tree well.

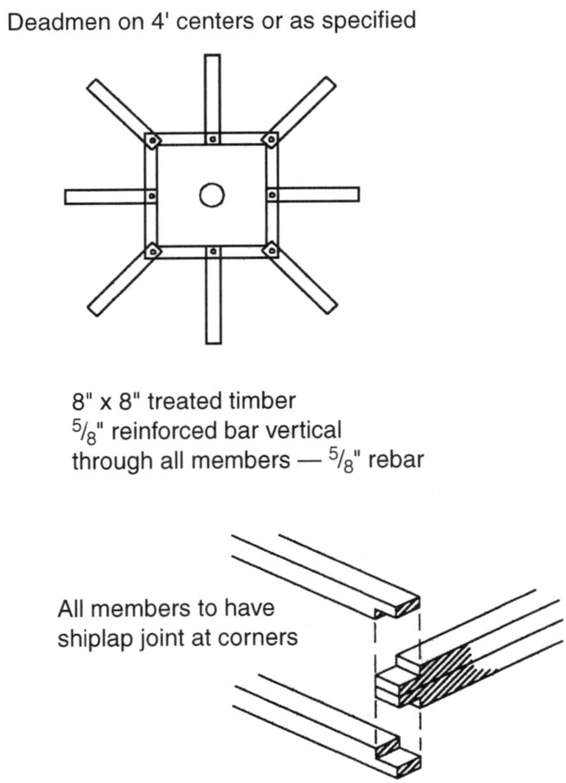

Figure 9.29 A timber retaining wall tree well.

Figure 9.30 Photograph of a timber tree well.

Once the drain tiles are in place, the well is constructed. The choice of material for the well can be varied. For shallow wells of 1 to 3 ft, bricks or stone can be used. These should be laid up in an open joint, that is, without mortar. This is sometimes referred to as a *dry joint.* A batter of at least 3 in/ft should be provided. It may be necessary to construct deeper wells with a greater structural stability. In such cases timber tree wells are often used. These structures allow the use of stabilizing features, such as a *deadman,* to be incorporated into the design. In either case, the well should be constructed allowing at least 2 ft from the trunk of the tree in all directions.

A means of drainage at the drip line is often provided (Figs. 9.32 to 9.33). This may be a series of drain tiles on end and extending into a gravel or stone bed or an actual gravel or stone channel provided to direct water to the root zone. Once the tiles and well are in place, a layer of stone 2 to 4 in in diameter should be installed over the pipe and cultivated soil. This layer should not exceed 18 in or 25 percent of the depth of the fill, whichever is least. It may be necessary to support the well or the drip line drain pipes with additional rocks.

The layer of rocks or stone must be of a material that will not react with the tree or soil chemistry in such a manner as to harm or inhibit the plant. The layer of rocks is covered with a finer "filter" stone to a maximum depth of 12 in or to within a foot of the ultimate grade. A layer of straw or filter fabric is installed on top of the filter stone. This prevents, to some degree, the soil fines from washing into the spaces between the stones and rocks, at least until the soil can begin to form some structure. Topsoil is then placed up to the finished grade. If vertical drain tiles were used at the drip line, these should be filled with small stone to prevent debris from filling and blocking the hole.

Figure 9.31 Photograph of a tree in a cut area.

Trees in cut

It is more difficult to protect a tree from a change in grade that involves removing soil from its base (Fig. 9.34). If soil is to be removed, it is probable that some root damage will occur, including the removal of some roots in the process. The roots most likely to be damaged or removed are the smaller roots on which the tree relies for feeding. The rooting characteristics of a tree will have some bearing on the degree of impact the disturbance will have. Elms, for example, are deep-rooting trees and will tolerate a modest change in grade. Shallow-rooted trees, such as conifers, are difficult to save and protect in cuts. It is key to the success of removing soil that the operation be done by hand to minimize damage to the roots. Steps can be taken to reduce the damage by promoting a new root growth at a lower level, but these efforts require at least one full growing season before the removal takes place and so are rarely used.

Construction site management is important to ensure that the decision and efforts to save a tree are successful. Care taken during the construction

Figure 9.32 Photograph of a tree in a tree well.

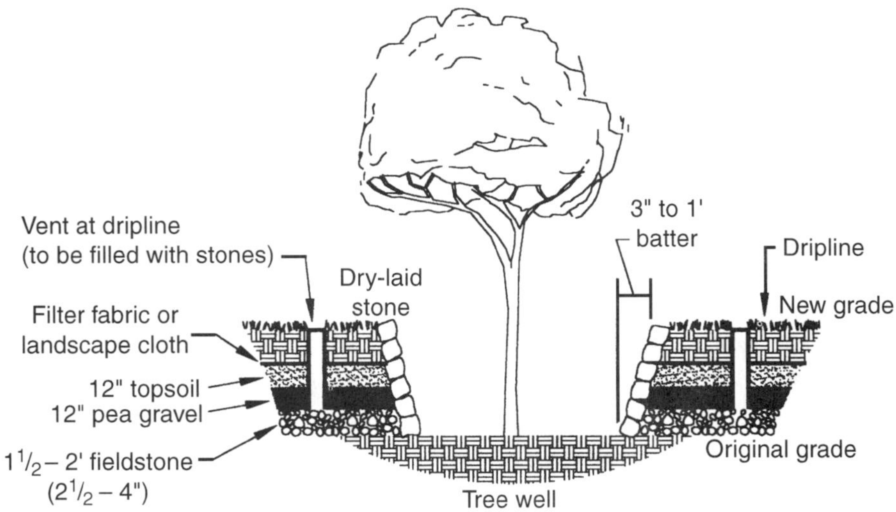

Figure 9.33 A dry-well tree well.

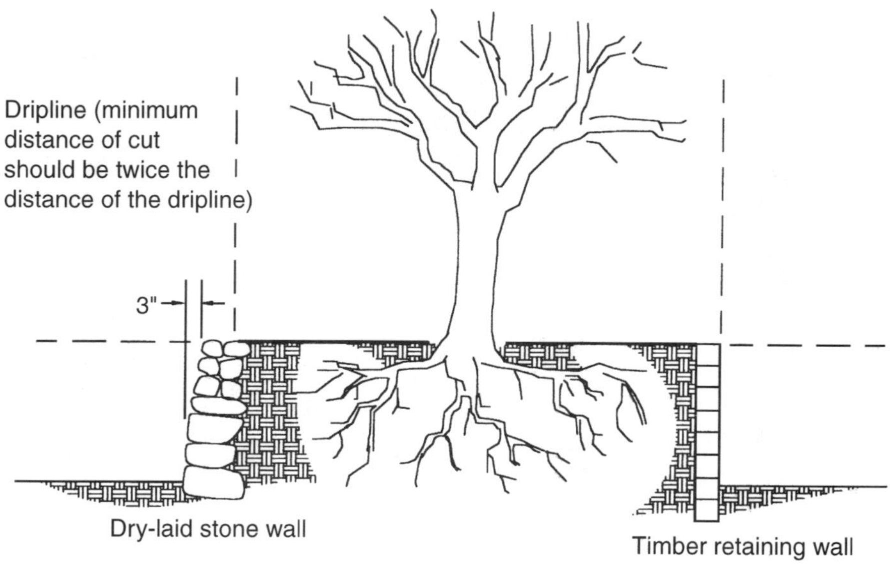

Figure 9.34 Tree in cut area detail.

process can minimize the risk of damage by subcontractors or careless operators. The procedures to protect selected trees would begin by clearly marking or identifying the specimen in the field. This can be accomplished by simply marking the trees to be saved with surveyor's tape and marking the trees to be removed with paint on the trunk. After marking or identifying the trees in the field, the next step is to communicate the plan to save the trees to the field crew so that everyone knows the plan.

Other steps that can be used to protect the trees and implement the plan include protecting trees from traffic by installing temporary fence around the root zone or better yet routing site traffic away from the specimen. Trees with low-hanging branches that are likely to be damaged should be pruned or the protective fence extended to encompass the low branches. Disposal and storage areas should be kept at least 50 ft away from the root zone.

Trees and Carbon Management

As concerns grow over the consequences of global warming, trees may emerge among the strategies to offset the continuing increase in carbon dioxide in the atmosphere. As trees grow, they fix carbon dioxide in their biomass, and early studies indicate that planting trees contributes significantly to accumulating carbon dioxide. If carbon management emerges as a strategy in the United States as it has in other countries, tree planting might represent a source of income in the form of pollution credits. A study of a tree planting program in Kichener, Ontario, indicates that between 34.5 and 68.8 million pounds of carbon dioxide is removed by its trees each year. This represents from 20 to 40 percent of the carbon dioxide load of Kichener (Doherty et al.). The importance of the use of trees and planned woodlands might be expected to grow in the next 20 years as a strategy to offset the impacts of development.

There is however research that indicates the use of forests as a long-term carbon management strategy may be overestimated.

Phytoremediation

Phytoremediation has become more common as part of a brownfield remediation strategy. In principle, *phytoremediation* is the use of plants to clean up site contamination. It is actually an umbrella term that includes several different approaches to cleanup. Phytoremediation is popular primarily because it is relatively inexpensive, and it can be both attractive and effective. The principal disadvantage lies in the time it may take to be effective.

Phytoextraction, sometimes referred to as *phytoaccumulation,* is a type of phytoremediation that relies on the plant's natural capacity to absorb and incorporate specific materials into their tissues. Some plants are efficient collectors of metals such as lead, mercury, or nickel. These plants are periodically harvested and either incinerated or recycled. Incinerated material may be recycled or placed in a landfill. The ash from incinerated materials represents a significant reduction in the volume of material that must be disposed of.

Phytodegradation, on the other hand, relies on the metabolism of plants to decompose certain contaminants once they are absorbed by the plant. Plants have been found to produce various enzymes and acids that cause the decomposition of contaminants. Still other plants are able to absorb organic materials and volatalize the material into the atmosphere through the processes of respiration. *Rhizosphere biodegradation* occurs around the plant roots rather than within the plant. Substances released by some plants into the soil around their root systems encourage the growth and development of communities of microorganisms, which in turn biodegrade the contamination. This approach has been shown to be effective against petroleum contamination in soils.

Some 400 plants have been found to be *hyperaccumulators,* that is, they will absorb and store contaminants, particularly metals. Mustard (*Brassica juncea* and *Brassica carinata*) has been found to be an effective accumulator of chromium and lead. Trees have found particular use in managing water tables on impacted sites. Poplar, cottonwood, and willow trees have all shown promise in experiments in which the trees have been used to lower local water tables and reduce the degree of contact between groundwater and shallow contaminated soils. Hybrid poplar trees (*Trichocarpa deltoides*) have demonstrated the ability to absorb and break down trichloroethylene, as well as some other organic contaminants and some metals. In Europe poplars and willows have been successfully used as biofilters for various organic contaminants. While phytoremediation provides significant promise for site remediation, many of the applications have been studied under hydroponic situations and have not been studied under field conditions. Still other applications involve *transgenic plants* that have been designed for the application at hand, but this sort of technology may be beyond the typical site development project's budget and schedule. Phytoremediation is an important consideration in those places where it will serve, but much remains to be learned. If a project requires remediation to be completed prior to the redevelopment phase, the rate of improvement using phytoremediation may be to slow. On the other hand, if redevelopment and remediation can proceed together, such practices may present an important low-cost strategy.

Bioremediation

Bioremediation is the use of microflora or microfauna to decompose or stabilize contaminants. Like phytoremediation, it is the use of living organisms to remediate a site. Significant strides have been made in the bioremediation field. In general, the most effective approach has been to identify organisms that exist on the contaminated site and that are already at work on the contamination. Some of these organisms are collected and brought into a laboratory to determine the ideal combination of factors such as moisture, air, light and nutrients that will facilitate the most ideal environment. Once determined, these conditions are re-created in the field. Since bioremediation may take 6 months or more, the site design may have to accommodate the necessary conditions during construction and perhaps even after construction.

Meadows

The popularity of native plants has resulted in an increase in growers and distributors of plant material so that adequate stock is usually available. Natural meadows are preferable to lawns for their function and for their low life cycle costs. Meadows are usually mowed only once in the fall. They provide important habitat and forage. The meadow functions as an important element in maintaining local water quality, and it is attractive. Nonetheless, sites using natural landscape have been singled out and criticized by some for the "wild" look, and some communities have even taken action to limit or even restrict native meadows. To avoid an unpleasant response, some education of local officials and neighbors may be required.

While many nurseries offer meadow mixes, it may be prudent to understand the seed mixture before specifying or using it. Meadows are complex plant communities. Seed mixtures should be evaluated for the number and type of specific species. The mixture should provide a combination of annuals, biennial and perennial plants, and native grasses. If a continuous blooming is desired, the species should be assessed for bloom time. Native warm season grasses make up from half to three-quarters of the plants in a natural meadow. These are clump-type grasses as opposed to the familiar cool season turf grasses. Warm season grasses tend to grow in the late fall and early spring and do not compete directly with summer germination and seeding of the wildflowers. Care should be taken to minimize the use of plants that aggressively spread and that might affect neighbors.

The first few years of the meadow are the most intensive and costly. Noxious weeds and exotics must be removed by hand during this period and continue until the meadow can establish itself. For the perennials and grasses, the first years are spent growing extensive root systems, and so the annuals will dominate the meadow. By the third year the perennials and grasses have established their roots and begin to flourish.

Meadow site preparation varies. Seeding meadow plants into small prepared areas within an existing landscape will allow the small site to be established and then naturally expand over time. Otherwise an area must be cleared of vegetation before it is seeded. The National Wildflower Research Center suggests, as an alternative to hand weeding, to water the area for a week or two before applying Roundup and then repeating the process to collect any resident but newly germinated weeds. After the second round, the meadow seed should be applied.

Toxic Plants

The trend toward specialized landscapes has increased in recent years, along with our appreciation of the natural environment. Specialized landscapes include therapeutic gardens, living laboratories on school grounds, scent gardens, designed for the elderly, and many others. Many of the plants commonly used in landscape planting pose some risk from toxicity. There are several poisonous plant databases found on the Internet. There are also Web sites that list plants of particular concern for various animals such as cats, dogs, or horses. Table 9.7 provides a list of common poisonous plants. The list is not comprehensive or species specific. Landscape designers should become familiar with the toxicity of plants commonly used in a particular area, especially if the designer is working on landscapes or sites for particular end users such as children in a day care

TABLE 9.7 A List of Some Toxic Landscape Plants

Common name	Species	Poisonous part
Autumn crocus	*Colchicum autumnalle*	Bulbs
Angel's trumpet	*Datura* (some species)	Seeds, leaves
Apricot	*Prunus ameniaca*	Stems, bark, seed pits
Azalea	*Rhododendron occidentale*	All parts
Baneberry	*Actaea spicata*	Berries, roots, foliage
Bleeding heart	*Dicentra* (some species)	All parts
Buchberry	*Lantana*	All parts
Buttercup	*Ranunculus* (some species)	All parts

TABLE 9.7 A List of Some Toxic Landscape Plants (*Continued*)

Common name	Species	Poisonous part
Calla lily	*Zantedeschia aethiopica*	Leaves, rhizomes
Castor beans	*Ricinus communis*	Seeds
Choke cherry	*Prunus virginica*	Leaves, seed pits, stems, bark
Daffodil	*Narcissus*	Bulbs
Daphne	*Daphne mezereum*	Berries, bark, leaves
Delphinium	*Delphinium* (some species)	Seeds, young plants
Eggplant	*Solanum melongena*	All parts except fruit
Elderberry	*Sambucus* (some species)	Roots, seeds (stones)
Euonymus	*Euonymus* (some species)	Leaves, fruit, bark
Four o'clock	*Mirabilis jalapa*	Roots, seeds
Foxglove	*Digitalis purpurea*	All parts
Hemlock	*Conium maculatum*	All parts, roots, and root stalks
Hens-and-chicks	*Lantana*	All parts
Hyacinth	*Hyacinthus orientalis*	Bulbs, leaves, flowers
Hydrangea	*Hydrangea macrophylla*	Leaves, buds
Iris	*Iris* (some species)	Rhizomes
Jerusalem cherry	*Solanim pseudocapscium*	All parts, unripe fruit
Jimson weed	*Datura stramonium*	All parts
Jonquil	*Narcissus*	Bulbs
Larkspur	*Delphinium* (some species)	Seeds, young plants
Lily family	(Many species)	Bulbs
Lily of the valley	*Convallaria majalis*	All parts
Lobelia	*Lobelia* (some species)	All parts
Lupines	*Lupinus* (some species)	Seeds
Mandrake	*Podophyllum peltatum*	Roots, foliage, unripe fruit
Mistletoe	*Phoradendron flavescens*	Berries
Monkshood	*Aconitum napellus*	All parts
Morning glory	*Ipomoea violacea*	Seeds
Narcissus	*Narcissus* (some species)	Bulbs
Nightshade	*Atropa belladonna*	All parts
Oak	*Quercus* (some species)	Acorns, young plants
Oleander	*Norium oleander*	All parts, including dried leaves
Poinsettia	*Euphorbia pulcherrima*	Leaves, flowers
Pokeweed, inkberry	*Phytolacca americana*	All parts
Potato	*Solanum tuberosum*	Green seed balls, green tubers
Privet	*Ligustrum vulgare*	All parts
Red sage	*Lantana camara*	Green berries
Rhododendron	*Rhododendron*	All parts
Rhubarb	*Rheum raponticum*	Leaves
Sedum	*Sedum* (some species)	All parts
Snow-on-the-mountain	*Euphorbia marginata*	Sap
Spindle tree	*Euonymus* (some species)	Leaves, fruit, bark

TABLE 9.7 A List of Some Toxic Landscape Plants (*Continued*)

Common name	Species	Poisonous part
Sweet pea	*Lathyrus odoratus*	Seeds, pods
Tansy	*Tanacetum vulgare*	All parts
Tulip	*Tulipa*	Bulbs
Virginia creeper	*Parthenocissus quinquefolia*	Berries
Wisteria	*Wisteria*	Seeds, pods
Yew	*Taxus*	Needles, bark, seeds

Chapter

10

Project Management Issues

One way of viewing planning and design is as a process of avoiding failure; design proceeds from weighing solutions in terms of what won't fail. Henry Petroski has observed that design is always evaluated in terms of failure and success is celebrated in terms of failure avoided. "Brilliant success avoids failure brilliantly," he writes (Petroski 2000). Failure is most often perceived as the antithesis of success, but in fact, it is sometimes a matter of perception—one person's failure is another's success. "The operation was a success but the patient died" is an expression of this axiom.

Sometimes failure is an event. The elevated walkways in the Kansas City Hyatt Regency were a successful design. When they failed because of an ill-advised contractor's modification, it was an event. The design was sound. A bridge can be said to be a successful design until the day it suddenly falls down. Success in this sense is a condition or a state, but once a design is seen to have failed, it cannot be seen again to be otherwise.

Success and failure in site development may be defined in many ways and have many causes. Economic failure may be a result of poor financial planning, a change in the marketplace, unexpected development or operating costs. The examples that are most often cited in discussions of engineering failures almost always are limited to projects or designs that were on the cutting edge of design and materials. Catastrophic failure in site planning is rare primarily because of the standards of care and practices that have been tested and refined over its long history. This vast experience has led to safety factors or practices of overdesign that are routinely employed to avoid risk and reduce liability. These safety and design practices have evolved from trial and error as much as from rigorous engineering and scientific study, and they may be so familiar to us that they are followed without question.

All responsible designs and designers proceed on the basis of what they have learned either firsthand or as a student. There are very few truly inspired and new solutions to problems; rather, solutions tend to be iterations of solutions that have worked in the past. Failure may occur because the design problem or circumstance changes, thus requiring new applications for familiar methods or an entirely new consideration. Still, as students or practicing professionals, we rarely study failure in any meaningful way. No one likes to dwell on mistakes, particularly one's own, but it is the lessons of mistakes that make our experience valuable. Failure is rarely discussed openly, and its causes are often dismissed as ineptitude, poor judgment, or misadventure. There is no body of design literature that explores the projects that have not worked except as a gloss before moving on to other topics. The literature of site planning and design is primarily a catalog of practical methods and a library of success stories. Expositions on failure are uncommon, but we readily acknowledge that we learn more from understanding failure than we do from mimicking past successes. To study failure gives us understanding of the underlying principles and the forces at work in situation, but it also gives us an appreciation for the choices that were made and why.

Design failure is usually divided into technical and nontechnical causes. Technical causes are addressed directly in the practice of overdesign and safety factors. Although it may currently be an unpopular notion, most technical problems can be discovered in the quality-assurance processes, primarily through repeated checking by multiple reviewers. As it happens, nontechnical causes of failure may be more difficult to identify or address.

The nontechnical threats to project success might be summarized as falling into one or more of the following categories:

1. Inadequate capitalization through design and construction phases and into the future
2. Regulatory resistance or apathy in the form of a lack of interest, a lack of authority, or a lack of will
3. Community resistance and image issues
4. Client infatuation or the honeymoon syndrome in which clients proceed with a vision but without a well-defined plan or the fiscal and management discipline required (Often the client's vision is visible only to the client.)
5. Designer infatuation or the Taj Mahal syndrome in which design professionals pursue their goals without regard for the project limitations

It is the role of the project manager to avoid all of the causes of failure.

Effective management of a project is as important as any of the other various design and planning skills. In the end, firms are often distinguished as much by their abilities to successfully manage the project as they are any of the other individual parts of the project. Experience suggests that more clients are lost because of management issues than design concerns, and so firms struggle to find, develop, and keep effective project managers. Many times project management is treated as a skill one is simply destined to acquire with time and experience. This is as true for management skills as it is for design skills; experience improves our skills and burnishes our judgment, but our experience is more productive if it is supported by training and a sound understanding of the underlying principles and practices. The fundamental objectives of project management are, simply stated, to manage cost (within budget), time (on schedule), and quality (meet specifications or expectations). Actually doing it, however, can be a challenge.

Many firms are organized around key people that direct the project and staff. Harold Kerzner makes an important distinction between "project managers" and "project champions" (Table 10.1). In most contemporary environments, working with qualified and creative professionals requires an interactive, fairly open, style of management. Most of us have attributes of both manager and champion. Certainly most organizations have people of both types and benefit from their relative strengths. Still, the project manager is better suited to the levels of interaction required to work with project teams and stakeholders.

The Project Manager

Firms are organized in many different ways so there is no way to identify a single role of the project manager, but there seem to be several principles that contribute directly to the success of the project manager. In general, project managers are effective communicators, good problem solvers, technically knowledgeable, effective advocates for their projects, and articulate representatives of the project, the firm, and their profession. There is a variety of ways in which firms organize the role of the project manager, and it is difficult to choose one as the most effective. Project managers in small firms may have

TABLE 10.1 Differences between Project Managers and Project Champions

Project managers	Project champions
Prefers to work in groups	Prefers to work alone
Is committed to management and technical responsibility	Is committed to technology
Is committed to organization	Is committed to profession
Seeks to achieve objective	Seeks to exceed objective
Is willing to take risks	Is unwilling to take risks
Seeks what is possible	Seeks perfection
Thinks in terms of short time spans	Thinks in the long term
Manages people	Manages things
Is committed to pursuit of material values	Is committed to the pursuit of intellectual values

SOURCE: From Harold Kerzner, *Project Management,* 3rd ed. (New York: Van Nostrand Reinhold,).

quite different responsibilities from project managers in large firms. Differences also exist between private for-profit organizations and nonprofit organizations. How the project management role is organized reflects the culture of the organization, and therefore, it is difficult to choose one form as better than another, but there are common strengths and roles required of the project manager (Table 10.2).

In general, project managers are responsible for the successful outcome of the entire project—they design quality, financial performance, and schedule compliance. Meeting these objectives requires the coordination of in-house resources, subcontractors, scheduling, client's needs, estimating, establishing and meeting budgets, and recognizing and resolving problems, in addition to site planning and design responsibilities.

TABLE 10.2 Scope of Project Management

Who is the client?
Is there a contract?
What is the objective of the project?
Who are the stakeholders?
Determining the scope of the work:
What is it you are going to do?
What is your objective?
Who are you doing it for?
What resources will you need?
Project organization
Subcontractors
Equipment
Where are the decision points in your project?
What regulatory involvement is required? Permits?
What are the deliverables?
When is the job finished?
Writing the project schedule:
How long will it take?
What needs to be done first?
What is the critical path?
When must it be done?
Who sets the pace and the standards?
What regulations will affect the project?
What problems should be anticipated?

Communication

The project manager's role is primarily one of communication. The project manager is the clearinghouse for information both inside and outside of the organization. To be effective, the project manager must be an advocate and balance the sometimes conflicting interests of the firm, the project, and the client. It is the project manager that articulates the project objectives and controls the design and planning process within the organization. To do this, above all other skills, the project manager must have effective communication skills.

Communicating effectively is always a challenge. The imprecision of language, the tendency of people to hear different things, and the ambiguity of memory all work to muddle the message and the information. It requires an effort beyond casual communication to be effective. The project manager must establish a record of communication from the outset of the project and maintain this practice throughout its life.

Project managers soon learn that communication requires preparation. Project information between staff and subcontractors should flow in a consistent form and pattern. The methods of communication are more effective if they are used in the fashion and in the same way. Although it is best to provide clearly written communication, if less formal methods are preferred, communication should still occur in a consistent format. The consistency alerts the listener that a particular communication is different—it is more formal, and it has more weight than casual or mere informational communication. To provide the most effective communication, managers should avoid verbal directions to staff. Verbal directions should be supplemented with written confirmations. In the modern office environment, there are numerous tools to provide such written interaction.

Meetings should be organized with a written agenda. Managers should control the meeting as to content and time, but they should be sensitive to new information that might indicate problems or unexpected concerns. All meetings should be summarized in a brief set of minutes, which establishes a formal record of communications. The purpose and the value of the agenda and minutes process is in the consistency and expectation created in those being communicated with. Project records that include meeting minutes are often valuable later in the project to re-create the decision-making process. It is also highly recommended that project managers keep a project journal of some form.

External communication to clients, regulatory agencies, and other stakeholders should be carefully prepared. Written communications often have a life well beyond the intention of the author and so must stand alone. Every written document must contain all the relevant facts discussed or provide a reference to another document where the facts are to be found. This provides a future reader with the ability to understand or re-create the writer's intent and purpose.

Public communication, especially presentations, requires even greater care and planning. Preparing for a public presentation first requires an articulation of the objectives of the presentation: What is the purpose of the presentation? What is to be accomplished? Articulating the objective provides clear guidance on how to prepare for the meeting and for how to assess the results. After establishing a clearly stated objective, the preparation should identify the key points and supporting arguments. Site planning is fairly unique in that although trained and experienced professionals are responsible for planning and design, project approvals are made by individuals who may have no training or experience. This requires that presentations include carefully prepared education of the public and public officials. In any case the quality of the presentation and the credibility of the presenter are critical components of the successful presentation.

Project managers may also consider the preparation of routine project reports. While these may not be appropriate for small or short-term projects, larger projects often benefit from periodic summaries. Quarterly summaries to the client and key staff provide updates as to progress, issues that require attention, and performance measures. They provide a reference for correspondence and establish a project history. For many projects these reports have little value once they are complete, but for those few times they are needed, they are very valuable. During the project they serve as important internal and external communication tools.

Finally, project managers should consider keeping a project journal or personal diary. As a practice, the diary provides an ongoing record of decisions and contacts. Project managers can use the diary to prepare summaries and letters to the file and similar documents. There are many tools that exist today to assist the project manager in meeting all these communication demands. For example, *personal communication assistants* (PCAs) and similar devices are compatible with desktop computers and provide valuable assistance to the project manager.

Leadership

Project managers must provide leadership in the form of setting objectives and motivating staff to meet those objectives, but leadership style is usually developed on an individual basis. Most leaders develop a style that is an amalgamation of their own personality, skills they have learned either formally or informally, and the circumstances in which they are working. That said, there are some characteristics of leadership that seem to be more effective than others in professional design environments.

Working with creative, highly trained people requires a leader to be able to direct people mainly through influence, perhaps even more than through authority. Often a project manager is responsible for outcomes that rely on the performance of people who do not work for him or her. Or a project depends on the timely and precise performance of people often working apart from one another and coordinated by a project manager who is not their supervisor. In such cases the project manager must find an effective method of communicating and influencing that relies as much on informal influence as it does on formal lines of authority.

There is an ongoing need to build consensus in multidisciplined project teams. The project must proceed on many fronts with confidence that each team member is progressing and that each role is properly informed and coordinated. The project manager must find the means to generate consensus through leadership, facilitation, and influence. The design professional in today's marketplace is highly trained and motivated and requires a participatory role or understanding in their work. It is unrealistic to try to build a staff of competent and reliable professionals without developing a sense of project ownership and responsibility in every individual. This sense of professionalism is a reflection of the individual character of each staff member and the degree to which he or she feels engaged by the organization. This feeling or perception of involvement can be difficult to establish and nurture, but it is a cultural element of most truly successful and innovative firms. It may be the firm owners and senior-level personnel who empower this creative culture, but it is the project managers who breathe life into it through their methods and behavior.

In the modern sense, then, leadership is most effective when it acts to empower others—when followers feel they can use their own judgment and initiative within the context outlined by the leader. Leaders must therefore be patient and persuasive, and they must recognize, by being active learners, the value of followers' contributions.

Managing change in the organization

There are four ways to consider the future: (1) Things will go on as they have been but get better. (2) Things will go an as before but get worse. (3) Something serendipitous and wonderful will happen, and things will be great. (4) We will think and plan to influence the future as best we can. Designers and planners intuitively value the last view since projecting a vision into the future is the essence of design. The historian Alexander Toynbee described what he called the "challenge-response formula of success." He observed that when a response is equal to the challenge, the response is defined as successful, and this remains true until the challenge changes. To some degree the effective leader is captured by the process or system in which he or she operates and so as Warren Bennis, a noted author on leadership, observes, success is by definition the seed of failure since the challenge will inevitably change. Project managers must provide leadership in terms of the project, but firms require innovators as well.

Change in the firm is inevitable. Resistance to change is natural, but it is the most significant roadblock to innovation. Within the professional environment, individual experience is composed of three aspects: skill, knowledge, and attitude. These elements act as both cause and effect; they are honed by experience, but they also influence behavior and perception. Abraham Maslow once observed, "A man that is good with a hammer tends to see everything as a nail." To become and remain successful, a firm and an individual must avoid the trap of success and intelligently embrace change.

True change occurs in organizations in one of three ways: It is either imposed by outside influence, it is the path of least resistance, or it is sought. Change may be imposed upon the organization from the outside by changes in regulations, new ownership coming into the firm, or a change in market conditions or competition. The path of least discomfort or pain may lead to change because it is easier or more comfortable to do something differently from the way it has always been done. Such change is undertaken to avoid something unpleasant.

The last form of change is embraced because it feels good—because it is rewarding or promises to be so. Of the ways in which change occurs, it is only this last one that is a positive experience and the only one that can be described as a function of leadership and management.

The Computer in Project Management

Relatively inexpensive and powerful computers have made significant contributions to project management. The contributions come at a price, however, and many project managers find the maintenance of project information systems to be a time-consuming effort, focused more on measuring outcomes than on managing projects. Project management software, however, should serve to manage projects—not just collect accounting information.

Good software programs are available to assist the project manager in the organization and tracking of project elements. Such programs use a high degree of integration and cross utilization of information, allowing multiple reports from a single set of data. In general, project management software should be able to track project performance in terms of costs, milestones, and schedules. It should provide forecasting and cost control tools, but it should not be limited to accounting functions or reports. Project management information systems should be designed and implemented to support project management and so should address the needs of project managers primarily. There is a wide variety of software programs for project management. They range from modestly priced software for creating schedules to expensive network programs. There is a package for every firm and situation. Project managers should evaluate software in terms of what they need or desire, not in terms of what the software company says it can do. Many firms have invested in elaborate packages only to find that project managers use only some of the capabilities of the software.

Contracting

The simplest contracts require only that there be an agreement for exchange of value for consideration. Formal contracts are documents that acknowledge the conditions and limits of the exchange, the nature of the value to be exchanged, and the nature of the consideration. The contract language and standard clauses commonly used by design professionals usually include a description of the limits of liability, and indemnification clauses, and they specify who owns the drawings, when and what reports are due, and how and what payments are to be made. Some contracts include schedules and specific descriptions of the scope of the work; some contracts might refer to other documents that contain this information.

Most design professionals include in their contracts a *limitation-of-liability clause* that limits the liability of the firm, principles of the firm, licensed design professional, or other staff to a sum no more than the sum of fees paid to the firm. In fact, most firms view this language as negotiable and are often persuaded to increase the limits of liability in exchange for increased consideration. Another form of liability limits is the exclusion of work done by others or effects caused by the actions of others. This limits the firm's liability to work actually done by the firm and excludes liability that might arise from actions of others such as labor stoppages or substantial changes to the scope of the work.

Typically contracts for design services include a statement or a clause that clearly states that the firm makes no warranty of its work, implied or otherwise. In general, this clause requires the work of the design firm to be judged by a *standard of reasonable care*—that is, the degree of care generally observed by other design professionals under similar circumstances.

Most design professionals have worked without a contract at one time or another. Over time many firms establish close working relationships with clients and subcontractors. In many cases this arrangement is mutually satisfying and profitable. In other cases, design professionals have experienced difficulties with getting paid or resolving unhappy situations. Many professional liability insurance carriers require project-specific contracts for projects that are of equal or greater value than the firm's out-of-pocket deductible would be.

Professional Liability

Risk management begins with knowing what the designer is being paid to do and what the client expects in return. One of the most important and necessary aspects of formal written contracts is the exercise of describing the work and the expected outcomes. Firms should never assume greater responsibility than they contracted for, and they shouldn't contract for work they are unprepared for.

Professional liability usually stems from negligence, omissions, and/or errors. *Professional liability insurance,* sometimes called *E&O insurance,* is purchased to protect firms and individuals from lawsuits and costs arising from omissions and errors. This type of insurance may not address negligence. Furthermore, professional liability insurance may have limits as to the activities that fall under the professional design umbrella. For example, a firm that chooses to act as a contractor, even for a moment, may not be covered by insurance for claims arising out of such action. Most professional design liability insurance includes a *pollution exclusion.* This clause has been interpreted differently by various companies but is commonly considered not to include the collection of samples and similar small quantities necessary for professionals to conduct good work.

Protecting yourself from claims is generally referred to as *risk management.* Some providers of E&O insurance provide training for staff to learn about risk management and to recognize problems while they are still manageable. Such firms sometimes offer discounts in insurance rates for firms completing such training. The basic principles of risk management are fairly straightforward, as shown in Table 10.3, but they are often found to be difficult to practice. Ambiguity in project scope or in objectives is a fertile ground for difficulties later on. A tight scope of work provides a clear set of tasks and deliverables and significantly less room for misunderstanding. Once the scope is established, the project manager must be on guard for "mission creep"—that is, taking on more work or tasks than the project originally called for. More work is usually welcome, but not if it is going to be difficult to get paid for it later or if it is the seed of client discontent. Additional work should be formally recognized and incorporated into the agreement between the firm and the client.

Professionals with a good relationship with their clients are often called upon to assist the client when trouble arises. Occasionally the professional may be asked to assume a management role in construction activities. While such opportunities may be exciting and demonstrate the firm's value to the client, the professional may be moving from a professional relationship to a contractor relationship in terms of liability. While there may even be business advantages, they are not without costs. Professionals must be careful to never unintentionally cross the line from recommending and observing to directing and supervising.

Finally, perhaps the most difficult part of risk management is to recognize and react to the earliest warning signs of a problem. This is difficult because few of us enjoy dealing with problems, especially if they result from our decisions. The fact of the matter, however, is that problems identified and addressed early are most likely not to progress to insurance claims or lawsuits. If nothing else, costs are contained by early action. In many situations recognition and swift action makes the relationship between the firm and the client much stronger. We all make mistakes eventually, but only those who manage the mistake effectively can actually profit from it.

TABLE 10.3 Basic Risk Management

Know what you are being paid to do.
Do not assume greater responsibility than you have contracted for.
Formally recognize changes in the scope of the work.
React immediately to problems that may be potential claims.

Quality Assurance in the Design Process

Quality assurance is a critical element of the design and planning process, but it remains an elusive and ambiguous objective; defining quality is perhaps more difficult than achieving it. There are many approaches to quality assurance, but in general quality is achieved through communication, technical competence, and a commitment to the work. These are the necessary attributes of the firm and the individuals that make up the firm. Quality is not accomplished through checklists, though for some people checklists are helpful tools. Pushing responsibility down to the lowest level of the organization does not achieve quality if the staff is unqualified or the project is poorly defined. A competent staff cannot produce quality work without effective leadership, communication, and resources.

Getting Paid

Getting paid can be a challenge. Unpaid invoices are important warning signs that every project manager should be aware of. A client that is slow to pay may be unhappy with some aspect of the firm's performance, or underfinanced, reevaluating the project, or just slow to pay. The project manager must determine why the invoices are not paid and what, if anything, to do about it.

In an unscientific survey of project managers, the most common reason for lack of payment was related to communication. Although other reasons may have been given by the client, the underlying reason was a perceived lack of attention on the part of the client. Whether there was an actual lack of communication was less important than this perception. Many project managers have discovered that an effective tool for getting paid is to make a special call to a client after the client has received the invoice to confirm that it has been received and to inquire if there are any questions about it. If there is a problem, it surfaces early instead of a month or two later.

Once a problem is identified, however, it is critical that some action be taken in fairly short order. In many cases the quick and effective response strengthens the relationship between the client and the professional. In cases of untrustworthy clients, the sooner the problem is identified, the sooner action can be taken.

For many project managers, dealing with unpaid invoices is among the least favorite tasks. Unpaid invoices place the project manager in a circumstance in which both the client and the firm are unhappy. It is important that the manager remain objective. The firm is entitled to be paid for work contracted and performed. The client is entitled to receive the benefit of work performed and fairly paid for. There must be a win-win, and it is the manager's role to balance these interests and solve the problem.

Payment terms and conditions should be part of every contract. Firms should bill and expect to be paid in accordance with the contract. Exceptions should be dealt with immediately and in person. Clients that refuse to pay or do not meet payment promises should not expect continued service, and the relationship should be formally terminated. Early and frequent contact with clients is the best way to assure payment or to identify a problem with a client.

Greening Up the Design Practice

The practices and suggestions in this book are standard procedure in some firms, but they will require new thinking in others if they are to be implemented. The reluctance to change in some communities and even in design firms is evident from the types of projects still being produced. Design professionals are not the only party in the design process but they are best suited to educating and bringing change to the built environment. Firms engaged in land planning and design should seek opportunities to provide access to public transit and to create linking pedestrian corridors and bicycle paths. The character of communities is more than the eyewash of clever plantings and curvilinear roads. Ultimately, their character is defined by how they work. Transportation is therefore a critical feature of the modern community.

Design professionals should encourage zoning ordinances to permit mixed-use development so that homeowners can walk to the store, to school, or to work. It is necessary to advocate for these changes when not representing the interest of a client so that the motivation is clearly the quality of the design. Buildings should be situated to maximize energy efficiency, allow for pedestrian access, and reduce development impacts. The cost implications are generally favorable because of narrower and shorter roads and reduced lengths of utilities.

Plans should incorporate the protection of trees and topsoil during site work. Specifications should require that when backfilling a foundation or grading around a building, construction debris must be removed rather than buried. Contractors should be encouraged to minimize job site waste by centralizing cutting operations to reduce waste and simplify sorting. Clearly marked bins for different types of usable waste (wood scraps for kindling, sawdust for compost, and so on) should be set up. Salvaged materials could be recycled or donated to low-income housing projects, community groups, and so on.

In addition to introducing more green design practices to the site, firms should consider finding ways to "walk the walk" to improve the environmental performance of their own business. Reducing the environmental impact of the firm contributes to the firm's credibility and serves as a model for other firms. Simple strategies such as monitoring the type of vehicle used in the company fleet and encouraging employee carpooling and the use of mass transit are important steps. The use of recycled paper or rebuilt office equipment and using recycled materials in office furniture and decorating speaks volumes to clients and the public. The well-designed and appointed office could serve as a calling card for the firm's values and commitment. The daily routine in the office should reflect the designer's values.

Dealing with the Public

The site design profession may be unusual among the professions because the work of designers is always under public scrutiny and subject to change and approval by nonprofessionals. This aspect of the profession is often cause for frustration borne of required changes to plans that contribute nothing to the project or, worse yet, undermine the design in some fashion. Nonetheless, it is a tenet in the United States that land use is a public process and locally controlled. Professionals must learn to work effectively with citizen planners and local authorities. Much of what an effective designer does is to educate the public citizen in the value and values of the project.

Presentations

Preparation is the key to every presentation, and preparation begins with understanding what the presentation is to accomplish. A clear and concise objective should be articulated, and the presentation should be built around it. It has been said that for every 5 minutes of presentation there should be an hour of preparation time. The fact of the matter is that some presentations take more preparation and others take less. The articulation of the objective should help to determine the preparation effort.

Every effective presenter eventually develops a style that works, but there are some characteristics of effective presentations that should not be overlooked.

Preparation. Once an objective has been stated, the next step is understanding the audience to whom the presentation will be made. In a public meeting, is the presentation being made to the members of the community or to the members of the planning commission? Depending on the circumstances, it could be one or the other or both. It is necessary to understand why the members of the public are there: What is their interest? How will they be affected by the project? By addressing the key issues in the presentation, the designer will be addressing their concerns directly.

The next step is to develop a strategy for the presentation. There is a tendency among some design professionals to effect a casual, what might be described as a laid-back manner of dress and behavior. This might work for some groups but not for others. If the designer's ideas are to be taken seriously, he or she must be taken seriously. If casual attire and manner are taken as a reflection of his or her work or the quality of his or her ideas, the presentation may have to overcome the designer's public image before anyone actually listens to it.

The designer should know ahead of time how much time is allotted to the presentation. He or she should plan to end early. By intentionally running short, he or she will allow time for questions and any overrun. No one enjoys listening to a 10-minute presentation for a half an hour. Also, the designer should know what he or she will leave out if pressed for time. Spending the bulk of the presentation on one or two points and forcing several more in the last 2 minutes undercuts the credibility of the presentation.

If the designer is going to use visual aids, they should be designed to support the presentation, not to replace it or compete with it. Every visual should have a purpose that directly supports what is in the presentation. There should not be too much information on the visual; it is not a good idea to have the audience reading while the designer is talking. If there is reading to be done, it should be provided in a handout, to be given out at the end of the talk. Every visual should have a purpose, but it should not contain too much illustrative information either. Data presented on visuals should be as simple as possible. Complex charts and tables generally do not work well and serve to confuse the listener and muddle the presentation. Some presenters use a rule of thumb that a visual should be up for a maximum of only 3 minutes. This keeps the information per view limited and keeps the presentation on a good pace.

Notes should be prepared to guide the presentation, but they should not be read directly off the note cards or the slides. The presentation should not be memorized; instead, the notes should serve as prompts. Any equipment to be used during the presentation should arrive early and be set up and tested ahead of time. The designer should always be familiar with the equipment. If there is an equipment failure, he or she should not spend time working through the failure while the audience waits. Instead the designer should continue the presentation without using the equipment. The key is to minimize disruption and display confidence and aplomb. The designers should remember that he or she is the presenter, not the equipment. Finally, the designer should always arrive on time to set up the equipment, learn the room, and get to know the members of the audience, and to know who the formal and informal leaders are.

Presentation. A presenter should always begin by introducing himself or herself and acknowledge an introduction made by another. He or she should demonstrate an open and objective approach to the subject. Posture should be comfortable. Sitting should be avoided, as should leaning on furniture, putting hands in pockets, or assuming any other casual body stance. The presenter should face the audience. The use of stories and jokes is generally to be discouraged. First, few people tell jokes well, and second, the choice of a joke or a story is difficult to manage without offending someone. The presenter should stand straight, emphasizing main points with natural gestures. An occasional big gesture may help in making a point, but it shouldn't be done too often because big gestures are distracting. Instead, the presenter should use his or her voice to make points, speaking louder or softer, faster or slower, to give emphasis or to underscore a point. He or she should use a pause to draw attention to an important point and give it a moment to sink in. The key to the presentation lies in a person's eyes. People are more likely to like and trust a person who makes direct eye contact. The presenter should look at each person he or she is speaking to and make eye contact at least once with every key person. The presenter should always make eye contact with a person who asks a question. In many public presentations the audience is sitting before a panel, and the presentation area lies between them, making it difficult to face the audience and the board. In such circumstances, it is best to assess the purpose of the meeting and the objective. If it is a public presentation, ask permission of the board to face the audience; if the presentation is to the board, too, request an action to face the board and apologize to the audience.

The presentation should open with a statement of the speaker's primary objective. Many trainers instruct presenters to always find a way to describe how the project benefits the listener. In many cases when a development project is being presented, the public is not interested in the benefits. It may be best to not antagonize them with benefits if the speaker is unwilling to address the negative impacts as well. Hard data should be given to the audience, as well as information on how these issues are addressed in the plan. The thinking and problem solving that went into addressing the public's concerns should be demonstrated as well.

The presenter should keep track of the time. When questions are asked, the presenter should be sure to listen and avoid shuffling papers, looking at his or her notes, or adjusting the equipment. The question should be acknowledged somehow and directly answered. If the presenter does not have an answer, he or she should say so but then promise to follow up. The language of the presentation should be conversational and simple. Technical jargon should be avoided unless the group is apt to be familiar with the jargon. Acronyms should be avoided.

Closing. The presenter should always have given thought to his or her closing in advance. It is better to leave out part of the presentation than to not deliver a closing. At the end of the presentation, the presenter should remind the audience of the original objective for the talk and recount the reasons that support that objective. If the speaker wants a specific action or decision, to occur right then, he or she should ask for it. It should not be left to someone else to parse together the speaker's request for an action; the speaker should simply and directly ask for what he or she wants.

Not in My Backyard

A common experience of developers and site professionals is community resistance to a proposed project; the so-called *NIMBY effect,* or *not in my backyard effect.* The typical NIMBY reaction is based on the community reaction to the project's negative impact on property values. A number of studies have been completed that speak to the negative impact of specific projects on property values. Some studies of the impacts of NIMBYs on neighboring properties provide policymakers with additional data with which to develop plans and public policy, but many consider only the negative impacts of the unwanted project and ascribe no value or benefit to the larger public or society. If in fact a power plant in a small town devalues property overall by $200,000 to $18 million, is there no offset in value to the town and its residents? Would the property in this small town be of greater value without electricity?

To deal with the expected NIMBY reaction, designers should consider how the project might be sited or prevented and offer three approaches to minimizing the impacts. Some NIMBYs might be prevented by redefining the parameters of the solution. For example, instead of building a large low-income housing project, why not offer housing vouchers to eligible citizens? (In fact, the Section 8 housing program of HUD does exactly this to eligible residents or eligible housing units.) Other options include the use of recycling to reduce the need for landfills and transfer depots. If the residents become more efficient or conscious of the costs of their behavior, they might see a benefit to the new project after all (if we were more energy efficient, we would not need another nuclear power plant and therefore eliminate the NIMBY) (Kenyon). In fact, however, if the residents perceive the cost of avoidance or loss of convenience to be greater than the negative impact, these approaches to preventing the NIMBY are unlikely to work. NIMBYs fail or are prevented by grassroots action when the costs (loss of property value) are greater than the perceived benefits.

NIMBYs are a problem to communities that fear a loss of property value and/or a loss of quality of life without receiving any offsetting value or return. The fact of the matter is that a contemporary community places a high value on quality-of-life issues that require a significant return. The "carrot" of jobs and lower taxes are not adequate offsets to these impacts. The paper suggests that siting issues should be left to the marketplace and that the NIMBY should have to survive on its own economic merit.

Since most land use decisions are ultimately local decisions, this consideration, however, fails to recognize the regional nature of many NIMBYs. A power plant or a landfill, or even a group home, might service an area and provide benefits to a much larger population than the local population within the scope of its immediate impact. The solution to the considerations of local populations is to provide compensation for negative impacts. Compensations must be in a form that will be capitalized into the affected community so as to support the continued viability of the community. These compensations may be in the form of tax relief, or host fees, or some other offsetting value that neutralizes the NIMBY for the locally affected population, so that it allows the benefit to go forward for the larger population.

Chapter

11

Historic Landscapes and Preserving Land

Historic preservation is an important area of practice with its own sets of concerns. The value of historic preservation has become more apparent to communities with the rapid expansion of suburban areas and the effort to "renew" older parts of cities. There are numerous laws, regulations, ordinances, and programs that address historic preservation across the United States. In general, these efforts are focused on the preservation of buildings; less often is there concern with historic landscapes in and of themselves. Exceptions, of course, are those areas of exceptional beauty or natural wonders that are preserved as parks or monuments, and places of note like the Gettysburg Battlefield or Seneca Falls, origin of the woman's suffrage movement. As with other areas of site planning and design, work with historic landscapes requires a familiarity with a set of underlying principles, and it is an area of expertise in itself.

Investigation of the Historic Landscape

The investigation and analysis of the historic site have some additional elements that are not normally part of the site analysis phase (Table 11.1). On a historic site the designer is trying to understand the limitations and opportunities of the site in order to modify the site for the intended purpose. The site analysis is an attempt to understand what has already been done to the site, how and why the site was altered in the past, what role the landscape played in the past, and ultimately how to preserve, restore, or rehabilitate the landscape.

A key difficulty in historic work is that historic values are often subjective. For example, finding compatibility between a historic site and a use proposed for it may be difficult to do when balancing the preservation concerns with the desire to maximize the use of a site. Even finding a clear "time context" can be difficult on a site with a great deal of history or with numerous periods of

TABLE 11.1 Fundamentals of Historic Landscape Preservation

1. There must be a compatibility between the historic and natural character of the site and the proposed use of the site.
2. A clear "time context" of the site must be developed.
3. A comprehensive identification and assessment evaluation must be completed to identify the distinctive elements of the site and how these elements are to be placed in the time context of the site.
4. The site assessment should note whatever changes have occurred over time and how the site is different now from the way it was during its time context. Care should be taken to determine if these changes are important elements in their own right.
5. Restoration is always preferred over replacement of an element, and therefore stabilization and repair should be undertaken carefully.
6. If new work is to be done, it should be designed so as not to interfere with the character of the site. In any effort to introduce or maintain out-of-context elements, the "fit" should be weighed and considered carefully.
7. Objectivity should be maintained about the site, the time context, and project.

development (Figs. 11.1 and 11.2). While it is recommended that the professional maintain an objective view and approach to the site, the time context, and the project, it is often difficult to do, and more often there is never one right answer.

Landscapes that are significant historically may range from the work of well-known designers retained by historical figures to places that are remembered for a significant event or person, from landscapes of cultural importance to places of natural beauty or ecological significance. Many organizations and agencies have developed guidance documents to assist and even to prescribe the evaluation and treatment of historic landscapes. Projects utilizing public funding, for example, may be required to use a specific set of guidelines or processes. Other valuable resources may exist in the offices of local and state historic commissions or societies. Many sites already have had cultural or historic landscape reports prepared for them.

A fairly thorough history of a site is necessary to understand the appropriate time context of a landscape. *Time context* refers to the historic period that is of importance for the site. While some landscapes may not be time sensitive, others are associated with a specific period or series of events. Establishing the appropriate time context is critical to assessing the condition of the site from a historical perspective. Some sites have fairly obvious time contexts, but others may have multiple contexts making the selection of just one difficult or even impossible. In such cases the selection of time context may be a matter of some disagreement and may involve ideas from a variety of sources. In any case, the selection of time context is rarely the business of the site design professional alone. The site professional, however, should be aware of the importance of knowing the site context before design work begins.

A complete inventory of the site is also necessary. The inventory planning should include reviewing all of the available records pertaining to the site,

Figure 11.1 The context of historic sites may be difficult to determine. This formal garden was constructed more than 150 years after the establishment of this historic site, and it is not consistent with the original design of the grounds.

including plots and maps, diaries, articles, journals, local records, or any other source of information about it. It is suggested that even unsubstantiated rumors be collected. A history of ownership such as a chain of title may be particularly important. On larger sites aerial photography may be a useful tool. Historic aerial photography may provide important information about the site in its recent past (Fig. 11.3). Current aerial photography sometimes reveals site features not seen from ground level (Fig. 11.4).

Using a topographic map as a base, a site should be prepared that includes the significant features (not related to size but to historic significance) and shows all the buildings, foundations, fences, stone piles, tree masses, major or important specimen trees, surface water features, outbuildings or appurtenances, road pathways, areas of fill or cut, and visible boundaries. The assessment is an effort to "read" the elements of lost landscape features, old roads, foundations, unexplained grading, and masses of ornamental trees, all of which tell a portion of the site's history. For example, in early America, trees were planted around houses to shade the house in the summer or provide a windscreen, but trees were also commonly planted in some places to commemorate special events in the life of a family, a community, or the nation. During the U.S. Centennial year 1876, many communities planted groves of "centennial trees." Landscape elements that sometimes appear disjointed or unorganized

Figure 11.2 Foundation plantings were not used at the time this house was built in the 1600s in Maryland.

by modern standards merely represent rational design choices of another time—the underlying form of the landscape hidden from modern sensibilities.

During the site investigation, photographs of conditions and features should be collected. Also, visiting a site in low light conditions (early morning or toward sundown) often reveals patterns in grading and other site features not seen under brighter light conditions. Exotic plants as well as important plant types and species should be noted.

The purpose of the site inventory is to identify the site features and to assess the integrity of the site: How much of the historic or important landscape remains, and how has it changed? Do the mature trees of today represent the character of the site in the past? Are the "old" features of the site in keeping with the time context of the site? Integrity must be defined by how much the site retains its historic, natural, or ethnographic character. Is the site significant? Are the remaining features significant?

Using Historical Photographs and Photography in the Site Analysis

Photographs are important to site analysis in a variety of ways. The use of historical photographs may provide particular insight to the past uses of property or help to locate features now lost. Old photographs may be found from a vari-

Figure 11.3 An early aerial photograph of a former Pierre L'Enfant garden.

ety of sources such as local historical societies, libraries, public historical records, private collections, or local government archives. The difficulty in using old photographs is usually the lack of information about the location being photographed or the site location of the photographer. Very often clues to such information can be found in the perimeter of the photograph. Also, physical features like hills, chimneys, or buildings in the distance may still be visible from the site and can be matched with the photography. Old photographs are particularly important if the site has important historical significance.

It may be helpful to use two cameras in the field—one using black and white film and one using color. If an attempt is made to compare current conditions and those shown in an old photograph, a current black and white photograph could be very helpful; however, the differences in camera equipment should also be considered. Old photographs appear much different because some older cameras provided a wider field of view, as much as 120 to 160°, a far larger field than modern cameras capture. When comparing new and historical photographs, it is important to remember the differences in the equipment as well as the time of the year and time of day the original was taken.

Figure 11.4 A later aerial photograph of a Pierre L'Enfant garden.

Landscape Style in the United States

Public landscapes in colonial United States were largely influenced by English and French traditions, but archetypal landscapes still reflected the interpretation of the character of their new land through the experience and traditions of the new immigrants. Almost without exception the planned landscape forms that emerged from the colonists reflected the new country and land, but they included formal and geometric gardens. The form of the archetypal gardens and landscapes reflected the straightforward requirements of the designers (Fig. 11.5). These early places represent much of the cultures the early settlers came from, but as generations came and went, new forms emerged that adapted to the demands and materials of the new land and climate.

For the most part, gardens were made to grow food. Land cleared from the wilderness or in town was at a premium, and simple forms were easiest to arrange and manage. More important than the esthetics of such a landscape were its orientation to the sun, quality of the soil, drainage, and proximity to water. Plants selected for such gardens were chosen for utility; what could the plant provide to the larder or the home pharmacy? The archetypal landscape of the United States can reveal much about the people who settled here—where they came from and how they adapted to the new land. Just as the forms and functions of buildings reflected the use of materials, adaptation to climate, and the esthetic tastes of their builders, the layouts of homesteads and villages also spoke for the settlers. Buildings and spaces under such conditions were not oriented toward the public highway but to the things that mattered most: sun, slope, water, and wind.

Early public landscapes such as parks and government properties tended to be small by today's standards and were generally formal. The civic landscape represented the power of humanity, its formality clearly setting the works of human beings apart from those of nature. The landscaped grounds on private estates ranged from the formal designs favored in the seventeenth century to the more naturalized landscapes of the eighteenth century. The English first parted from the formality of the seventeenth century through the creation of landscapes portraying an Arcadian ideal. These landscapes were designed to be sort of a three-dimensional painting of a supposed scene from Roman or Greek antiquity. This style abandoned the use of the features of formal landscapes, such as rectangular water forms and formal arrangements of shrubbery and flowers, to assume a more natural, although contrived, appearance. Water features were designed to have a natural, irregular appearance and to serve the landscape as an element incorporated into a whole design rather than as a discrete element as a formal pool or fountain by itself. This movement away from formal gardens and toward more natural appearing landscape was championed by Lancelot "Capability" Brown (1715–1783). In

Figure 11.5 Photograph of early American homestead.

Brown's vision, the landscape took on a pastoral image that blended with the natural landscape but was as much the product of design as the formal gardens he eschewed.

Thomas Jefferson and other wealthy gentlemen of the time introduced the European ideas of landscape design to the United States. Pierre L'Enfant (1754–1825) worked for many wealthy landowners of the time designing gardens and estate landscapes using both formal and informal styles and elements. Eventually many of the principles of "Capability" Brown's romantic landscape were adopted in the new United States. In England the new preference was for the natural look of the Gothic landscape, but during this period the English combined the Gothic with sweeping lawns and natural forms of lakes without attempting to create a single scene. Although some aspects of the Arcadian tradition, such as grottoes and ruins, were used, the most notable characteristic of the landscape of this period was its inherent naturalism. Generations of garden and landscape architects, including Frederick Law Olmsted, would eventually be influenced by the work of Brown. The most noted landscapes of this era were often criticized at the time for being indistinguishable from the countryside. Much of the "natural" movement, however, was lost on smaller urban properties where the opportunity for large expanses of lawn or artificial lakes and such did not exist. Small urban gardens tended toward simpler forms of the formal gardens of the archetypal tradition or earlier more formal period of design.

Andrew Jackson Downing (1815–1852) was among the first of the professional American landscape designers. Downing believed that the landscape should improve the house but not be too formal in itself. His romantic landscape designs attempted to integrate the house and the surrounding landscape by ensuring that flowers and ornamentals could be viewed from within the house and that fragrances from the garden could be smelled and appreciated from inside as well (see Fig. 11.6). Downing's designs were seen as being a blend of the formal and informal; he planted flowers and shrubs in circular, formal beds, but he designed lawns to be bordered by informal, curvilinear beds of shrubs and trees. He liked informal arrangements of trees that framed or enhanced particularly good views. Downing influenced landscape and garden design throughout the middle of the nineteenth century. His books *A Treatise on the Theory and Practice of Landscape Gardening Adapted to North America* (1841) and *Victorian Cottage Residences* (1850) were arguably the most important gardening and landscape design books in the United States of their time and are still occasionally reprinted.

In the nineteenth century, the design of large private and public landscapes and gardens moved away from the wild and natural forms of the eighteenth century and back toward some measure of formality. Where buildings and homes were placed in the landscape with little to separate them from the lawn beds of flowers, terraces and ballustrades were introduced. Downing continued to advocate the natural romantic style until his untimely death. Attempts to bring the romantic landscape into smaller and smaller properties eventually led to a disintegration of the style until it came to mean merely random

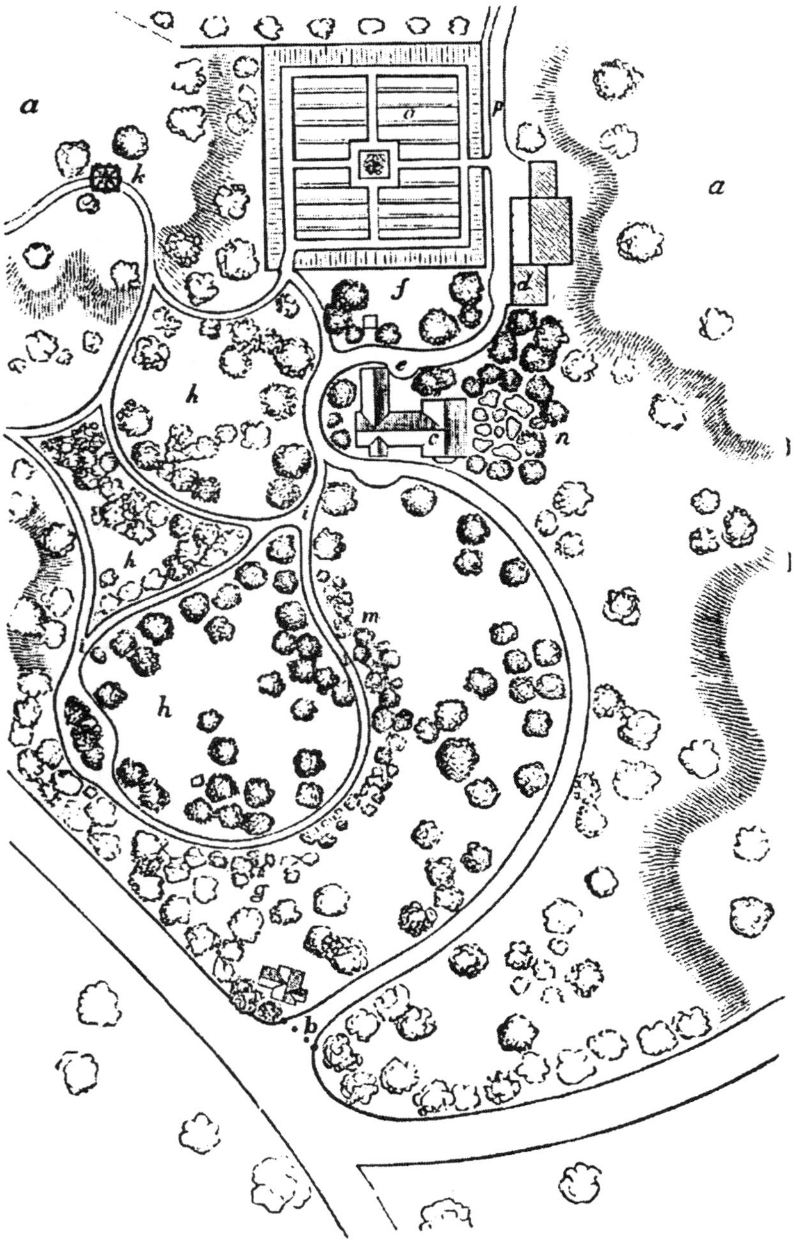

Figure 11.6 A site plan from Andrew Jackson Downing's *Victorian Cottage Residences* (New York: Dover Publications, 1980).

flower beds and specimen trees. About this time the availability of landscape plants increased significantly, and gardens were revised to include specimens from all over the world. It could be said that this period had no clear distinctive style of its own and was merely a period of transition from the romantic to what would become the "Victorian garden" some years later. To make the landscape more inviting, artifice and formality were reintroduced by designers as fountains, fences, and other similar features. Many large landscape and garden projects in the United States during this period called upon earlier European styles.

Frederick Law Olmsted (1822–1903) started his practice in 1856 very much a student of the romantic style but he was not limited to Downing's vision. Olmsted had studied the design of gardens and articulated his own vision, which was firmly grounded in art and science. Further, Olmsted was something of a social activist with firm beliefs in the value of public landscapes as enhancing the quality of urban life. The development of Central Park in New York City (1857), the Emerald Necklace park system in Boston, and similar park systems across the United States introduced important urban open space to the nation's cities. Olmsted's designs were more than the romantic image produced by the romantic movement. Instead, his were working landscapes so to speak, designed to perform certain functions such as waste treatment and flood control while serving as visual and social amenities. Public access to green space became an important element of city life because of the vision of many public leaders and the designs of landscape architects such as Olmsted, Calvert Vaux (1824–1895), and many others.

Among the important lessons of the best work of this period is that landscapes that are well designed and constructed are soon thought to be naturally occurring. It is forgotten that these great landscapes are the product of thoughtful design. With the passage of time, it is forgotten that as designs, they may require maintenance or that as working landscape features, they may need to be managed and protected. Many of the most important parts of these landscapes have been lost because they have matured with such a natural appearance.

The *Victorian garden* emerged to embrace elements of many styles. While rooted in the principles of Andrew Downing, the Victorian garden was an expression of its time—exuberant, often fanciful, busy, even cluttered. The Victorian era of the late nineteenth century had the resources and the materials of the industrial revolution at its disposal. Victorian gardens might be fenced using one of many commercially available wrought- or cast-iron fences, though wood remained the most popular and least expensive fencing material. The style of Victorian architecture was to set a house high on a foundation, and so it was that *foundation planting* first became popular during this time. The period is also known for the introduction of many exotic species of plants. Plants were chosen for their textures and colors, and eclectic mixes of plants were common on the borders of lawns and in more formal beds. Exotic plants became fashionable—the more exotic, the more fashionable. Formal *parterres* (gardens with walking paths among the plantings), *knot gardens* (very complex planting designs), and rose gardens were also popular features. In essence, the Victorian garden was, like so much of the era, a statement of wealth and class more than anything else.

As with earlier gardening and landscape styles, most homes could not afford the embellishment of a fully developed Victorian garden; however, many homeowners adapted aspects of the fashion so that small formal gardens, statuary, exotic plants, and even fountains or gazing balls might be introduced into the yard. The common practice of shrubbery foundation and border plantings edged with flowers is from this era. During the Victorian era and well into the twentieth century, Gertrude Jekyll (1843–1932) conducted her horticultural practice. Much of Jekyll's work reflected her interest in wild self-sustaining landscapes. She contributed much to restoring the connection between the garden and nature that was lost in the Victorian and similar styles of her day. Her influence was felt for many years, and it is felt again in the current move toward self-sustaining landscapes and the use of native plants.

In 1917, *An Introduction to the Study of Landscape Design* was published. Its authors, Hubbard and Kimball, observed that landscape design at the time could be divided into one of two types: romantic or classical. *Romantic landscapes* were informal and full of contrasts whereas *classical landscapes* were formal and restrained. The *eclecticism* continued into the early twentieth century, but eventually a variety of styles emerged from the enthusiastic Victorian landscape. In general, the variety of styles might be described as occurring along a continuum ranging from the natural or English landscape elements seen in some of the designs of Beatrix Ferrand to the "architectural" garden styles employing patios, gazebos, ponds, fences, outbuildings, and other structural elements. In the post-World War II era, suburban gardens and landscapes flourished, but there were few important styles or trends that emerged from this era. Garrett Eckbo's work extended indoor living into the landscape, but for the most part suburban landscapes tended to mimic the most common aspects of the Victorian or Downing gardens. The estate lawn of the nineteenth century became an icon of the suburban good life, and entire industries have grown up in response to the desire for this type of lawn. As the new century begins, the design of lawns is being reconsidered, and more regional and native landscapes are emerging as truly esthetic and important forms.

Landscape architecture continued to mature, with its practice ranging from suburban homes and residential developments to commercial and public spaces and landscape restorations. Building on its rich past, landscape architecture developed to freely employ formal and informal elements, and natural and more architectural styles, as suited the designer and the site. Site design has become a broad area of professional practice including engineers, landscape architects, and others. The range of the profession has also grown from traditional site development to new areas of practice and art. As both the design community and the public have become more aware and concerned with environmental issues, landscape architects and site designers have expanded the scope of professional practice to include landscape restoration and sustainable development in their many forms.

Planning for the Historic Landscape

Planning for the historic landscape involves many tasks, and site design professionals will most likely be working closely on a team with other specialists and professionals (Table 11.2). In addition to the tasks that any site planning effort requires, there are the concerns for the importance of the site as a subject in itself. The historical research may or may not directly involve the design professional, but the inventory of the site features, site analysis, and site plan will. The project may include maintaining substantial documentation as the site is disturbed as well as accounting for and avoiding the sensitive elements of the site in the plan.

Developing plans for historic sites will require a fairly thorough understanding of the site history and time context. For site design purposes, a map of existing features is important. It is suggested the designer confirm the feature map in the field to gain a first-hand understanding. Addressing the physical characteristics generally flows from understanding the intended use and the philosophy of the project. Preservation projects will necessarily proceed differently than rehabilitation projects. It is critical that the underlying philosophy is understood at the outset.

The difficulty in assessing an existing designed landscape is that the design itself is often blurred over time. The designs of many of the Olmsted firm's parks, for example, are largely seen as natural rather than as designed landscapes. Unfortunately, as some of the parks became overgrown, the landscape features that were designed to carry out functions in the parkland were lost. An important example is the Olmsted design of the Fens in Boston's Emerald Necklace. The original design successfully addressed and accommodated sewage and stormwater flows in a beautiful park setting. Users and planners alike forgot the function of the design over time and came to see the Fens only in terms of its beauty so that when the Charles River was eventually dammed, the function of the Fens as a working landscape was destroyed. Thus, an existing designed landscape must be investigated as more than a planting plan.

All too often the landscape around a historic building or monument is not treated as part of the historic character of the site or is sacrificed to parking and providing convenient public access. However, such landscapes can sometimes be modified to reflect contemporary tastes and standards. The new treatment of the site includes several challenges such as parking, lighting,

TABLE 11.2 Treatment Types for Landscapes

Preservation. Involves protection and stabilizing of site features rather than replacing them.

Rehabilitation. Involves repairing and/or altering a property to be compatible for another use while preserving features of historical or cultural significance.

Restoration. Involves re-creating an accurate depiction of the historic form or features of a property by reconstructing missing features or removing features added later.

Reconstruction. Involves depicting through new construction the form and features of a nonsurviving landscape or object.

way finding, accessibility, liability, and the like, which were probably not addressed in the original site plan. Finding a balance between these practical necessities and the integrity of the site is challenging. Many of the elements of the restored or rehabilitated landscape are both intrusive and necessary. It is necessary to provide for parking, and to stabilize pathways for safety and wear-and-tear. It might be necessary to introduce lighting and paving to provide for pedestrian or motor vehicle safety. Signs are both intrusive and necessary. The designer is faced with a challenge to balance these opposing forces in a design that serves the mission and the character of the site.

Creating navigable pathways

Lighting and signs provide both for safety and enhanced use of the site. An important means of integrating the historic character of the site and a new use is to educate people as to the history of the site. Signs are usually of two types: fairly simple signs providing direction and somewhat more detailed signs providing historical information. Signs providing directions should be simple and direct with clear and easily seen lettering or symbols. It is sometimes recommended that such signs use a minimum of letters or characters, but off-site signs should use abbreviations or acronyms only after careful consideration.

Signs can provide a simple and direct way of communicating history and increasing the site visitor's appreciation and awareness. Signs should be designed to minimize intrusion by being both noticeable and subtle. This is accomplished by carefully selecting the size, color, and location of signs. Placing signs very low to the ground, for example, keeps them out of the views but makes them easily found and read. Once visitors or users are aware of the location, they quickly adapt. The use of color or symbols increases visitor familiarity and maximizes the value of signage. It should not be surprising that designing content and sign communication is a specialty in its own right.

Adapting the historic landscape for modern use

Whether the intention is to adapt a historic site to a new use or to restore or preserve the site as it was in an earlier time, the site must accommodate contemporary use by visitors and staff. Walkways may be constructed where none existed, and the use of cars and perhaps buses must be addressed. In both cases automobile traffic should be kept a distance from the key elements of the site. Important views or features should be respected and maintained. It may be appropriate to construct unloading areas nearby, but parking and general vehicle circulation should be directed away. The use of plantings and grading to screen parking areas might be considered as well.

Parking areas should be designed to minimize their intrusion and impact on the site. Parking lots may influence the quality of the site visually, audibly, and environmentally. As addressed in previous chapters, there are a variety of methods for screening and mitigating all these effects. Vehicle circulation should be simple and smooth. The use of one-way circulation patterns and traffic calming strategies allows for narrower cartways, which in turn tends to reduce vehicle speed and noise. The narrower cartway also minimizes storm water runoff and maintenance issues.

Walkways should be planned to meet expected peak loads. Where narrow walks may be less intrusive, if the walk is underdesigned, damage from foot traffic may result. The impact of wider walks may be minimized by careful selection of paving or path materials. Wherever appropriate, the layout of walks should anticipate the "rounding" tendency of pedestrians and provide for natural and comfortable changes of direction rather than always using intersections at right angles. Circulation might benefit if walkways are slightly wider at intersections, rest areas, and points of interest where people might linger or slow down.

Accessibility on the historic landscape

Walkway accessibility is also an important consideration. Frequent rest areas and varying degrees of difficulty contribute to the site user's experience. Providing accessible routes in historic landscapes must also be considered in the design process. While it may not be necessary to provide accessible routes that would threaten or destroy the significance of the site, alternatives need to be considered. The determination as to whether an alteration would destroy the significance of the site or feature may require the advice of a local or state official. Americans with Disabilities Act (ADA) regulations require that at least one accessible route be provided. The regulation states a preference for a route that is used by the general public, but it allows for an alternative if no entrance used by the general public can comply with the accessibility requirements. An alternative access route must be open (unlocked) with signs indicating its location during hours of operation. The alternative location must also have a notification system or remote monitoring system. If the site has rest rooms, at least one accessible toilet must be provided along the accessible route. At a minimum, an accessible route should provide access to all of the publicly used spaces on the level of the entrance and access to all levels whenever practical. All displays, signs, and other graphic or written information should be positioned where they can be read or observed by a seated person—that is, no higher than 44 in above a floor surface.

Meeting environmental challenges in landscaping historic sites

Today the historic character of a community is just as likely to be defined by its former factories and mills as it is by the homes and estates of the former owners of those mills. With the expanded scope of the types and character of sites is a greater risk of encountering some form of environmental impairment. Among the challenges to be addressed in the preservation and rehabilitation of historic landscapes is the potential for environmental conditions created by past practices or activities. These conditions range from relatively simple issues such as difficult soil quality to an uncomplicated underground storage tank removal to more involved projects involving contaminated soils or active remediation activities. With the new focus on risk-based strategies for dealing with site environmental problems, the site plan may have to accommodate mitigation strategies into a final design.

Redevelopment of urban areas presents new design challenges and opportunities, but urban areas generally are host to a variety of confounding problems that are not present on new land, or green fields. In recent years brownfields have emerged as one new marketplace, and urban undercrowding is also an emerging challenge. The redevelopment of urban waterfronts has been an important area of practice for many landscape architects, but extending the success of the waterfront into the fabric of cities remains a challenge that clearly has a design component to the solution. Many urban landscapes have suffered from neglect or what has been called *deferred maintenance.* In general, urban historic landscape projects are one of four types:

1. Restoration or rehabilitation of existing urban open space such as a park, campus, or estate
2. Conversion, reuse, or redevelopment of existing buildings or features of a site from one use to another, such as conversion of former factories to a retail or residential use
3. Restoration of environmental function, such as urban stream restoration or urban forestry projects
4. Redevelopment of new urban spaces by razing existing structures to ground and beginning with new construction

Each of these has levels of concern that must be balanced throughout the design and construction process: the character of the site as it is and the risks to neighbors, occupants, or others; the risks to workers during construction; and the risks to end users. Urban redevelopment requires the design professional to have an understanding of the environmental issues on sites and the implications of design solutions. The historical context of sites in the process of a site analysis and an assessment of the environmental legacy of the past uses should be included in the typical site analysis. Some of the many tools available for this analysis are listed in Chap. 2. These are important sources of environmental information as well. For example, using the USGS quad map for an area over time can clearly demonstrate significant changes or land uses over time. These maps can show areas of fill or cut and the locations of buildings, roads, streams, and other surface features. Aerial photography is also an effective tool when used as a historical reference. Again the location of features such as buildings, storage areas, streams, former woodlots, tree rows, alleys, or other features obliterated in later development are often clearly visible in aerial photography dating back to the 1920s and 1930s. Most urban areas have been mapped using aerial photography at least several times since the 1940s.

Soil maps are valuable in terms of identifying the native soils in areas subsequently developed or the soils at the time the area was mapped. In urban areas soils maps assist in identifying former wetlands by soil type that did not exist at the time the wetland inventory maps were produced. Soil maps are also a source of historic aerial photography, though the scale and clarity is sometimes difficult to work with.

Local sources of information such as maps, business directories, photographs, libraries of historical societies, and living memory are often important. Though it is often necessary to find corroborating sources of information, sources of data such as newspaper stories, local letters to the editor, or travel letters that were popular in the late nineteenth and early twentieth centuries are important and, if not included in the site analysis, would otherwise be lost. These sources are often preserved in local historical societies or by the historic preservation agency in the county or state. These records are saved and catalogued, but often they are not used by designers in the assessment process. These collections should be used as they often include photographs, paintings, and even site maps and diaries that are invaluable in the assessment of a site. An interest in the environmental conditions of a site requires an additional level of effort and new sources of information during the assessment. If an environmental site assessment (ESA) has been performed, it should be viewed as an important source of information. The ESA should include a look at the historic conditions to determine the presence of any indicators of contamination or conditions that would result from contamination. Appendix A is a more detailed discussion of the ESA process.

While no list can cover all of the possible sources of contamination or activities of concern, there are certain rules of thumb that can assist the designer in making this assessment. Table 11.3 presents a short list of the general types of past activities that should raise concern. The presence of such an activity is not a definitive indication of environmental trouble, but it is an indicator that a closer look is warranted.

Tables 11.4 and 11.5 present lists of risks and contaminants that should be considered carefully by site designers and planners. The "Priority Pollutants" is a collection of materials used as a general screening tool. In the absence of specific knowledge of contamination, a priority pollutant scan or screening may be done. A knowledge of the types of contamination that could be reasonably anticipated on the urban historic site would assist the landscape architect in the assessment and design process as well as assist the client in determining effective strategies for addressing the mitigation concerns. The most probable contamination issues for a given area are best determined by the local history and industries. Among the more common issues found on the redevelopment site could include: underground storage tanks (USTs) and related petroleum pollution: asbestos-containing building materials (ACMs); lead and other metals; polychlorinated biphenyls (PCBs); polycyclic aromatic hydrocarbons (PAHs); and volatile and semivolatile organic compounds (VSOCs).

TABLE 11.3 Past Activities of Environmental Concern

Printing operations or photography processing
Dry cleaning operations
Service stations, body shops, fuel depots, and so on
Industrial operations, particularly plating, chemical manufacturing, metal refining, oil refining, equipment manufacturing, tanning, battery manufacturing, plastics manufacturing, machining, electrical equipment, dye shops, pre-1980 electrical substations, transformers, capacitors (regardless of size)
Old building materials
Industrial suppliers, maintenance shops
Solid waste collection, transportation, or disposal sites
Areas of fill, including "made land" on soils maps
Transportation-related activities, including rights-of-way and railways
Wood preservation operations
Boiler rooms, steam generation plants, coal gas plants
Nearby activities (particularly up gradient or upwind)

TABLE 11.4 A Summary of Priority Pollutants and Typical Sources

Pollutant	Typical sources	Risks and negative impacts
Asbestos	Building materials (see Table 11.5)	Asbestosis, carcinogen
Lead	Paint, building materials, fuel additives, old pesticides	Chronic health effects, may be found in soils and/or water
Heavy metals	Metal refining, fuel additives, industrial operations	Chronic negative health impacts
Hydrocarbons	Storage tanks, fuel storage, improper disposal	Carcinogenic
Solvents	Degreasing operations, metal manufacturing, painting, semiconductors, dying operations, printing, dry cleaners	Carcinogenic
Polychlorinated biphenyls	Pre-1980 transformers, capacitors, mastic, hydraulic oil	Suspected carcinogens
Semivolatile organic compounds	Chemical plants, coal tar by-products, insecticides, solvents	Possible carcinogens, skin irritants
Pesticides, herbicides	Farms, roadsides, railroad siding, golf courses	
Polycyclic aromatic hydrocarbons	By-products of incomplete combustion; coal plants, and so on.	Possible carcinogens

TABLE 11.5 Asbestos-Containing Building Materials

Material	Use	Period of use
Surfacing material	Sprayed or troweled on	1935–1970
Thermal insulation	Batts, blocks, pipe coverings	1949–1971
Textiles	Fire blankets, lab gloves/aprons, rope, fire cord, tubing, tape, curtains, building felt	1920–1981
Concretelike products	Flat, corrugated roof tiles, shingles, pipe	1930–1981
Roofing	Asphalt materials, caulking, mineral surface of roofing shingles	1910–1980
Asbestos-containing compounds	Joint compound, dry wall, wall board, wallpaper adhesive, mastic, tiles, plaster, putty, stucco, spackling	Unknown to 1978

Land Preservation

The conversion of productive and valuable farmland into residential projects and associated commercial development has been accomplished in many regions throughout the United States. In general, the loss of farmland is associated with changes of use to urban centers and transportation corridors that allow workers to travel to and from their jobs. A ring of growth—that is, the suburban growth—occurred on the fringes of most cities after World War II. Beginning in the 1970s and blooming during the 1980s, another surge of growth occurred. The earlier suburbs were tied to the original urban fringe because of an undeveloped highway network and a reliance on traditional means of transportation such as urban mass transit and walking. Without the means to quickly and conveniently travel, people chose to live near centers of shopping and work.

The second wave of suburban growth occurred largely on the strength of a developed network of highways and a reliance on the automobile. The development of regional shopping malls and suburban office parks in conjunction with well-developed highways provided the impetus for a new growth wave that spread much farther into the countryside than its post-World War II predecessor. Coupled with the significance of the numbers of baby-boomers, the development of the countryside outside the traditional urban fringe was unparalleled. The phenomena known as "edge cities," or "exurbs," were manifestations of this new ring of growth.

The preservation of farmland and the loss of farmland to development is largely a regional issue. In Iowa, farmland preservation programs are geared toward community zoning issues and providing *right-to-farm ordinances* and the like while still encouraging local development. In primarily rural states and regions, the preservation of farmland is less of an issue than the development of rural communities and local economies. The pressure to preserve farmland increases toward the large urban centers. In the fringe of the eastern urban band, traditionally rural areas have seen increasing development pressure as transportation and technology have enabled people to live farther from work and work farther from cities.

Between 1982 and 1992, the amount of developed land in the United States increased by nearly 18 percent, from 78.373 to 92.352 million acres. This increase was an unprecedented surge of expansion and prompted, among other things, the concern for our rural heritage and the preservation of farmland. In this same period the number of family farms decreased, and farm foreclosures increased dramatically. The amount of cropland in the United States dropped from 420.96 to 380.64 million acres, a decrease of nearly 20 percent. In spite of the decrease, food production continued to increase. The total developed portion of the continental United States in 1992 represented only 4.9 of its total land mass.

The losses of cropland included the conversion of cropland to other uses, such as commitment to the Conservation Reserve Program (31.8 million acres), or other agricultural activities such as pasture (14.8 million acres), forest (3.1 million acres), or range land (2.1 million acres). Of the 59.7 million acres taken out of crop production, 4.2 million were converted to developed land uses (Kellogg, TeSelle, and Goebel 1994). The amount of land developed from other agricultural activities in addition to the 4.16 million acres of cropland include 2.4 million acres of pasture, 2.1 million acres of range land, and 5.58 million acres of forest.

In 1988, the Council on Environmental Quality (CEQ) and various departments of the U.S. government jointly published the *National Agricultural Lands Study*. This comprehensive study looked at development trends and land conversion throughout the United States, and it reported various methods by which remaining agricultural land could be protected from development. The report was intended to serve as a resource to local officials in the effort to manage the development of agricultural land in their jurisdictions. Although the report has been criticized because of errors contained in the calculations that influenced the conclusions, the methods described in the report have been implemented in many places throughout the country. These methods were classified according to the approach by which they would affect farmland preservation.

The first approach was to reduce the attractiveness of agricultural areas for development through comprehensive regional planning and zoning. By identifying areas for growth and developing infrastructure in those areas, farmland would be less desirable for development. The second approach was to create programs that would reduce the burdens felt by farmers as development encroaches on them. The establishment of agricultural districts and right-to-farm legislation, coupled with tax relief for farmland and deferred taxation programs were among the methods used to reduce the influence of developers. Providing financial and estate planning to assist farm families with the management of their assets was also part of this effort.

The last methods were focused on preventing changes in land use through land use regulations, zoning restrictions, the purchase or transfer of development rights, and the use of restrictive covenants or easements. It is the effect of this class of methods on neighboring properties with which the research reported in this paper is concerned.

Preventing changes in use

Property rights in the United States have been described as a "bundle of sticks"—that is, an owner of property accrues rights, which are discrete from one another but are taken as a whole or a bundle. These rights would include the right to restrict access, the right to sell all or a portion of land, or the right to use one's land for some economic end. To decide on the disposition of land is one of the "sticks" in the bundle of sticks often used to describe the natural rights of landowners. The development of farmland can be prevented if an owner agrees to give up or modify his or her right to develop the land. The owner can decide to encumber his or her land with a restriction that will "run with the land"—that is, it will transfer with the title. By so doing, the landowner can reasonably assure that the use of the land by future owners will be constrained in accordance with the restriction while none of the other rights in the bundle are lost.

The difficulty of an institutional approach to encouraging landowners to give up this right is that the greatest asset held by many landowners is the value of the land itself. Most farm owners cannot afford to simply give up the value of the land since it is their largest asset. There are a wide variety of programs available to landowners across the United States to restrict the use of land in the future. These programs generally follow or are permutations of one of the following types of programs: transfer of development rights, restrictive covenants or conservation easements. The purpose of each of these is to preserve the farmland either as agricultural land or as open space.

The *transfer of development rights (TDR),* or the *purchase of development rights (PDR),* is perhaps the most important of the programs because it involves an exchange of comparable value. The general principle of TDR programs is that undeveloped land sold as farmland has a value something less than undeveloped land sold for development. The difference in these values is identified as the *development value of the land*—the price to purchase the development rights stick from the landowner's bundle. In actuality, there is no difference between the PDR and the TDR, but the terms are used exclusively. The TDR programs involve the transfer of development rights from an owner to another party. In some cases this transfer is made through zoning, growth control restrictions, or the outright purchase of these rights. In either case, an owner is paid for his or her rights, and the rights are transferred to another person or body.

The method of determining the value differs somewhat from program to program, but in the end the landowner receives cash for his or her right to develop the land. It is important to note that this right does not vanish but rather is transferred to another "owner"; the right to develop the property belongs to someone else, usually the state or county. In some programs the rights are merely leased for a specified period of time, and in others the rights are sold outright.

Restrictive covenants are a simpler approach to preservation. A landowner may place a restriction on his or her land at any time and cause that restriction to "run with the land." In a case in which a restriction is placed on land to restrict a future owner's use, the right remains with the land but it is encumbered. With no other party involved other than the owner, these restrictions are unenforceable by any third party. If the restrictive covenant is shared because of subdivision or multiple ownership, the restriction is enforceable by any of the other parties that have standing.

The last major type of program is a *conservation easement.* In this case, a landowner assigns an easement to a third party. The conditions of the easement require that activity on the land is usually restricted to exclude development. The easement is enforceable by the third party in protecting its interest. The easement usually allows some manner for the third party to confirm that the landowner complies with the easement agreement. The easement is usually given to a private nonprofit organization and involves some cash payment on the part of the landowner to support the organization's efforts. The easement does not restrict the owner's use in any other way and does not provide the easement holder with a right of use, although some limited right of entry is usually included to allow the third party to confirm compliance. In general, the only financial inducement of landowners to participate in such an arrangement is certain tax advantages available to them though some limited cash-for-conservation programs.

In each of these cases the landowner retains all of his or her rights as an owner except for the development rights. The land may be sold, and the restriction passes with the title. The owner may use the land in any manner other than to develop it. Each of these approaches is intended to work best in conjunction with other programs and growth controls, although in practice there is a wide variability in the administration of the programs and controls.

The *preservation of isolated farmland* refers to the practice of restricting development on farms that are outside of agricultural zoning districts or are not contiguous to other preserved farms. In these cases the land adjacent to the preserved area is still subject to development. The CEQ study is interested in the impact of preservation on these adjacent lands.

The impact of growth controls on land use and value has been extensively studied. The role of agricultural zoning is significant in the preservation process for several reasons. The most obvious, and most important, of these is that agricultural zoning would reduce development pressure on farms by restricting development. Second, agricultural zoning acts to reduce the value of farmland. It is popularly believed that agricultural preservation raises the value of farms because it eliminates the "interferences" of development, such as complaints from nonfarming neighbors. In fact, Vaillancourt and Monty (1985) discovered in a review of more than 1200 land sales that land in agricultural preservation zones lost between 15 and 30 percent of their value compared to the sales of nonrestricted farmland. In this way agricultural zoning would also help government TDR programs by reducing the development value of properties and therefore reducing the cost of purchasing development rights. In practice, it is not clear that this effect is factored into the price being paid to landowners for development rights in areas with agricultural zoning. This may represent something of a windfall for landowners and a programmatic inefficiency in the purchase of development rights.

Similar results were observed in land sold outside the outer growth boundary of Portland. The Oregon land plan called for a 20-year delay in approving development beyond a specific growth boundary. This delay was perceived in the marketplace as a binding constraint on the development of the land and significantly influenced land sales (Knaap 1985). Some studies indicate these restrictions may also raise the value of unrestricted properties, and in particular the prices of single-family dwellings.

The value and effectiveness of growth controls are the subject of wide-ranging and heated debates throughout the United States. While the scope of that discussion is well beyond this study, agricultural preservation in general is well received by the public and landowners, although for different reasons. Farmers and landowners resist farmland preservation to the extent that zoning and other growth controls are seen to reduce the market value of their assets. Purchase of development rights programs address that concern by paying for the privilege of transferring the private landowner's right to develop his or her land to the public. Ideally, there is no loss of value in this arrangement. Inducing landowners to voluntarily restrict their property without compensation can be more difficult. Such landowners often incur significant expense to preserve their land and are motivated by more altruistic purposes. The tax incentives and other programs do not offset the costs and reduction in asset value.

Conservation easements and similar approaches are the tools of private organizations. Public programs usually use cash buyouts or leases of development rights, or they try to restrict development through growth controls. Private organizations, such as land trusts and conservancies, approach owners of properties that are not eligible for public programs. In this way, landowners who are interested in preserving the farmland or open space, whether or not payment is involved, have a means of doing so and the public's interests are served. These are farms that did not score high enough to be considered for the purchase of development rights because they are not prime agricultural properties or because they do not fit in the profile of a sustainable agricultural operation. The latter might be the result of proximity to existing development or infrastructure, or a zoning district that encourages development. Private groups are under no restriction or imposed guidelines as to the properties on which they accept easements.

The effect of preservation on local revenue

The American Farmland Trust (AFT) and the Lincoln Institute of Land Policy have both recently looked at the impact of preservation on local revenue generated from land taxes. Communities often look toward the development of open land into commercial or residential uses as a means of increasing their tax base. The American Farmland Trust conducted a study on the *cost of community services* (COCS) in the municipalities of Agwam, Deerfield, and Gill in Massachusetts to assess the impact of preservation (AFT 1992). The COCS is a study technique that is used by communities to assess financial impacts of various land uses, assuming current standards of infrastructure and services are applied. The purpose of the COCS is to compare income to expense for different land uses.

The AFT study found that while farmland did not contribute as significant a return in taxes as the same area did as a developed parcel, the farmland also required fewer services. For example, the study reported that a community can spend as much as $1.15 in services for every new dollar in additional revenue received from residential development. On the other hand, farms were found to require only $0.33 in services for every dollar paid in taxes; a net gain, according to the study (AFT 1992). Although the developed community functions as a whole, such commercial and residential properties are inexorably linked through different uses. The study considered them as distinct and separate functions. The study did not consider the cost of a community but rather specific elements within the community; perhaps there is a balancing of costs between the uses that the COCS method does not identify. The study did reveal that the cost of commercial or industrial development was nearly as significant a gain as farmland (Lincoln Institute of Land Policy 1993).

The effect of preservation on growth controls

Mill (1990) describes zoning as a means of assigning development rights to a community; the right to develop land is a collective right of the community that a potential developer must purchase or be given. This approach essentially circumscribes the Lockean theory of property rights as a "bundle of sticks" and natural rights, but it may have some value in explaining actual behavior. Essentially, under Mills's theory, a landowner must have the agreement of the community before developing his or her land. This approval occurs in the form of plan approval and the payment of fees. In effect, the zoning takes a right from the bundle of sticks and then returns it to the developer once the developer agrees to the community's conditions as to how it will be used. This restriction on the use of the land serves to devalue the land to some degree.

The adoption of agricultural preservation districts is an adaptation of the classic *Euclidian zoning model* of establishing zones or districts in which specified activities or uses are permitted. By applying the urban formula for development to rural communities, these communities provided the framework for the urban sprawl. Most studies in land use and development behavior use models that assume a certain efficiency in the marketplace, and that efficiency is usually based on the *urban center model*. In this case *efficiency* refers to the prudent and effective application of the zoning ordinances to apply the "highest and best use" to land, usually taken to mean the most profitable. The urban center model assumes that there is a central business district and that land uses are driven from the efficiencies gained by proximity to that urban center. In the simplest form, the highest value and most efficient activities are located closest to the center, and the least valuable activities are located progressively farther away. Nearly every study on zoning and land value assumes this model to be representative. However, the model is clearly not representative of what our experience tells us is true.

Current development trends reflect our reliance and preference for the automobile. The automobile and communications technology have eliminated the central business district as a factor. Physical proximity is no longer a primary issue to our culture; it has been supplanted by time. The development of highway systems, originally conceived after World War II as a national defense system, allowed us to travel farther and faster when and where we wanted. The development of networked computers and faxes has accelerated this ability. The need for a central business district is disappearing at an accelerated rate. Public policy and development planners cling to the old center-city planning models and struggle to find the prescriptions to save "downtown." In fact, planners have little influence over growth and the distribution of development in Pennsylvania or in the United States in general. Studies show that the restriction of development does not direct development into the urban center but into the edge cities. Developers are motivated by the ability of a project to pay for itself, and their location decisions reflect the desires of the marketplace rather than the public planning models. The growth of King of Prussia, and hundreds of similar centers, does not reflect a model of planning; rather, it reflects the market forces.

Traditional studies also have limitations when applied to rural circumstances; however, there are general trends and behaviors that have relevance. The value of agricultural land does reflect its development potential and expectations. The determination of the purchase price of development rights is essentially one manner of accounting for this value. Any influence, particularly zoning, that restricts development on a parcel of land in fact lowers the market value of the affected parcel but concomitantly raises the market value of adjacent parcels (Hennebery and Barrows 1990). In one of the few studies concerned only with agricultural issues, Hennebery and Barrows found that agricultural zoning had both positive and negative effects on land prices. Among the measures they addressed were the externalities of agricultural zoning. *Externalities* are the effects of growth control or land use restrictions on adjacent parcels of land. Their study found that agricultural zoning decreased the price of land in the district but that the values of large parcels of land were only nominally affected. They believe this finding reflects the value in removing development pressure from farmland; farmers are willing to pay more for large parcels that are not threatened by nearby development.

Many studies have considered the effect of growth controls on land and housing values. The externalities of zoning and growth controls are well documented. While growth controls may limit growth and appear to be a positive effect, the spillover effect of restricted growth is often observed where growth is simply squeezed into another area. This has been described as a *balloon*; if you squeeze it in one place, it simply pushes out somewhere else. The growth restriction will lower undeveloped property values and raise developed property values. In effect, growth controls are a windfall for current homeowners. Likewise, it will raise the value of undeveloped lands adjacent to the controlled area (Pollakowski and Wachter 1990).

William Fischel (1990) conducted a review of the literature on the subject and draws three conclusions from his study:

1. Land use controls—in particular, controls that attempt to guide overall growth in an area—are an important constraint in the marketplace.
2. Land use controls do provide some benefits to the public that no other mechanism available can provide.
3. The cost of these controls is probably greater than the local benefits they provide.

In effect, Fischel observes that while growth controls impose a greater cost, in terms of reduced land values and increased housing costs, than they do a benefit, there is no other mechanism available with which the public can pursue these ends.

Another notable element of Fischel's study is that the power to create these controls generally rests with the people that will benefit the most from higher housing prices. As growth controls are implemented, it is the price of existing houses that rises and the value of undeveloped land that falls. As rural areas become developed, eventually a critical mass is reached, and the local power shifts from the traditional owners of large tracts of land to the more numerous owners of smaller tracts. The newcomers wish to "protect" the rural character of their community and are willing to do so at the expense of owners of undeveloped land.

The Effect of Land Uses on Property Prices

There are numerous studies that review the effect of a specific land use on land or housing prices, but the results of these studies are inconsistent and divergent. The lack of concurrence is a result of the various models used. There is debate in the literature as to what the elements of the model should be. Without any consistency, it is difficult to draw any conclusions from these studies (Pogodzinski and Sass 1990). In general, these studies are concerned with the *disamenity effect* of presumably undesirable land uses on neighboring properties. The most notable result of these studies seems to be the unexpected minimal effect of undesirable land uses.

These studies attempt to identify the *disamenity margin*—that is, the distance from an undesirable land use where the effect of the land use on land prices is zero. Studies have looked at power plants, nuclear power plants, group homes, solid waste landfills, strip mines, and industrial facilities. The results have often been surprising: Undesirable land uses often had a very small effect on housing prices and sales. Two independent studies reported that the Three Mile Island nuclear power plant did not have a statistically significant effect on property values. Even after the accident, property prices of homes sold nearer the plant did not vary significantly from homes farther away, nor were the frequencies of sales notably different (Gambe, Nelson). Even solid waste landfills had a more limited disamenity margin than anticipated. Shortle et al. (n.d) found that well-managed landfills had only a very weak effect on local property values. A correlation was noted between the volume of waste accepted and the effect on prices. Landfills accepting less than 300 tons per day were found to have nearly no effect on housing prices while landfills accepting more than 500 tons per day were found to demonstrate a weak effect on housing prices (Shortle et al. n.d.). The effects of a given land use on adjacent lands is not always an intuitively obvious effect. Facilities feared to be a disamenity may have a negligible or even positive effect. The presence of group homes in a lower-socioeconomic neighborhood was found to actually favorably influence housing prices. An *amenity effect* was discovered that attenuated over distance (Gelman et al 1989.) Clearly these results indicate the possibility of the unintended consequences of land uses.

The literature of professional journals of real estate appraisers, planners, design professionals, and real estate developers failed to identify more than a few studies and references on the subject. These references were primarily in the form of questions from the professional society to its members rather than actual studies. An exception was a study reported in 1968 regarding the effect of scenic easements on property. This study found that properties encumbered with a scenic easement lost value (Williams and Davis 1968). This loss was believed to be the direct result of the owner's loss of his or her right to use his or her land unencumbered. The study indicates that these losses were limited to the portion of the land affected by the easement. The unencumbered portion was not observed to suffer a loss. The study did not explore if the unencumbered portion of the land may have increased in value as a result of the easement. Coughlin (1991) reported that while the use of agricultural zoning programs to protect farmland was a departure from the general use of zoning as a guide for development, agricultural zoning was more effective if used in conjunction with other programs such as TDR. Agricultural zoning, purchase or transfer of development rights, and an agricultural infrastructure are all different but complementary elements to successful farmland preservation. Coughlin's (1991) criteria for evaluating the effectiveness of agricultural zoning did not extend to the consideration of externalities or spillover effects.

The Lincoln Institute of Land Policy published Jay E. Closser's *Assessing Land Affected by Conservation Easements, A Resource Manual* in 1993 to be used in conjunction with a training seminar for real estate professionals. The manual observes that sales of land abutting "conservation land" appreciate more quickly than other property. While the land under conservation is reduced in value by as much as 90 percent, the adjacent land will increase in value faster and greater. The manual does not offer a method of evaluating this increase in value.

The American Society of Appraisers (ASA) has also observed, and recommended to its members, that the preservation of land could have "considerable value to an abutter for the additional privacy and seclusion it would afford" (Czupyrna 1993). To date the ASA does not offer specific guidance or models for making that determination. Federal tax law does consider the positive impact on the value of neighboring properties. The tax code requires that any deduction taken for a conservation easement must be reduced by the amount of increase in value experienced on property owned by the donor or a relative to the donor (Small). The tax law includes what is known as the "enhancement rule," which is described through the use of examples. The text in the code states, "By perpetually restricting development on this portion of the land, [the landowner] has ensured that the two remaining acres will always be bordered by park land, thus increasing their fair market value" (Lincoln Land Institute 1993).

There was significant agreement among those surveyed that the presence of open space or preserved farmland would be expected to raise the value of adjoining properties if the adjoining property were to be used for residential uses (Russ 1995). There was also agreement that open space would have greater value to adjoiners than would preserved farmland. The presence of open space or preserved lands is viewed as an amenity to potential home buyers for which they would be willing to pay a premium, but it would have little if any value in commercial, industrial, or public properties. There might be some additional value of land adjacent to preserved farmland to farmers as well since the presence of the preserved parcel ensures that the land will remain in agriculture use indefinitely. This assurance offsets the "impermanence" concern farmers must weigh in purchasing land or improving leased land. The *impermanence factor* refers to the risk of a parcel of land next to or near a farm being converted to residential or other nonagricultural uses.

The fact of the matter is that preserved land, whether open space or farmland, may serve to attract development because of the amenity of the view shed and the certainty of future land use. The value of the amenity, however, is insufficient to attract development without other factors such as location of the parcel, access, or available water and sewage disposal (on site or public). The presence of agricultural security areas and agricultural zoning would also serve to discourage development. Based on the survey results, it can be expected that the preservation farmland or open space will increase the value of adjacent land that could otherwise be developed. Further, this value would be based on the amenity value of the permanent open space or farmland, and home buyers would be willing to pay some premium for permanent visual access to the preserved land.

Chapter

12

Landscape and Culture

Land is important as a manifestation of power and wealth, but the landscape is more than simply the space itself. It is perceived emotionally, as a source of beauty, and of solace, as well as economic well-being. Land and the landscape are also a system—a series of intertwined related flows of energy and materials upon which we depend in the most fundamental ways. In most discussion, the concept of landscape is confined to areas of gardens or the plants and constructed surface features in a place, but this definition clearly limits the landscape to something superficial—only those places that reflect the direct, intentional effects of human modification. This idea of the landscape as a reflection of positive human influence is an old one.

Landscape is more than just space. For example, it is four dimensional, it has history, and it is time that is the measure of all landscapes. The common view of landscape is of an unchanging constant. Unlike hardscape elements of design and the built environment, the biotic elements of a landscape are dynamic, and they grow and mature. Olmsted once observed that it is the vision to see a landscape in its future mature state that will occur long after the designer has passed that is the genius of the profession of landscape architecture.

Landscapes are also the media of cultural experience. Landscapes are altered by societies in many ways. Impacts of humans on the environment occurred even before the first humans chose to change their environment. Cultures and societies subsequently made huge impacts—a 4000-year-old canal system in China, Mayan irrigation systems, Syrian and Roman water systems. It is accepted fact that Native Americans periodically burned woodlands and prairie to manage the environment. Roman aqueducts that conveyed cool mountain waters to the urban center changed the landscape in appearance and made Rome a different kind of city. The damming of American rivers bears witness to the will and attitudes of our society as a culture and as political will. The current environmental restoration efforts also speak of our attitudes. These changes reflect the qualities and values of a culture that is judged best in historical contexts with the clarity of time.

As professionals, we know that all of our expressions of landscape are simplifications of the natural world. The natural landscape is complex and diverse. We cannot perceive the complexity of the landscape, nor can we duplicate it in design, so we direct our attention to what is important to us by eliminating from consideration facets of the landscape we consider unimportant or peripheral to our interests. In this paradigm we assign pieces of the environment values based on our interests: We keep items of high value to us, perhaps even enhance them, and we exclude or ignore—sometimes we don't even see—items of low value to us.

The low value or expectation of environmental and landscape quality can be measured in the lack of visibility low-quality circumstances have. Many complain about the boring sameness of the modern built landscape, perceiving that one place is all but indistinguishable from the next place. But the universality of the modern landscape, the loss of local significance, seems to be practically invisible to most people. Degraded urban streams are invisible to most city dwellers. John B. Jackson (1986) wrote of Pausanias' observations of the Greek countryside in which he reported only the human-made elements he saw. Jackson referred to this as "political observation," saying that Pausanias "did this because these were the only objects with what to him was visibility." As it is, most of us in the modern society have no direct frame of reference to nature or the environment. At best we see nature through other activities or interests, and we value it according to those interests. Hikers and snowmobilers both value trails but have very different frames of reference. For many, possibly most, nature is observed vicariously, primarily through the media such as magazines and television and possibly billboards (Fig. 12.1).

To the degree that the land is influenced by people, it is a landscape distinct from wilderness, and perhaps, the best definition of *landscape* is eventually found in this human aspect of land. Influence can be intelligent—that is, by design—or it can be unintended. In either case, land that is influenced by human activity is the landscape. It has been estimated that 60 percent of the world is "managed landscape"—that is, that it is managed because the decisions of people have the greatest day-to-day systemic influence. We overcome aridity with irrigation, chemically enhance soils for greater fertility, change the shape of land to facilitate our activities, remove forests for suburbs, and divert runoff from soils and fertilize our surface waters, just to offer a few examples. Impacts are not new, but while humans have been impacting on the environment for thousands of years, it is the character and scope of the modern industrial age impact that is harmful. This has been called "deficit spending in the environment"—the degradation of natural systems, air, water, soil, species diversity, and so on, is usually externalized from the market costs of products and services. We measure our competence by how fast we move information and transfer energy. Most Americans—as many as 76 percent by some accounts—consider themselves "environmentalists," but the average American spends 3 percent of his or her time out of doors.

Figure 12.1 Photograph of a billboard.

Landscape values are learned behaviors, passed from generation to generation, usually by example. Professional designers usually produce works that are based on a cultural ideal or a refinement of traditional values. Exceptions are noteworthy primarily in their differences from the traditional models or their expanded views of the landscape. Often the degree of change is quite small and can be seen as incremental evolutionary steps toward new values, new cultural norms.

Settlers of America attempted to bring what they knew with them as they moved to new places. Within 20 years of landing in Massachusetts, early English settlers were already abandoning the familiar central construct of the "ideal town" for outlying areas. Single-family dwellings were developed without regard for "political" designs or layout. These homesteads were instead built with an orientation to nature, along contours, relevant to streams, and so on. Clergy and village leaders discouraged "outliving" as forsaking God and church and civilization. Within four decades of the original planned towns, plans were abandoned and development began to reflect the abundance of cheap land and rich agricultural opportunity. The old world culture adapted to the new world conditions. The abundance of space in the new world skewed traditional values and roles. Few professional or educated designers worked in early America. Land was developed relating to the landscape, economy, and a practical aesthetic. James Stilgoe observed that the archetypal landscape and land use "objectified common sense, not the doubtful innovation of professional designers" (Stilgoe 1982). In 1875 a survey of six counties in New York State reported that about 21 percent of the total cover was timberland. In 1990 a survey revealed that more than three times that area was now in timberland, more than 1.8 million acres. Words changed, meanings changed, and new social roles were created. The New England idyll of a village was transplanted, with varying degrees of success, across the entire United States. Even today, architect Andres Duany and his innovative "New Town" is an admitted return to these traditional ideals. Perhaps the appeal of this traditional village is its image of hosting a simpler way of life.

Economists, on the other hand, tend to view the landscape as a universal flat plain on which features are obliterated and only markets matter. Nature itself, beyond its resources or effects on resources, is removed from consideration because it is not valued. Simple surface features such as water expressions or topography may be considered only to account for patterns of distribution. Resources such as minerals or materials are included but only as they regard markets and distribution of wealth. Economic principles are guided on the assumption of "free goods"—that is, natural resources that have no inherent cost. Only the costs of extraction or acquisition and distribution are used to determine market value. In this way economies have been subsidized by nature; yet few economic models have ever included the negative impacts of the markets on the landscape. The negative impacts of exploiting natural resources are distributed in the marketplace without regard to benefits—that is, the costs of exploitation are often not borne by those who benefit from the exploitation. The absence of environmental values in the economic system is both a reflection of larger cultural values and in part, the cause of them. The impacts of degraded environments and damaged landscapes are always felt locally, but the benefits of the destruction may be distributed far away. Forests are dying; most of the river pollution in the United States is from nonpoint sources, agriculture, and urban runoff; we are mining prehistoric groundwater; and we are disturbing basic life systems, the atmosphere, water quality systems, soil systems. These impacts are the externalities of our economic behavior—costs that are not calculated into the market prices of products and services.

The Use of Land

The Use of Land is the final report of a task force created by the Citizens' Advisory Committee on Environmental Quality (Reilly 1973). The citizens' advisory committee was created in 1969 by a presidential executive order. The mission of the task force was to investigate the laws, policies, and trends that existed at the time, to characterize the effectiveness and impacts of these public actions on past and current land use, and finally to make recommendations as to how land use might be better regulated. The task force undertook its mission with a presumption that land use trends evident at the time of the study were unsatisfactory and unsustainable.

The report observes that "unrestrained, piecemeal urbanization—supported by a value system that has traditionally equated growth with the good life—

has produced too many dreary, environmentally destructive suburbs, . . . too many bland indistinct city centers; extensive mismanagement of the earth's resources and rising popular discontent." The task force studied how the process of development might be organized, controlled, and coordinated to protect the environment as well as the cultural and esthetic characteristics of land while still balancing the needs for housing, manufacturing, public access to open space, and infrastructure. The investigators recognized that the laws and institutions, as well as the national psyche, had begun and mature at a time when development was encouraged.

The seemingly limitless expanse of colonial America resulted in a legal bias in favor of development. American landowners have an expectation that they might use and develop their property as they deem fit, even at the expense of the environment or the interests of the public. The task force came to the conclusion that the existing laws and public policy frameworks are inadequate for effective planning and control of development—if only because of the pro-development prejudice built into our legal doctrines. The conclusion is qualified, however, by the task force's belief that "with property rights go obligations that society can define and property owners should respect" (Reilly 1973).

This statement would appear too simplistic a response to such a complex issue, but the report goes on to provide numerous instances in which landowners' uses of land have been successfully restricted by law or policy. In fact, the authors devoted an entire chapter to that subject: "Adapting Old Laws to New Values." A variety of case law is reviewed in a case study format to illustrate how environmental protection and the interests of the public have been asserted over the rights of the individual landowner. The importance of state laws and constitutions that do not require compensation for "takings" is discussed. The wide variance in state and federal approaches to the regulation of land complicates the issue even further. In spite of the differences, there are many examples of land use restrictions that have been sustained in case law as well as common law. These include restrictions on nuisance—activities that clearly endanger the public health and welfare—and the right of eminent domain, and the use of zoning ordinances to regulate the use of land for the benefit of the public.

The greatest value of the task force's report might be in its exploration of the various methods and approaches to how development might be controlled and coordinated. Beginning with the obvious need to restrict development of lands that serve some sort of valuable function (groundwater recharge areas, particular wildlife habitats, steep slopes that might be unstable or subject to erosion, and so on), the report identifies a hierarchy of sorts for identifying areas that deserve protection from development. After the environmentally sensitive areas, areas required for future public recreation and areas that would serve as buffer zones between urban areas are seen as most important. Unique or highly productive farmland is the last of the identified important areas.

The methods identified for establishing protection for these areas included requiring dedication of open space by developers through the use of density bonuses and open-space ordinances, outright purchases of property by government or private entities, requiring environmental impact statements for all "large" developments, and encouraging the use of open-space easements and transfers of development rights. In the time that has elapsed since the report was published, requiring the dedication of open space (or cash in lieu of land) and the use of conservation easements have become relatively commonplace. Density bonuses and transfers of development rights have been used extensively but with mixed results.

The report goes on to make a series of recommendations that encompass activities by various levels of government, and private nonprofit organizations, and private individuals. The essence of these recommendations are the following:

1. A national land use policy would serve the planning and development community in a manner similar to the laws and policies that have addressed air and water quality issues. The task force saw the National Environmental Policy Act (NEPA) as the proper root for the development of this policy. A national policy or legislation would link federal funds to land use actions and regulatory programs at the state and local levels.
2. Restructuring tax policy at the state and local levels would encourage effective planning and quality development.
3. Noneconomic incentives would promote stewardship such as the granting of open-space easements or land trust creations among private parties.
4. Innovative planning considerations such as density bonuses or density transfers would encourage the clustering of housing and commercial development and the protection of larger tracts of open space.
5. Land use education of the public, public officials, and institutions would focus on the important links between development, land use, environment, and quality of life. The task force observed that public policy and the opinions of lawmakers and the courts often follow public opinion; therefore, an educated public will "lead" the lawmakers to new interpretations of the Constitution and legal doctrine or encourage the creation of new law.
6. Continue to urge the participation of citizen-planners in the land use planning process, and encourage the involvement of neighborhood groups as well.
7. Encourage the broader use of environmental impact statements as part of the development and planning process.

The value of *The Use of Land* as a contemporary reading is its role as a benchmark. The report is a fairly in-depth review of land use trends and planning policy as it had developed up to 1973. From the vantage point of 2002, we may evaluate the impact of the report and the effectiveness of its recommendations. For example, the task force clearly identified the creation of a national land use policy as a priority, but no such policy has been developed. In fact, since the report was published, a policy of the Reagan administration encouraged the sale and development of publicly owned lands. This policy was carried out without the benefit of any development guidelines or restrictions on the buyers.

In spite of the lack of national leadership on these issues, significant grassroots—albeit often unorganized, unsophisticated, and underfunded—efforts have sprung up all over the country. These local efforts have influenced the actions of elected officials and caused the development process to include consideration of the negative impacts on the community. The creation of laws that allow for the transfer of development rights, conservation easements, and preservation of farmland and sensitive areas are largely due to the efforts of private groups. Public policy and regulations have followed the interests and direction of the organized "environmental movement." Whether by design or by accident, the environmental and preservation interests have implemented through private funding or caused public action to implement, in varying degrees and success, aspects or parts of all of the remaining recommendations. The task force was unable to identify any direct method or means by which the federal law could restrict landowners beyond the limits described by the Constitution and defined by the Supreme Court. The recommendations rely on guidelines and compelling landowner responsibility as a reaction to financial incentive and public opinion.

Public Land and Private Land

Private property is among the most fundamental principles of a free nation. The rights of the landowner are important considerations in any action that affects private property. Public lands, on the other hand, are viewed quite differently. Although important distinctions are supposed to exist between public and private organizations, many of them begin to blur under examination. In spite of at-large perceptions, public and private landowners may be more alike than they are different. This is especially true as the size and scope of organizations increase. Public and private land uses share the same objectives of management, and in this way the similarities of the land use—its management and stewardship—may ultimately be more important than the differences that exist between specific landowners and land users. The private and public sectors are constantly interacting; there is no private-sector organization that is not touched in some way by an element or extension of government. These interactions extend into the use of land and land ownership. In the United States, it is private individuals and businesses that most regularly use public lands.

By way of a comparison, these relationships are fundamentally the same in Canada and the United States, but in Canada the government's role tends to be more direct and centralized. The Canadian government is an owner or partner in many business organizations. In the United States government-private partnerships exist but usually in a more indirect fashion in the form of financial or other material aid rather than actual ownership. Despite the indirect nature of the relationship, however, the goods and services of the federal and state governments in the United States account for about 25 percent of the U.S. gross national product (GNP) and so management decisions within the government have an impact on the private sector.

The private sectors in both countries are affected by the activities of the government regulations and policies, adopting what might be viewed as a "public component" into the behavior of the private organization. Landowners and operators in both countries adapt to social demands, or public expectations, to comply with or to avoid new regulations or the expropriation of rights or use of the land. Interaction between the two sectors is increasing, and often each can resemble the other in practice. Governments are judged more often on their ability to respond to constituents and are often held to almost businesslike standards of performance. Private landowners and users are held more often to a measure of public accountability for their activities and stewardship.

Ownership of land is a limited concept. Private property in both countries is based on the English common law and feudal doctrine of "fee simple." Under this concept, land is held in ownership at the sufferance of the government. Landowners are subject to the payment of taxes, obligations of use, and "eminent domain." In the United States landowners are protected to some degree by the Fifth Amendment to the Constitution, which requires due process and compensation before the government may "take" public property. In Canada the rights of landowners are protected by statute only, which is subject to change. The rights and extent of ownership are complicated further by the fact that ownership and interests in the land may be separate. The separation may be by the choice of the owner (as in the case of a lease, in which the owner assigns use to a lessee) or by law. An example of the latter would be wildlife that are held in trust by the government for the people. Under normal circumstances a landowner may not have the right to take wildlife without the permission of the government. Mineral rights are examples of an interest in land held separate from title, as are water rights.

The patterns of ownership in Canada and the United States are different in several ways. The land in the United States is 58 percent privately held whereas in Canada the government (federal and provincial) owns 90 percent of the land. A more important and fundamental difference exists in that Canadian property owners do not enjoy vested property rights. The government retains property rights despite the transfer of title. This important difference provides the government with greater latitude in developing and implementing policy. In the United States, the Fifth and Fourteenth Amendments assign and guarantee property rights and access to due process to private landowners. In light of these fundamental differences, it is surprising that the policies and results are not more divergent between the two countries.

Government control is more fragmented in the United States than it is in Canada as well. This fragmentation is considered to significantly inhibit successful management and planning of metropolitan areas or regions. Political economy in the United States and the popular commitment to local autonomy precludes any significant changes or reforms to this situation. Comparisons of planning and managing land use and development in selected American and Canadian cities reveal that in spite of the basic differences in ideologic form, planning theories and approaches to development are similar. After World War II the public planning process emerged in a leadership role, directing economic development in the established city centers in both countries. This new role occurred in response to the vacuum left by private enterprise during the war years. Prior to this time economic development had been almost exclusively a private-sector undertaking. With the rise of public planning came the involvement of the public in the planning process. Citizens and neighborhoods became part of the process of development and land use (Feldman and Goldeberg).

The highly centralized form of decision making in Canada did not allow for the process to be changed as easily as in the United States. Although the tools and instruments of policy are similar, the processes of implementing the tools remain fundamentally different. The control of growth and land use is managed by government through the use of zoning laws and the extension of infrastructure—most important of which are sewage, water, and transportation. In the United States, the presence of multiple municipalities often with conflicting interests and objectives has served to limit the ability of regional planning to constrain urban sprawl. Development on the edges of cities is beyond the control of those cities in most cases. Cities and those wishing to influence growth have a limited palette of choices with which to control land use. The first and most effective way is to buy land. By purchasing land, the government, as land owner, has the rights guaranteed by the amendments to the Constitution and can control the use of land. The second way to control and influence growth and development is through the political planning and development approval process.

In Canada, the process tends to be less fragmented. Government is more centralized, and decision making is less politicized. Although attempts to measure planning and land management success are at best of limited value, Canadian cities seem to be considered more "livable" than their American counterparts by their residents, but the Canadian results are mixed. The management of Canadian forests and rural land development are not the equals of their urban counterparts. The uncertain results of the planning and management processes are clear indications that the public's interests may not be best served by government's management and control of the process as it stands in either country. Answering the questions of who should decide how land is used and what community is to be considered (a very localized view of community or a more regional view) requires careful review. The consideration of a landowner's rights and the rights of the public must also be balanced.

The study of the attempts to control urban sprawl and development in the United States and Canada reveals that there are as many similarities as there differences between the two countries. In both countries growth and development are controlled more through political maneuvering and through controlling the approval process than through a deliberate planning and management process. The areas of greatest concern tend to be the areas of weakest control—points of municipal boundaries and edges of growth areas or transition zones.

Recent experiences with siting hazardous waste disposal facilities can serve as examples of how the planning process might be used and adapted. In comparing six case studies of sting projects, the common elements of the two successful projects were the "package" the project came in. In these cases the disposal facility was only an element—albeit the central element—of a larger package of development including infrastructure and industrial and business development, which had value to the community. The "enlightened" self-interest of communities overcame the fractured nature of the process and the organized objections to the projects.

In spite of ideological and constitutional differences between the United States and Canada, the similarities, especially in outcomes, appear to be greater than the differences in the two countries. The differences between public and private enterprise are also blurred across international lines. The process of land development and control may be rooted differently, and the processes may differ in context and form, but the outcomes share similar problems and results. The study of successes and failures indicates that different planning and control systems may be appropriate for different regions and cultural areas. The tradeoffs made to satisfy one set of political conditions will differ from the tradeoffs made in another. Political influences on the use and management of land are greater than the processes intend, or imply, and ultimately it is the political process that dictates much of the actual control.

Growth Controls

Between 1982 and 1992 the amount of developed land in the United States increased by nearly 18 percent, but at the end of this period the total developed portion of the country remained less than 5 percent. Culturally, North Americans have conflicting views of land and property. On one hand, we celebrate and place a high value on natural beauty and the integrity of the environment, and on the other hand, we tend to view land as "property" or as a commodity. These conflicting aspects are built in to the system of land use approval and regulation. Although there is a great body of law, public policy, and regulations that is meant to govern the use of land and particularly development, under the United States Constitution, the landowner-developer has generally been given the benefit of the doubt in the process. Government tends not to lead on issues of land use but seems content to follow, reacting to events and trends in the marketplace. While planners, economic development and redevelopment authorities, and federal and state programs all attempt to direct development, ultimately builders respond to their own perception of the marketplace and challenge the public planning effort to meet their needs.

Throughout the literature the message is the same: the final authorities in determining the nature of land use are the community and the consumer. Political influences on the use and management of land are greater than the formal process recognizes or intends. Case study after case study shows the power of public interest in determining the short-term decisions governing land use and the long-term influence of the marketplace in defining "public interest." Examples of successful government-directed development are not common. The lesson appears to be that government cannot compel a project to be successful; ultimately government projects must meet the same tests as private projects to be successful. Housing developments to replace substandard housing fail because they are located in undesirable sections of the city. No private developer would build there because *no one* wants to live there.

Communities are motivated by the interests of the individuals that make them up. The interests of the majority often are in conflict with public interests, and in these cases the interests of the majority tend to prevail. In the Mount Laurel, New Jersey, cases, the Court allowed the town to resolve its unfair housing practices in one of two ways: Build low-income housing throughout the city (including the upscale areas) or provide density bonuses and incentives for the development of better housing in the urban areas to be paid for, in part, by the more affluent areas. The city chose the latter. The power of community influence in preventing the development of projects that serve a broader base than the "host " community (NIMBYs) clearly demonstrates the power of a community's interest over a larger public interest. Equally important is the ability of municipalities to affect adjacent communities through the use of various growth management efforts. By restricting development, communities force development and the extension of infrastructure to adjacent communities.

Beginning in the 1960s and 1970s, a surge of activity in zoning and land development regulation has swept across the United States. As sprawl has continued, so have the attempts to control it. Local and state governments have proceeded with presupposition that growth controls do work. It is widely accepted that the effect of regulating growth through local regulations is to increase the value of developable land and the prices of existing housing. Obviously, owners of property that is valued higher after growth controls would tend to favor such controls, and buyers wishing to acquire property would not favor such controls. Still, does controlling growth through regulations have the desired effect?

Higher prices for existing housing could simply be a reflection of supply and demand. Regulations restrict or reduce the supply of available housing; therefore, the cost rises in response to an unfulfilled demand. Another explanation for the higher prices is the buyers' willingness to pay for what may be perceived as an amenity resulting from the restriction (Fischel 1990). The amenity may be in the form of preserved open space or reduced density, which result in a more attractive environment.

Many studies have been conducted to ask various questions pertaining to the effect of zoning on land use and value. Unfortunately, they have not produced a conclusive answer as to whether growth controls work. In fact, the complex nature of the relationship between a zoning ordinance and a property makes the design of a comprehensive analysis a mind-numbing proposition. Variances in location, political influences on the formation and administration of the ordinance, as well as the necessity to combine land uses into "common" types, all serve to reduce the resolution of the studies. For example, in one study open spaces included swamps and floodplains as well as golf courses and public parks. Evaluating the impact of zoning on property values without making a distinction in the qualities of the properties is an exercise in futility.

Generally studies seem to confirm what might have been intuitively obvious. Land located near public sanitary sewer collection systems is valued higher than land farther away. A moratorium on sewer connections increases the value of sites that have secured the right to connect to the publicly owned treatment works. Land zoned for higher-density residential development is valued greater than land zoned for lower-density residential development. It is interesting to note that the rate of increase in value is not linear—that is, the doubling of the density allowance increases the property value but does not double the value. One study noted property located next to city limits and zoned 20 units to the acre was valued as seven times greater than similar property zoned for only 1 unit to the acre. This same study looked at a site 15 mi from the city limits with the same density allowances and found only a threefold difference between the 20-units-per-acre site and the 1-lot-per-acre site.

In his work Professor William Fischel observes that of all the studies done, "almost all of them conclude that growth controls raise housing prices." One effect of controls also noted was the spillover effect. The *spillover effect* is simply that a zoning or other land use restriction has an impact on the areas outside of the prescribed boundary. Studies in Montgomery County, Maryland, found in the 17 planning districts in the county, that areas adjacent to districts with more restrictive controls had higher property values. It was concluded that the property near these restricted areas was seen as more valuable because of the restrictions. The impact of the growth management program, where it was coupled with conventional zoning restrictions, established a maximum number of new units each year.

It is not a surprise that one effect of this arrangement is higher property values of existing housing. In this way the zoning is seen as conferring a benefit on owners of existing homes in the form of higher property values. Another perceived value is the amenity value of a growth-restricted community. The cost of these values is shifted almost entirely to owners of undeveloped land and ultimately to buyers of new homes. A spillover effect of these restrictions was found in areas adjacent to the areas of restricted development where property values were "slightly, but significantly" increased. In his paper, Fischel notes that growth controls are in effect a transfer of "entitlements to develop from the landowner/developer to the community." Extending this thought, Fischel then suggests that fees charged by the community for reviews and impacts are really the developer buying back the entitlements from the community. In this argument, then, high-impact fees merely represent a community's "unwillingness to accommodate development, or it may represent a change of heart, . . . allowing some development if it pays its own way (and perhaps a little more)" (Fischel 1990). In fact, if this were true, under federal legal guidelines, this might be a taking since no overriding public health or safety interest was being protected, and the individual's right to "use" his or her land is completely compromised without due process.

Developers do not like impact fees, but they reluctantly support them as a cost of doing business, even more so if they are tied to performance criteria. Impact fees must be supported by justification and a fair-share measurement to be legal in more than 20 states in the United States. In general, the research reviewed by Professor Fischel indicates that growth controls impose a net cost on society. He concludes: (a) Land use controls impose constraints on the land/real estate marketplace, which in turn affects the cost of housing and land; (b) land use controls do provide a benefit to the public; and (c) the cost of growth controls are greater than the benefits that they provide.

The development of growth controls is a reflection of the attempts by legislatures and the courts to limit local exclusionary zoning. Affluent communities have not wanted to accept low-income housing, for example, have been able to avoid it by restricting all housing. These types of controls have not been undone through the courts presumably because they are "democratic" since they exclude everyone, not just low income residents. Another circumstance that has encouraged growth controls has been the use of citizen ballot initiatives. Many growth control ordinances are the result of such "direct democracy." In these cases the normal give-and-take of compromise is not built into the system, and the resulting controls tend to restrict a voting minority (owners of undeveloped land and business interests) in favor of providing the gains of restricting land use to the majority of voters. Could it be argued, in such cases, that the growth control is in fact a transfer of land rights from the owner to the public? If the answer is yes, then at what point does the loss of private rights, in this case the right to develop or use property, become a public taking of private land for a public benefit (the amenity of no growth)?

Growth controls also cause development to spread out; as developers and buyers seek opportunity, they move to new areas. The net effect is to encourage the urban sprawl that the growth control efforts were organized to combat. A municipality that passes strict development restrictions causes at least a portion of the development that would have occurred in its boundaries to be pushed into the next "ring" of municipalities. As these outlying areas become more populated, the political power of the community shifts from original residents to newcomers. The newcomers adopt growth controls to restrict further development and to maintain the character that attracted them in the first place. The long-term effects of this cycle have been observed to be a lower standard of living (more time spent commuting, more income required to sustain lifestyle, and more pollution and attendant costs), continued flight from the central business district, and a loss of the economies of established "agglomerations" of businesses.

Instead, businesses leave the urban center for the developing rings around the cities. In his 1991 book *Edge City,* Joel Garreau recognizes the appearance of commercial centers on the fringes of "traditional" cities as a logical extension of city-building and as a reflection of modern transportation (automobiles) and the limiting factor in modern lifestyle (time). Garreau pronounces the edge city to be the third wave of urban development after the development of central business districts. The first wave was the post-World War II flight to the suburbs, when "home" was removed from the city. This new freedom was made possible by the affordability of the family automobile. The second wave was the move of the marketplace from the central business district to the suburb in the form of shopping centers.

Garreau observes that the third wave is a predictable response to the success of the first two: If the home and the marketplace are in the suburbs, it is only a matter of time before the workplace will join them. Although cities evolved as a response to the need for protection, they remained largely independent marketplaces. Even within cities trade occurred throughout the city, without a single business center or district. This remained true, according to Garreau, until the industrial revolution when inexpensive power and mass production brought a number of changes. In the early days of industry, factories required large numbers of workers. The available transportation at the time was either foot power or, somewhat later, steam locomotives. The natural limits of the mode of transportation led to high concentrations of residential areas around what became central business districts.

Ironically, the author points out, the first predictor of the edge city was probably General Motors. In 1919 construction started on "New Center," which at the time was to be the world's largest office building. This huge complex, however, was not built in downtown Detroit but rather at a busy intersection about 10 mi from the downtown area. This new complex was accessible but only by "personal" transportation—the automobile. Ford soon followed suit—after all, the downtown location did not provide enough parking for all of Ford's employees. In 1928, the year Henry Ford began to build the Model A, and only 2 years after building the millionth Model T, Ford moved its first plant out of Detroit, to Dearborn. According to Garreau, the edge city was born.

Like all trends, the edge city has certain characteristics that are replicated, albeit in different degrees, in most other edge cities. Edge cities are developed along or near important intersections of highways, they are "anchored" by a regional shopping mall of at least 600,000 ft^2 (one that serves a population of at least 250,000 people), they have 5 million square feet of office space, they have more jobs than bedrooms, and they have an identity as a place. The underlying influences driving the growth of edge cities are the effects of technology. The automobile and the information age (computers, communications) have combined to shift the limitations of urban development from issues of distance and proximity to primarily issues of convenience.

According to Garreau, an unintended consequence of the development of edge cities has been the lack of recognition given to them by the design and planning community. It would be fair to say that the edge city is viewed as a blight upon the landscape by many architects, planners, and citizens. In the 1980s building boom, 80 percent of the development dollars were spent outside of the old central business districts, going instead to building elements in the new edge cities. The growth in the edge cities has been led by developers, and the planning and design community has largely resisted recognizing these developments as any form of urban development at all. Developers, on the other hand, responded to an almost intuitive sense of what the public wanted and the most desirable form of what "should" work. The developers viewed the edge city as a product of the culture and society and themselves as only providers of what people desired. The edge city is a result of the forces of the marketplace.

Designers and planners continue to believe that the central business district must be the focus of our planning efforts and that, in spite of all of the evidence to the contrary, the public really wants to be "downtown." Garreau provides ample argument to demonstrate that, in fact, the public does not want to be downtown and despite the best efforts by planners, is not going to change its behavior. Examples of reinvented downtowns actually serve to support Garreau's observations that these revitalized and saved central business districts have actually been remade into tourist and entertainment nodes that serve in the center of several outlying edge cities or into an edge city themselves. The requisite 5 million square feet of office space required to be an edge city, according to Garreau, is larger than all but the largest central business districts.

Accordingly, while the professional planners and designers continue to decry the form and trend that is the edge city, the developers and marketplace are going about building the urban form that best reflects our society and technology. The form is in its youth and not without its own problems, but it has been legitimized by the marketplace. It would seem it is here to stay. The disagreement over the "sprawl" of edge cities is rooted in the two faces of the American people: On one hand, our history is a history of building and developing the wilderness, and we have a constitutional predisposition toward a landowner's right to use property. On the other hand, as a people, we value the land for itself, and we identify with natural beauty and the land. Garreau observes that to preserve the land, conservationists would best be served not by fighting all development but by working with developers and the public to do two things: First, establish important specific features or attributes of an area that should be protected and then work to protect them, and second, work to improve the quality of life in existing urban centers. Until people see no advantage to living "out" and no disadvantage to living "in," they will continue to move out according to Garreau.

An interesting aspect of this is the concern that edge cities are developed indiscriminately and without the benefit of good planning. In fact, the underlying design and development criteria are very specific and concrete. Based on the scales of economy, as well as the expectations and limitations of the users and residents, the development of the edge city is a very deliberate undertaking. Garreau observes the sense that they are uncontrolled stems from their unfamiliar form. We simply are not used to cities with such a low density of buildings, ample parking, and greenery. The classic planning expectation that the right plan or right restrictions will bring everyone back downtown is not probable. The desire of preservationists to stop all development is equally unrealistic. Developers, on the other hand, must recognize the importance of landmarks and appropriate environmental planning, as part of the edge city development. In fact, viewed from another perspective, the edge cities often include the open space and opportunity to introduce—even retrofit—sites to incorporate most of the principles of sustainable design. Instead of resisting the trend, we should find ways of incorporating new design principles into it.

Land Takings

The political theories of John Locke were important to the framers of the U.S. Constitution. Lockean theory holds that human beings have certain "natural rights," that is, rights that are intrinsic to the individual and are not derived from the sovereign or the government. Representative government, in this view, is based on the premise that the state's rights are no greater than the sum of the rights of the individual. In other words, the state has no independent set of entitlements or rights. The right to hold and own property is considered a natural right. In these matters, the natural rights of the individual are equal to the rights of the government; the government is not entitled to any privilege or greater consideration than the individual. John Locke believed that such natural rights include an individual's right to "own" himself or herself—that is, one owns the fruits of one's labor and is free to determine the disposition of such fruits. In those cases where the interests of the people, many individuals, encroach upon the rights of the individual, the individual is entitled to compensation. In fact, the Lockean theory holds that the return on any government encroachment or taking must be greater than the cost to the individual. In such a situation, the value of the public good must be greater than the value of the right to the individual, and the compensation paid to the individual must also be greater.

Takings are public or private encroachments on the property rights of a landowner without the agreement of the owner. Takings are fairly complex legal issues and subject to ongoing definition between legislative and judicial actions. Common law establishes four conditions for establishing a taking:

1. The property must have equal value to the defendants and the plaintiff.
2. The taking is conscious and deliberate.
3. The taking is accomplished without a third party or natural events.
4. The entire thing is taken, not just a part or a portion of the thing.

If any of these elements is missing, then the event is not a taking. The individual may have recourse under tort law or under strict liability, but there is not, under common law, a taking. In any case, the plaintiffs' option for recovery is in court.

Takings may be public or private. In private takings, the "taker" is liable for the cost of replacement or the cost of mitigation for the taking. In public takings the government pays for the value that has been transferred to the government, not necessarily the value to the owner or even the market value. This is illustrated in the case *Kimball Laundry Co. v. U.S.* (1949) in which the government, under eminent domain, condemned property that included the Kimball Laundry Company business. The owner of the business sued the government for the value of his business. The government had paid the owner for the value of the real estate (the laundry was in a modest residential area). The compensation paid by the government was insufficient to relocate and reestablish the business. The Supreme Court ruled that the government must pay only for the value of the resource that it has taken, and it does not have to pay for attributes that the government does not use, such as the laundry business.

The U.S. Constitution is not designed to protect the thing that is owned, but rather the owner of the thing. The rights associated with ownership of property have several aspects: possession, use, and disposition. These elements are fully and completely the rights of owners. There can be no "degrees" of ownership, and no abridgement of the rights of the owner. Within the limits of public protection, police powers, and common law principles, any encroachment on the rights of ownership is a taking, according to the author. The U.S. Supreme Court has established a test to determine if an action constitutes a taking. For it to be found to be a taking, it must be shown that the following is true:

1. There is a *nexus,* or direct relationship, between the action and a legitimate public interest.
2. There is an invasion of the property or an owner is denied the right to control access to it.
3. The legitimate public interest has a greater value or benefit than the cost to the individual.
4. The individual is denied the use or other economic value of the property.

Property rights have been described as a "bundle of sticks"—that is, property ownership consists of a number of rights ("sticks") that are unique. For land ownership, such rights might include the right to control access to one's land or the right to sell a portion of it or to lease it to another, the right to make improvements on it, or the right to leave it fallow or unimproved. The loss or modification of one stick does not affect the others.

In *Penn Central Transportation v. City of New York,* the owner of the Grand Central Terminal building was denied a permit to build an office tower over the top of the terminal because of the historic value of the building. The Supreme Court held that the state might restrict the use of a portion of a property or a particular use of the property. It is a "fallacy" to assume, according to the Court, that the loss of any particular right of easement constituted a taking of property. This has stood as a cornerstone of preservation efforts for many years.

Takings by regulation are common. Although the state has the right, and arguably the duty, to regulate land use under the police powers, many land use regulations are not justified under the police power and might therefore be considered takings. The central purpose of government is to maintain peace and order. To do so, the government must have certain powers; the rights retained by the citizens must not be so broad that they prevent the government from discharging its responsibilities. These justified encroachments on to the rights of the citizen are the police powers. The form and structure of government is designed to limit the police powers and jurisdictions of authorities within the government.

Regulations that go beyond the justifiable police powers in controlling land use are subject for consideration as takings. Much environmental and land use regulation is within the police power. Where the justification is not clear, we simply need to answer the question, Is the regulation an attempt to control the defendant's wrong or to provide a public benefit? In some views if a public benefit is the goal, the regulation is a compensable taking of private property but there is a need for police powers of local government and the forced exchange of property (eminent domain) only for public use. In any case, the result of the taking should be for everyone's use or enjoyment, displaced owner and public alike, and each must receive something greater than the cost of the exchange. A "public good" must not be merely a benefit.

This approach to the police powers issue raises several interesting questions. The U.S. Constitution does not include explicitly any environmental rights, but many state constitutions do. In Pennsylvania, for example, Article 27 of the Commonwealth Constitution identifies a right of the citizens of Pennsylvania to have the protection and enjoyment of environmental, cultural, and historical features of the state. This explicitly extends the police power and responsibility of the state to include the protection of the environment and other entities. In Wisconsin similar constitutional language has caused the Wisconsin Supreme Court to hold that the environment is clean and pure and that pollution is a result of human activities. Based on the environmental rights of Wisconsin's citizens, therefore, an individual's property rights do not automatically include the right to change the land from its natural use or state. Use beyond these natural uses are subject to regulation under the state's police powers (*Just v. Marinette County*). A similar argument could be made (although I am not aware that it has been) under Pennsylvania's constitution. In effect, what is a taking under federal law may not be under state law.

Sustainable Development

Evidence of human activity's significant and far-reaching impact on the earth's environment continues to mount. It has been said that although just 60 percent of the earth's land surface is under the management of people, 100 percent of the earth's surface is impacted by the practices of that management. Whether we are aware or not, our activities have an effect on the world in which we live. Throughout this book practices and methods that contribute to reducing our impact have been discussed. In February 1996 the President's Council on Sustainable Development published *Sustainable America, A New Consensus for Prosperity, Opportunity and a Healthy Environment for the Future.* The report adopted the definition of *sustainability* as written by the

World Commission on Environment and Development: "The ability to meet the needs of the present without compromising the ability of future generations to meet their own needs." The president's council outlined ten goals for the nation to meet to become sustainable, but these goals could be summarized aptly in the first three, which concerned health and the environment, economic prosperity, and equity. Beyond the rhetoric, meeting these goals represent a substantial challenge because in our culture we have tended to see these things as antagonistic or mutually exclusive. To become sustainable, we must create a society that embraces all three.

In a classroom exercise, I ask my students to close their eyes and envision a high-quality environment, whatever that means to them. Then I ask, "When you form a picture in your mind of a high-quality environment, what do you see?" After the students write down what they see, I ask, "When you think of a prosperous economy, what image appears?" Then we discuss the different images. From these discussions, and others, I have found that we tend to think in extremes. The student images usually describe the worst of economic activities compared to the best of the environment. We tend to think of a healthy environment and a prosperous economy as mutually exclusive. Fortunately, it is possible to attain both good outcomes. But the challenge we face is to reconcile our economic interests with our environmental interests. As design professionals, we have particular challenges before us.

One theory of sustainability holds that as long as there is a fair exchange of value, the loss of environmental quality is acceptable. We can afford a less diverse environment, or a lower air or water quality, in exchange for an improved infrastructure. Future generations are compensated for lower environmental quality by this same improved infrastructure. This is known as *weak sustainability,* or the *constant capital rule.* On the other hand, there are those that believe there can be no exchange, that environmental values or quality cannot be offset by infrastructure, that the functional aspects of the environment cannot be equated to the values of, say, a mile of new road. In this view, functional values must be preserved and development must occur within the constraint. This is known as *strong sustainability.* Actual practice will necessarily lie somewhere in between, and we will learn to make decisions on a continuum between strong and weak sustainability.

Building the postindustrial landscape

The move from builder and expansionist thinking toward stewardship requires a shift in values, not just the adoption of new techniques or standards. Embracing stewardship is an intellectual shift away from the fiction of landscape control, from simply stabilizing the land, to visualizing a natural social satisfaction and fostering conservation just because it makes sense. Ultimately we need to, and we will, place greater values on nature and environmental quality, and these become part of our sense of satisfaction.

On March 26, 1996, Interior Secretary Bruce Babbit turned a valve on the Glen Canyon Dam in Page, Arizona, to release a flood of water from the artificial Lake Powell, into the Grand Canyon. The flood was intended to mimic the natural floods that can no longer occur seasonally because of the dam. Periodic flooding is believed to be necessary for the renewal of soil nutrients and temporary pools that serve as wildlife habitat. The experimental flood was to determine if natural flood cycles can be mimicked to a degree. Perhaps turning that valve remains one of the most tangible examples of the subtle changes taking place in the attitude of people in their interactions with the environment. The world-renowned historian Thomas Berry has correctly observed that the argument is usually framed as the good of nature against the bad of society, or the bad of society against the good of society. Berry suggests instead that each side has good and bad points and we should compare the good with the good and the bad with the bad.

Aldo Leopold (1887–1948) wrote that the first ethic guiding the relationships between individuals—the golden rule—had no counterpart, "no ethic dealing with man's relation to land and to the animals and plants which grow upon it." Aldo Leopold is best known as the author of *A Sand County Almanac* (1949 and 2002), which is a collection of his writings on ecology. He noted that the relationship between people and the land was strictly economic. People saw no obligation to the land or to nature. Since Leopold's death, the population of the world has more than doubled, rising from about 2.75 billion to more than 6 billion today. Our impact has grown even faster. For humanity to cope with this surge in population, part of the move toward sustainability must be the recognition that with property rights go obligations that society defines and property owners must respect. Aldo Leopold is famous for writing, "A thing is right when it tends to preserve the integrity, stability, and beauty of the biotic community. It is wrong when it tends otherwise" (Leopold 1949). Over the course of his professional career, Leopold experienced a profound change in his personal values through his fieldwork experiences and observations, realizing a dynamic shift in his perception of quality.

The industrial-age landscape must embrace the processes of nature as the models on which we base our activities and as the context in which we carry them out. This requires a new expression of landscape. The picturesque landscape based on the idyll English pastoral estate has devolved into strange configurations of single-story offices amid huge lawns, and $^3/_4$-acre exurban pseudo-colonial villages. We know inherently that there is an imbalance afoot, but we feel that there is little as individuals we can do to rectify it. In spite of our technology and economic prowess, we have a vague sense of dissatisfaction as we learn there is a limit to what we can consume of the natural world around us. The cycle of production, consumption, and disposal that has come to characterize modern living is not satisfying, and there is no end point to it. And there can be no end point as long as we continue in this direction.

We have become so competent and numerous that our day-to-day activities have global impacts. If in the past we worked to survive, then today we work to consume. While we resist the tyranny of nature for survival, we cannot deny its value to our own existence. Jay Graham, member of American Society of Landscape Architects (ASLA) wrote, "Society as a whole—not just design professionals—is looking at the physical environment in a more holistic, integrated fashion . . . The change in our environment in the past forty years alone should be enough to convince us of the impact of future decisions."

Landscape ecology and people

Landscape ecology is an emerging biogeographic science that seeks to understand the relationship between humans and their environment and the results as expressed within the human ecosystem. This discipline recognizes the integration of humans into the environment as both agents of cause and recipients of effect. The discipline has grown from work undertaken primarily in Europe to understand the ecological role of humans in shaping the land itself but also the relationships that occur within the ecosystem. The social human aspects of the ecosystem compound the complexity of modeling the relationships between people and the environment.

While science pursues a definition for the impacts and relationships, in fact, much of our interaction is guided by a mixture of values which varies widely in application. Attempts to quantify the impact of values on individual and social actions and decisions have improved our understanding of motivations and self-interest, so that we understand decision and action as underlying causes in the cause-and-effect view of human impact. Little research has focused on the relationship between the environment and values, in which the environment is the cause and the value is the effect. If humans are indeed a part of the landscape ecology, then it must be presumed they are influenced by that same environment.

Humans act and interact based on complex relationships that have evolved to promote the development and survival of the group. Individual actions are directed and assessed by the culture at large: In general, actions that are good for the individual and good for society are encouraged, while those that are good for the individual and bad for society are discouraged. Cultural and political infrastructures develop to support these tendencies. For example, the society requires children for its long-term survival; therefore, elaborate mechanisms evolve to encourage the nurture and survival of families. Some of these values are basic and perhaps biologically based (the nearly universal sense of protecting children from harm), and at the extreme are more elaborate, purely political constructs (tax benefits to married couples). In fact, these actions simply reflect the values of the culture at large.

Values are in one important sense a recognition of an important commitment and responsibility to the future. As guidelines, they serve to modify individual and cultural behavior in ways that contribute to the good of the group but do not benefit the individual directly. These values are often unrecognized as discreet moral choices, but they stand as a bundle of underlying—usually unarticulated—principles that brace the foundation of both social and individual behavior.

Change occurs incrementally, and it is typically undertaken in the pursuit of what is deemed a circumstance of higher quality by the members of society at large. Individual values tend to reflect the larger social fabric in terms of values and the definition of quality. As individuals form opinions as to the form and content of quality, their choices came to reflect the principles they have adopted to guide their own behavior. Though perhaps not a cognitive process, each person adopts a model of quality and the underlying framework that supports and advances him or her toward this model.

As circumstances change, individuals may test their principles against new information and adapt to the new information. Change may be expressed as small "course corrections" and, albeit less frequently, fundamental changes in an individual's principles. As the number of individuals with like feelings increases, the expression of these new values becomes entwined in the culture at large at the political and social levels. Predictably such changes are slow to occur; however, evidence abounds that the change is in fact occurring.

In 1991, the 22 largest environmental advocacy groups in the United States reported 11,533,818 members, an increase of more than 100 percent over their rolls in 1980 of 5,716,242. More Americans identify themselves as "environmentalists" as time passes. Surveys since 1990 have reported that from 58 to 85 percent of those asked considered themselves to be "environmentalist" (Hendee and Pitstick 1992). This finding is important because it means that individuals are identifying with the principle of being an environmentalist, rather than identifying with a particular environmental issue. In a survey taken within the Chesapeake Bay watershed area, 85 percent of those surveyed said they were somewhat or very concerned about the pollution in the bay, 68 percent said they were active or somewhat active in helping to reduce the pollution in the bay, but fewer than 10 percent could identify the most significant sources of pollution to the bay.

A study conducted by the Times Mirror organization in 1992 also found that most Americans cared strongly about environmental issues. When asked which was more important, 71 percent felt that protecting a wetland was more important than developing 50 new homesites on the wetland site. In the same group, however, only 48 percent felt that protecting the wetland was more important than protecting a broke homeowner's right to sell his or her land for new-home construction. While generally encouraging, the surveys conducted all show similar kinds of uncertainty in the public mindset. In the Chesapeake Bay survey, for example, while most respondents identified themselves as "concerned and active," only 10.5 percent said they belonged to an environmental group. While agriculture, construction, and suburban and urban runoff represents the greatest source of pollutants affecting the bay, those surveyed did not recognize that fact in their ranking of the top four sources of pollutants. Only 8 percent of those surveyed thought that farming was the most serious source of pollution. Rather, 7 percent thought individuals were, and only 2 percent considered construction to be a major culprit. The overwhelming majority believed that business and industry (32 percent), treated sewage (16 percent), and commercial shipping (15 percent) were the most serious sources of pollution. Their responses are in keeping with the fact that, in general, most people have little difficulty assigning problems to large corporations or the government for solution; they are less committed in expecting the same level of involvement from individuals.

Individuals acquire their values through cultural and personal experiences in both formal and informal circumstances. Formal experiences generally establish a standard of behavior, concepts of right and wrong, and cultural constraints based on fact and myth. These are the concepts that underlie the social contract and political fabric of the community at large. To varying degrees, each individual is a microcosm of the larger culture, which is very much a "work in progress" that fluctuates in response to the growth of its members.

We recognize intuitively that values do not operate in isolation from other influences and that most choices are an amalgam of factors and motivations. This is particularly true with environmental values. High environmental values can be expected to meet a high degree of conflict with competing motives such as economic opportunity, personal advancement, and lifestyle enhancement. The sources of environmental values have been studied at length. A study by Kempton, Boster, and Hartley (1996) identified three sources of environmental values: (1) religious or spiritual beliefs, (2) anthropocentric, or human-centered, beliefs, and (3) biocentric, or life-centered, beliefs. Although there is debate among scholars as to the amount of influence generated within each of the different sources, the study concludes that the nature of the source is less important than the cultural model through which the values are expressed.

The cultural model is the order in which people rank their values. If most people identify themselves as having "high environmental values,"—that is, most people consider themselves as sharing similar values toward the environment and nature—then the differences among them are the order in which they rank the shared environmental values. This order can be construed as the cultural model of the individual. Each individual thinks his or her order is the correct one. The cultural model thus reflects the individual's understanding of how nature works. As people have come to understand the interconnectedness of nature and the impact on nature of human activity, their sense of responsibility has increased and their need to protect nature has grown. The Kempton, Boster, and Hartley study looked at how different groups in the United States responded to questions regarding environmental values and how they ranked these values. In their study they found a shared set of beliefs among a diverse set of study groups ranging from Earth First! and Sierra Club members to laid-off sawmill workers, dry cleaners, and the general public.

When differences are noted, the researchers found them primarily in the cultural model—that is, the ordering of the values. For example, while, as a group, laid-off sawmill workers (70 percent) expressed that caring for their families was more important than caring for the environment, few Earth First! members (13 percent) did so. It is interesting to note, however, that more Sierra Club members felt this way than did sawmill workers. The values of the individuals range from the religious and spiritual ("The environment should be respected and protected because of God and/or Nature"), the human centered ("The environment should be protected and preserved for our use and the future of humans"), to the life centered (Nature has intrinsic value and should be respected and protected for that reason alone). The study concludes, among other things, that while the sources of values may differ, the values themselves are very similar. Although the source of the motivation may be different, the expressed environmental values have more in common than they do differences, and it is through the cultural model that these differences are expressed.

Most people consider the environment to be very important and consider themselves to be environmentalists, and the differences between them are matters of degrees. Individuals tend to hold values that reflect experience they view as high quality and that are not in conflict with other learned values. Values are interpreted and applied via a cultural model that is a paradigm of ranked values through which circumstances are viewed and decisions processed. In general, things valued as high quality are sought out, and things of low quality are ignored or discouraged. To some extent low environmental values or expectations of environmental or landscape quality can be correlated to the corresponding poor environmental and landscape quality of places. The sameness from place to place, the universality of the American City, and the loss of environmental quality in the built environment is a reflection of the low value and expectations of builders and users in the past. We tend not to see things of low value, and we tend not to provide for them in our construction.

If we are to embrace sustainable development, then the value of the environment must be appropriately reflected. Moving toward a model of sustainable development is not simply a matter of adopting new design standards, because if that is the extent of the commitment, sustainability will be a luxury, something only high-end projects can afford, the first thing lost when budget is an issue. Developing a sustainable culture will happen along with a rise in awareness of the components of a high-quality experience, which can be brought about with a more sophisticated approach to design. This sophisticated approach needs to be applied universally, to low-end as well as high-end projects.

Science and design

To begin with, it is important to understand what is meant by "design." Victor Papanek's definition "Design is the conscious effort to impose meaningful order" seems to express it best. Papanek gives a telesistic view of design—that is, a view in which the processes of nature are put to good use to meet a social or individual goal. Design is a process oriented toward an desired outcome. It is more than pattern and distinct from nature in this view. Nature does not design in this way. Form in nature is a reflection of dynamic processes that follow certain immutable laws. Thus snowflakes, the shapes of leaves, and ripples in sand are not design, they are pattern. People design to achieve some outcome. Design is intelligent, conscious, and intentional. Design is the application of knowledge, insight, and intuition. Designers work on several levels at once, but good design is always thoughtful. The creative design process proceeds on three fronts at once: the *objective* (technical knowledge, geometry, and the laws of nature); the *subjective* (individual insight, creativity, and experience), and the *intersubjective* (one's cultural view or social paradigm). It could be argued that design that proceeds on only one or two fronts is incomplete.

Science is one way of knowing. Other ways are knowledge stemming from traditions, faith, intuition, and experience. Our objective view is firmly rooted in science. From the fifteenth century until well into the twentieth century, western science proceeded on what has been called the *Newtonian paradigm,* the objective science of Newton and Bacon. In this view, sometimes called *reductionist science,* things can be known by reducing them to their essence and understanding, through observation, the parts and relationships between parts. Science understands nature as process. Where our understanding has been incomplete or we have not had the means to measure, it has been the role of science to push and expand our capability. Things are said to be "known" if they can be observed to be. The resulting objective knowledge is not good or bad; it is simply knowledge. The objective view has larger implications for our view of nature and how we have perceived our role in the environment. Design of all types is more or less applied science, and until recently the reductionist objective view has been the only significant view of science.

The twentieth century has been described by some as the "century of the engineer." It is an apt description inasmuch as modern society has placed much of its faith in the ability of engineers to solve problems. Beginning in the nineteenth century and through the twentieth century, engineering design was reduced to its most objective form. Engineering economics emerged as a means of studying and expressing objective efficiency in design. In this discipline, efficiency is narrowly defined in economic terms. Much good was accomplished with this objective, efficient approach, but much harm as well because the Newtonian paradigm and engineering efficiency were simply unable to consider other values.

Science is a way of knowing, but it is only one way. Science has moved past the rigidity of *reductionism* in which chaos theory and complexity consider the implications of simple actions within complex systems. Science has discovered *synergy,* processes in nature that are greater than the sum of their parts. As Frank Lloyd Wright once observed to Louis Sullivan, "Form follows function." Wright later revised this maxim to be "Form and function are one!" (Wright n.d.).

As the twentieth century closed, we had already started to appreciate the complexity of the world and the need for a deeper, more thoughtful approach to design of all types. The preservation of landscape function and the creation of beautiful and functional places are complementary. A sustainability paradigm will emerge from what is learned and produced by professionals working today. Ongoing works are variously called *sustainable design, regenerative design,* and other things, but what they all embody is a deeper understanding that comes from knowing both how things work and how they relate to other things. Sustainability, however, cannot ignore esthetics. Human beings desire beauty. To be sustainable, the environment must offer us the opportunities for individual as well as cultural fulfillment. In his book *Zen and the Art of*

Motorcycle Maintenance, Robert Pirsig contrasts what he calls the "romantic" and "classical" views—essentially, it comes down to, Do you want it to look good, or do you want it to work (Pirsig 1974)? The classical view is concerned with functionality while the romantic view is concerned with the appearance of a thing. Sustainability requires both.

Today it is inadequate to simply collect storm water and convey it away in the most efficient manner possible. We must consider its role on the site and place it in conjunction with a broader region. We must begin to design within the natural system where precipitation is an asset and a resource. The definition of *quality* in design has expanded to consider the natural system. As this recognition continues to mount, these parts of the landscape will become more visible because they will begin to have value for us. The role of science is to help us understand, but in the end it is only one part of the designer's voice.

Principles of sustainability

Nature evolves to become more diverse and more complex. The complexity arises as living things adapt to many niches, creating pathways and links through which energy and materials pass. The human approach to living, however, has been to focus on efficiency and uniformity, on concentrations of energy and materials—by definition, reducing the pathways of energy and materials. In contrast, nature is redundant, finding many ways to accomplish energy and material transfer. Where nature recycles everything, modern industrial systems are one-way systems of consumption. Although there have been significant improvements in pollution control and prevention, the quality of air and water in the United States continues to decline because the underlying system of cleansing and assimilation has been minimized while the generation of wastes has increased. If *pollution* is defined as the presence of a substance in quantities that exceed nature's or the ecosystem's ability to process it, then we have been inhibiting natural systems in two ways: increasing the presence of pollutants while decreasing nature's capacity to absorb them. Sustainability will require replacing or improving these processes of assimilation and natural cycling of materials.

The strategies for sustainable design are fairly straightforward. Site design professionals should be students of nature and look for ways to let nature do the work on a site. By using existing landscape elements or enhancing them, the development of a site can occur while the landscape functions are preserved both on the site and as part of a regional fabric. For example, expanding existing on-site wetlands to treat the anticipated runoff from a development preserves the integrity of the landscape system and reduces the development's impact on the site. By preserving or developing riparian zones in greenways, designers can preserve the hydrologic, habitat, and assimilative capacities of the landscape and simultaneously provide a site amenity.

By looking to nature and understanding how nature works, improvements can be planned in such a way that maintenance and life cycle costs are reduced and environmental values are increased. Natural communities are diverse because in diversity communities thrive. Instead of a few species of plants, there are many. Through diversity, plants are better able to resist disease and predation. Planning for diverse plant communities may involve some greater initial costs, but operation and maintenance costs are much lower as the community is established and matures in comparison to a more formal and monotonous plant scheme. This is not to say that less natural places are less worthy but rather, to suggest that natural plant communities also belong in the developed site. The best models for designing a functional site are found in nature.

Design is efficient when it minimizes development costs, and costs are kept low by minimizing the amount of material and labor required to develop a site. Through development, we decrease the infiltration on sites and increase runoff. We then collect and concentrate runoff into pipes designed to maintain an effective scour velocity and so deliver all of this sediment- and pollutant-laden water in a concentrated dose to a collection facility or worse yet directly into some natural surface water. Nature also concentrates runoff, but water in most overland flow accumulates on the surface only after much has soaked into the ground or the infiltration capacity is exceeded. Surface runoff is filtered as it courses through and around vegetation and is absorbed into riparian and littoral zones. By observing nature, we have learned to encourage infiltration rather than concentrating runoff into pipes, and we have also learned that the use of many small infiltration systems is preferable to using one large system.

We should use information and intellect to replace power. Throughout the world there are examples of architecture and site planning that were based on understanding the character of the place in which they were built. Designers, for the most part not professionally trained designers, accomplished design goals by understanding and taking advantage of the character of the places in which they lived. In New England, buildings were located on their sites to take advantage of the winter sun and to shut out the cold winds from the north and west. In the southeast, houses were built with tall ceilings and large windows to encourage air movement through the house. Eaves were built to shade the house from the summer sun, and houses were oriented to collect the evening breezes. In the southwest, Native Americans built thick-walled houses to insulate from the heat and cold, and they developed xeric forms of agriculture.

The availability today of inexpensive energy and building systems has resulted in a lack of sophistication in modern design, and time-tested models have been ignored. Buildings once reflected an elegance of design, a thoughtful construction based on an awareness of the environment. Buildings in this tradition were active working machines. But today, with cheap energy and systems engineering, buildings have become mere decorative boxes, containers for people and activities and managed by complex systems. Complexity, however, does not imply sophistication or elegance. There is a price for this lack of sophistication, beginning with higher energy and building operation costs. In a deeper way, such boxlike buildings contribute to our separation from the natural world; we do not have to understand the seasons if all we have to do is push a button for heating or cooling. Intelligently designing buildings and sites to reduce energy requirements is an important strategy for sustainable design.

As natural biotic systems are complex, so too are the abiotic, nonliving part of the environment. We should look for ways to use resources for multiple purposes. Wet ponds may be used to enhance cooling systems or provide habitat as well as treating and storing runoff. Parking lots might be designed to be shared between buildings whose uses are complementary. Building materials could be selected with recycling in mind, and so on. The key to sustainable design is to incorporate the principles of sustainability in every design decision (Table 12.1).

Emerging Trends

Much is happening in our transition toward sustainability. It is sometimes difficult to have a sense of progress, but a designer by nature must be an optimist and believe what may be done will make a difference. Smart growth has become the focus of much attention in recent years. The essence of smart growth is to slow down suburban sprawl by encouraging development in places where it can be best supported by existing infrastructure. Many special-interest groups have latched onto smart growth and tried to refocus its intent, but for the most part smart growth appears to be remaining well attached to its initial and fairly effective approach to addressing development concerns. Smart growth promotes environmentally friendly and community-oriented development. Among its concerns are transportation, community equity and health, environmental quality, and economic prosperity. It is difficult to determine how many individual smart growth community organizations there are in the United States, but the network of interested people and organizations is large and continues to grow.

Smart growth at the state level usually involves directing public funding to identifiable growth areas that have existing utilities, adequate public and highway transportation, and civic infrastructure. In this way the public funds (taxes from those living in the existing communities) are used to support existing communities rather than having to subsidize communities that do not exist. Growth is directed to contributing to the quality of life by encouraging the use of public transportation, the development of open space, and the stabilization of communities.

States are necessarily going beyond smart growth as growing populations and development begin to exhaust limited resources. In the arid Southwest,

TABLE 12.1 Elements of Sustainable Economy

A decrease in our reliance on nonrenewable energy
An increase in material efficiency through recycling and reuse and better design (doing more with less)
Improved functionality through design and a reduction in the use of energy
Reduction in the use of toxic materials
Recognition of environmental cost in accounting procedures
Repair of damaged or lost environmental functions

for example, the population of Nevada, the fastest-growing state in the nation, has doubled in the last decade. Arizona, second only to Nevada in pace of growth, has seen its population increase by 40 percent. California, the most populated state in the nation, expects to add 15 million more residents to the 34 million that already live there. Water has long been a contentious matter in the West, and with such increases in population, it promises to be even more so. In the fall of 2001 California law began to require that any residential project over 500 dwelling units must conclusively demonstrate that a sufficient supply of water exists for at least 20 years' use. Failing such a demonstration will prevent the project from breaking ground. Limiting development may be one component of smart growth, but ultimately it cannot be the only solution for the problem. It should be expected that people will continue to locate in the desirable sunshine states. Stopping development for whatever reason will serve to limit the housing supply, resulting in higher prices and limiting housing affordability still more. Development in some form will continue because it must. It is the form and related practices that will change.

The National Association of Home Builders (NAHB) has made a series of commitments to Smart Growth and to improve the practices of home development. Some have complained that NAHB brings a self-serving view to its articulation of smart growth and green building practices, but perhaps there is a similarity like the one noted in the Kempton, Boster, and Harley (1996) study: The differences are a matter of degree, not of substance. There will always be room for disagreement in sustainable development practices. Sustainability should not be anticipated as a one-size-fits-all or The Unified Field Theory of Development; rather, it is a new paradigm replete with its own problems, trends, and rewards.

The Chesapeake Bay Foundation announced its "Builders for the Bay" program in December 2001. This program brings environmental activists and builders together in an effort to establish sustainable building practices in the Chesapeake Watershed. Another encouraging trend has been the shift in the purchasing practices of the federal government in the United States. The Building Environmental Education Solutions (BEES) system was discussed in Chap. 1. It provides a method for evaluating the embodied energy and life cycle impact of materials. The federal government is also the largest buyer in the U.S. marketplace, purchasing more products, materials, goods, and services than any other single entity. In requests for proposals (RFPs) for buildings and site work, various federal agencies have started to include sustainability in the evaluation criteria. With the purchasing power of the federal government, design and construction firms have had to become more expert in sustainable and green development practices as a matter of business.

The private sector has begun to respond. The Leadership in Energy and Environmental Design (LEED) program is a method of building assessment developed by the U.S. Green Building Council that provides professionals with a consensus-based set of practices to guide design decisions. It is a self-assessment process based on life cycle considerations and accepted energy and environmental design principles and practices. The LEED system is a point-based rating system that recognizes the values of redevelopment and steering development to existing urban areas. Designers are encouraged to meet fairly specific performance standards with regard to storm water management, minimizing disturbed areas, reducing the amount of water required for the landscape, reducing the heat island effect, minimizing light pollution, and encouraging alternative transportation. In addition to the LEED program, other important sources of information have emerged such as the Center for Watershed Protection and the Green Building Network.

Industry has also started to change behavior. Many large corporations have started to rethink their approach to the environment. In the past, environmental compliance had been the target and consisted largely of treating or managing wastes after they were produced. The costs of compliance was significant; the consequence, if not the objective, of many environmental regulations was to increase the costs of producing wastes, particularly hazardous wastes. In the 1980s businesses began to realize that cost avoidance was a better strategy: If they never produced the waste, they would not have to manage it, thereby avoiding the costs and liabilities altogether. The so-called Pollution Prevention, or P2, movement was the result. By the early to mid-1990s, business began to move past P2 toward an industrial ecology. Bradley Allenby, a vice president of AT&T, has been an important promoter of this concept.

Industrial ecology is an approach to conducting business that uses the model of an ecosystem as its guide. Recognizing the principles nature employs such as recycling, using multiple pathways for energy and materials, and system redundancies, industries have started to look for ways of developing similar relationships among themselves. In such a system, one firm's waste might become raw material for a second firm, waste heat from one operation may be used to power the process of another, and so on. In principle everything is used, and there is no waste. While real-world obstacles must be overcome, successful efforts have demonstrated that such an approach is more than simply cost avoidance—there is a positive contribution to the bottom line.

Challenges

Although there is much already underway, much remains to be done. This period of change provides the opportunity for design professionals to assume leadership roles not only within their professions but within their communities. As noted before, the site planning and development profession is one of the few if not the only profession in which trained and experienced practitioners find their reasoned and well-founded work subject to review by people with little or no training. Sustainable design issues present a challenge to the design profession to lead and teach through the design and planning process. More design professionals should participate on local planning and zoning commissions. There should be greater political involvement on the part of design professionals, beyond simply protecting or renewing practice acts every so many years. In the absence of such leadership, we must be content to be led by others.

Sustainable design implementation will require a greater awareness of the underlying science of site design. It also will probably require working in teams with greater interdisciplinary makeup. There is not a significant existing body of research or criticism that addresses the practices of landscape and site planning and design. With notable exceptions, most of the literature for planning and design is directed toward recounting historic or contemporary successes, and most criticism among professionals is focused on esthetics. Effective means of criticism and review between professionals would strengthen the state of practice. Study of what has not worked and why are at least as useful as celebrating another success. Colleges and universities would seem to be the best places to promote this type of thinking.

Finally, although much research may be done, there are few places where this research is made available or disseminated within the profession. Much of the research that is promoted tends to be conducted by manufacturers, and while some of it is interesting, there is rarely any review or criticism. Site planning and design is an interdisciplinary profession of applied science, technology, and art. This should serve to increase the opportunities for research.

Appendix

A

Environmental Site Assessments

Brownfields are abandoned or underutilized lands and/or buildings for which redevelopment or reuse is complicated due to the perception or presence of environmental contamination. Estimates of the number of brownfield sites in the United States range from just tens of thousands to more than 450,000. The number of acres of impacted land is uncertain. The industrial heritage of many cities has left them with acres of properties deemed valueless because of the potential environmental liability associated with former industrial practices. These sites are largely the unintended consequence of federal and state legislation that was aimed at cleaning up contaminated industrial sites. Federal and state laws have often required significant cleanup and mitigation efforts to be undertaken before these properties could be reused to their fullest potential. Rather than assume the expense and environmental risks, would-be buyers have walked away from these suspect properties within once-thriving industrial centers. The consequences have been far reaching: deeply eroded tax bases, pockmarks within urban development projects, lost employment opportunities, underutilization of existing infrastructure, and the sprawl of industrial development into suburban and exurban greenfields. The resulting spreading out of industrial and commercial sites has had its own consequences from increased use of automobiles, reduced environmental quality in outlying areas, and increased pressure on the limited resources of suburban and rural communities.

At the heart of the issue is the liability and uncertainty created by the Comprehensive Environmental Response, Compensation and Liability Act (CERCLA) passed in 1980, widely known as the "Superfund." CERCLA mandates among other things the cleanup of contaminated sites. Sites are evaluated for cleanup and given a score and assigned to the CERCLA list with a ranking. Sites that rank high enough on the point system are placed on the National Priority List (NPL). The EPA administers superfund programs through the Comprehensive Environmental Response and Liability

Information System (CERCLIS) list, an extensive database containing information on contaminated sites.

Liability at the Federal Level for Cleanup Costs

Under the Superfund law, liability is assigned according to the concepts of *strict liability* and *joint and several liability.* Simply put, strict liability is the assignment of liability without fault, and joint and several liability asserts that any responsible party can be held liable for the entire cost of a cleanup. Under the Superfund law, a *responsible party* (RP) may be liable for costs incurred by the EPA and state agencies for assessment, removal, and remediation; costs incurred by other persons for assessment, removal, and remediation, costs for damages to natural resources, and costs of health and environmental assessments. The Superfund law goes on to extend liability to any of the following:

Current owners

Current operators

Any prior owner or operator who owned the property when hazardous substances were disposed of

Generators of hazardous substances

Transporters that brought substances to the site

Protection from Liability for Cleanup Costs

There are several ways by which a landowner or prospective buyer may protect himself or herself from the onerous liabilities that might be imposed under CERCLA. The only defenses recognized by the statute are *acts of God or war*; *a substantiated claim of minimal involvement*; or an *innocent landowner defense.* To be considered an "innocent landowner," a person must be able to demonstrate that he or she acquired the property through no fault of his or her own (inheritance) or he or she acquired the property in an action to secure a loan or other fiduciary action (but did not act as an operator). To be "innocent," he or she must further show that he or she did not know or have reason to know of the contamination on the site. To meet this test, a purchaser or lender must show that he or she attempted to protect himself or herself through the exercise of "good commercial or customary practice." A purchaser who acquires property or becomes an operator of the property after contamination is present is not liable if, at the time of acquisition, he or she made a commercially reasonable investigation and as a result had no reason to believe the site was contaminated [42 USC sec. 9601 (35) (A) and 9607 (b)].

Further, CERCLA requires that property buyers make "all appropriate inquiry into the previous ownership and uses of the property consistent with good commercial or customary practice in an effort to minimize liability" [42

USC sec. 9601 (35) (A)]. When a court is evaluating the sufficiency of a person's inquiry, the judge is to consider "any specialized knowledge or experience on the part of the defendant, the relationship of the purchase price to the value of the property if contaminated, commonly known or reasonably ascertainable information about the property, the obviousness of the presence or likely presence of contamination by appropriate inspection" [42 USC sec. 9601 (35) (B)].

The congressional history of CERCLA has been important and used by the courts to further expand the definition of acceptable "due diligence." In part, the congressional history reads:

> The duty to inquire under this provision shall be judged as of the time of acquisition. Defendants shall be held to a higher standard as public awareness (grows), as reflected by this Act, the 1980 Act, and other Federal and state statutes. Moreover, good commercial or customary practice with respect to inquiry in an effort to minimize liability shall mean that a reasonable inquiry must have been made in all circumstances, in light of the best business and land transfer principles. Those engaged in commercial transactions should, however, be held to a higher standard than those who are engaged in private residential transactions. (H.R. Rep. no. 962, 99th Cong., 2nd sess. 187, 1986).

In other words, lenders, commercial real estate developers, and the like will be held to a higher level of inquiry. The "higher level" is undefined. Over the years various efforts have been made to provide a definition in law or regulation, but to date none has survived. "Good commercial or customary practice" has come to mean a Phase I environmental site assessment, at a minimum.

Liability at the State Level for Cleanup Costs

In addition to the concern about cleanup liability, potential buyers and existing owners were faced with uncertainty because there were no guidelines for the cleanup of non-CERCLIS sites—that is, hazardous material generation and/or disposal sites for substances that were permitted at the federal level but not the state level. Since the early 1990s more than 32 states have initiated their own programs to provide cleanup guidance. In this book, state brownfield programs are discussed in general terms in Chap. 7. More than 20 states have entered into *memorandums of agreement* (MOA) with the EPA. These agreements provide some assurance that the EPA will not override the states' programs. There are, however, other important federal initiatives that may affect the state programs that are discussed in the following paragraphs.

Federal Initiatives That Impact Liability at the State Level for Cleanup Costs

The EPA issued a policy statement in 1995 that said it would not take action against the owner of a property where hazardous substances are present solely as a result of subsurface migration in an aquifer from sources outside of the property. This is commonly referred to as the *Offsite Contaminated Aquifers Policy.*

The *Comfort/Status Letters Policy* has been developed in the absence of a federal brownfield statute. Under this policy the EPA has established a practice of providing "comfort letters" to eligible prospective owners. The purpose of these letters is to provide information about the site and about the EPA's intention to exercise Superfund authority. These letters do not provide a liability release or assurance that the EPA will not pursue the landowner for cleanup costs if additional information or contamination is found. The "comfort letters" are based on existing and known information only.

Also in 1995, the EPA issued guidance for entering into agreements with prospective purchasers of contaminated properties if there is to be a substantial direct benefit to the agency in terms of cleanup or a substantial indirect benefit to the community. Essentially the agreement is a covenant between the purchaser and the EPA that in return for the benefits, the EPA agrees not to sue the purchaser.

In January of 2002 President Bush signed the Small Business Relief and Brownfield Revitalization Act into law. The law provides specific relief from some liability under Superfund for small businesses and nonprofit organizations. The law also provides grants and loans for site assessment and cleanup similar to the Brownfield programs administered by the EPA since the mid-1990s. Perhaps the most significant element of the law is the change from the strict liability provision of CERCLA. Section 222 of the law exempts prospective purchasers and their tenants from the owner/operator liabilities of the Superfund law so long as they do not interfere with cleanup or restoration activities. The law also provides definition for the "all appropriate inquiries" test necessary to qualify as an "innocent landowner" under Superfund. The new law directs the EPA to develop regulations defining the standards and practices which will comprise all appropriate inquiry by January of 2004.

Environmental Site Assessments

There are any number of site-assessment protocols published in the marketplace. Various professional associations, private interests, and public agencies have published different versions of site-assessment methods; however, the predominant protocol and, ostensibly the measure of professional practice most widely used, is the *ASTM Standard Guidelines*. The American Society for Testing and Materials (ASTM) is a volunteer organization, made up of representatives from government, industry, and academia, whose purpose is to review technical subjects so as to establish standards and protocols. The ASTM E.50 Committee is the committee responsible for environmental assessment. The ASTM standards are very carefully crafted so that language is important. At various times the ASTM has been criticized for developing "standards" that limit the discretion of practicing professionals. The ASTM E-1527 standard identifies its purpose as follows:

> To define good commercial and customary practice in the United States for conducting an environmental site assessment of a parcel of commercial real estate with

respect to the range of contaminants within the scope of CERCLA and petroleum products. As such this practice is intended to permit a user to satisfy one of the requirements to qualify for the innocent landowner defense to CERCLA liability.

Terminology

The language of the ESA is fairly precise, and users should be aware of the meanings of particular key words and phrases as they are used in the reports. For example, a *recognized environmental condition* is a condition in which there is a presence or likely presence of any hazardous substances or petroleum products on a property under conditions that indicate an existing release or a material threat of a release of any hazardous substances or petroleum products. The term includes hazardous substances or petroleum products that have been used, stored, and disposed of in compliance with the law, but it is not intended to include *de minimis* conditions. Conditions that might be considered *de minimis* are very limited incidents of contamination like oil stains from parked cars—that is, small quantities that do not represent a material risk of harm to people or the environment.

ESAs are conducted to a large degree on the basis of existing information. The protocol says the information used in the ESA is limited to what is *practically reviewable.* In other words, it is limited to information that is available in such a form that substantive analysis of the data is not necessary. Records that cannot be retrieved by reference to the property location or geographic area are not generally considered practically reviewable. All ASTM standards include a section called "Significance and Use," which describes the application of the standard and its limitations. Section 4.5 of the E-1527 standard guideline lists the principles under which a site assessment is to be undertaken and used. The E-1527 standard clearly acknowledges that Phase I ESAs are not sufficient in scope to eliminate uncertainty and that by definition, the Phase I is not an exhaustive investigation. The nature of the Phase I and the range of properties that might be subject to an ESA require that the practice be flexible.

Phase I ESAs

Section 6 of the protocol describes the scope and depth of the Phase I investigation. The typical Phase I process consists of four parts: a review of existing records of activity and site conditions, interviews with knowledgeable persons, a firsthand site reconnaissance, and a report that summarizes these activities and the relevant findings. The typical Phase I does not include the collection or analysis of samples. Section 6 also describes who may conduct an environmental site assessment under the ASTM protocol and that person's duties. While the professional is required to conduct the interviews and the site reconnaissance, he or she is not required to independently verify all information but rather may rely on information provided by others unless there is actual knowledge that information is incorrect or unless it is obvious that it is incorrect.

The objective of the *records review* is to obtain and review records that will help identify recognized environmental conditions in connection with the property. Information pertaining to environmentally significant permits, actions, activities, or events may be found in a variety of databases. *Minimum search distances* (MSDs) to be used in accessing these databases are described in the standard as "approximate" and may be reduced at the discretion of the environmental professional in consideration of the local density—that is, the distance hazardous substances might be expected to migrate based on geology, hydrogeology, or other conditions. Some distances are qualified as minimum search "property only." The recommended minimum search distances in the ASTM standard are the following:

EPA CERCLA National Priority List (NPL), 1 mi

CERCLIS list, 0.5 mi

Resource Conservation and Recovery Act (RCRA) Treat, Store, and Dispose (TSD) list, 1 mi

RCRA generators list, property and adjoining property

Emergency Response Notification System (ERNS) list, property only

State NPL/CERCLIS lists, 1 mi

State landfills, 0.5 mi

Leaking Underground Storage Tank (LUST) list, 0.5 mi

Registered underground storage tanks (USTs), property and adjoining properties

Accuracy and completeness vary from record to record and site to site. The user or environmental professional are not obligated to identify mistakes or to confirm information provided except to make a reasonable effort to compensate for mistakes or insufficiencies that are obvious or in conflict with other information. Further, the search is limited to records and information that are "reasonably ascertainable" and from standard sources. The environmental professional is not required to identify, obtain, or review every possible record that might exist; only sources and data that are reasonably ascertainable from standard sources. The standard defines "reasonably ascertainable" as that information which is (1) publicly available, (2) obtainable within reasonable time and cost constraints, and (3) practically reviewable. "Reasonable time and cost" are considered to be a nominal cost within 20 days of a request. "Practically reviewable" means that the information pertaining to the site is accessible without excessive analysis. Information should be accessible for a limited geographic area. Information should be accessible by address, Zip code, county, and so on. If information is not reasonably ascertainable or available, the environmental professional should attempt to find information from other sources such as by *interviewing* owners, neighbors, or local officials. Sources of information shall be documented—name, date, last update of the database, and so on.

In addition to interviews, existing records are used to provide *historical information.* Typical sources may include the following:

Department of health or public safety

Fire department

Planning department

Zoning and/or building permit

Water department

Local utilities

Lists of local landfills, hazardous waste sites, spills, fires, public and/or private wells

USGS (U.S. Geological Survey) maps; groundwater, bedrock geology

USGS (U.S. Geological Survey) topographical maps

Soil Conservation Services soil maps

Department of Interior wetlands maps

Historical USGS topographical maps

Aerial photography

Recorded land title records

Standard historical sources

Insurance maps

City directories

Property tax files

Street directories

Building department records

Zoning and/or land use department records

The degree to which historical information is sought and used is usually based on what is reasonably ascertainable and appropriate. For example, land records may go back to 1700 for a property, but if the property was undeveloped until 1960, there is usually little point in going back that far. Long periods of the same use do not need a further investigation (for example, if an apartment building in 1940 is still an apartment building in 1960, no interim check is required).

The purpose of *firsthand site reconnaissance* is to visually and physically observe the property and any structures to the extent that the investigation is not obstructed by bodies of water, adjacent buildings, or other obstacles. Exterior areas are viewed from all adjacent thoroughfares. The purpose of roads or access points is determined. Areas of stressed vegetation, stained soils, areas of fill, disposal areas, material storage areas, adjacent property uses, and general housekeeping are observed and noted. Building interiors, including accessible common areas, hallways, utility areas, boiler rooms, maintenance and repair areas, and a representative sample of occupied areas are inspected.

During the site visit proceedings, the current use of adjoining properties and, to the extent possible, the past uses of subject property and adjoining properties are determined. The physical characteristics of the site such as topography, aspect, drainage, existing water bodies, roads, descriptions of structures, paved areas, water supply, sewage disposal, existing material storage area, presence of drums or tanks, odors, pools of liquid, heating and/or cooling methods, stains or corrosion, floor drains, sumps, pits, ponds, lagoons, solid waste disposal facilities, and wells are observed and noted.

The purpose of the *interview* is to acquire more information about current and historic uses and conditions observed in the site reconnaissance or found in the record search. Interviews may be conducted in person, over the phone, or in writing, and they may be done before, during, or after the site visit. Typical candidates for interviewing include the site manager, occupants, owner, local officials, and neighbors. Persons interviewed should be asked to be as specific as possible to the best of their knowledge, but, of course, they are under no obligation to answer questions. If possible, the environmental professional should acquire copies of documentation that may support the interview. Typically such documents include the following:

Previous site-assessment reports

Environmental audit reports

Environmental permits or records

Registration of tanks

Material safety data sheets (MSDSs)

Right-to-know or emergency plans

Site-specific reports, drawings, and so on

Correspondence

The recommended *report format* is described in Appendix X2 of the ASTM Standard Guide. Briefly, the report should consist of an executive summary, supporting documentation (including references, copies, records of conversations, photographs), a professional résumé and/or the credentials of the environmental professional(s), opinions, findings, and conclusions.

Phase II ESAs

The Phase II environmental site assessment is intrusive, and samples are collected and analyzed. Phase II assessments are performed when common sense or a Phase I indicates the potential for or an actual presence of environmental contamination. Phase II is usually directed to a specific finding, for example, an area of stained soil, a known underground storage tank, a chemical storage area, or an area of suspected contamination. The purpose of the Phase II is usually to confirm the suppositions of the Phase I and to identify the limit and extent of contamination usually in three dimensions. It may be necessary to conduct an expanded Phase II investigation to meet the requirements of a state voluntary cleanup program. It is by definition also a reiterative process.

Sampling plans and methods for Phase I and Phase II ESAs

Some suspected areas can be preliminarily identified using historic topographic maps, aerial photographs, or firsthand observation. These resources assist the environmental professional in developing a sampling strategy and plan. The collection of samples should be undertaken only after an understanding of the objectives of the sample collection are established and the limitations of the plan are agreed upon. Samples have limited value. Depending on the nature of the site and the objective, the sampling plan can vary widely from the taking of a very few samples to a complex three-dimensional sampling plan. A sampling plan may include a list of the sample collection methods, equipment to be used, the rationale for each choice, the number of samples to be collected, the depths and location where they will be collected, and a decision as to whether the samples will be composited.

Compositing samples involves thoroughly mixing individual samples of equal weight and performing only one analysis. For the sake of comparison, the results are typically multiplied by the number of individual samples. The advantage of using a composite is that doing so reduces the sampling cost substantially. The disadvantage is that in the case of a positive result, there is no way to know where the problem may be.

Other elements of the sampling plan include a site plan showing the sample location (approximate), the equipment decontamination procedures to be used, the analytical methods to be used, the name and location of the laboratory, the quality-assurance and quality control (QA/QC) methods, the chain of custody, a demonstration of compliance with regulatory requirements or practices, a plan for the disposal of wastes generated by the sampling plan, and a health and safety plan for the field staff.

Where to take samples on the site may be obvious from the Phase I findings; however, there remain protocols to be observed. Sampling programs may be divided into one of four groups *judgmental, random, systematic,* or a *combination of the first three.* Within those groups, samples are of three types, *grab, composite,* and *integrated. Judgmental sample programs* are based on specific information about a release or contamination such as observed areas of stained soils or areas known to have been affected by a spill. The advantage of this approach is that it tends to be cost effective; however, it is definitely biased toward the worst-case situation, and marginal areas may be missed.

Random sample programs are developed to evaluate all areas of a site by dividing the site into a grid and using a random selection of grid intersections to choose sample locations. Each location has an equal chance of being selected for evaluation, and results are therefore said to be indicative of an average condition. This method is used in those situations in which statistical analysis is to be used and is appropriate or information is to be used for "legal" purposes. Such sites might include areas of widespread contamination. Advantages of this approach are that it is a low-bias, sitewide approach. Its disadvantages are that grid design is often restricted by inaccessible areas or areas where sampling isn't feasible (street surfaces) and that random selection of locations can result in clustering of samples. The number of samples to be collected in a random sampling process is determined using the formula provided in EPA SW-486. In general, a grid is established that will provide enough sample locations so that 10 percent of the site is covered by the requisite number of samples.

Systematic sample programs are similar to random programs to the extent that the site is divided into a grid, but in this case samples are collected at each grid intersection. Grid patterns do not have to be square. They could be rectangular or circular (around a tank), depending on site shape and geometry. Difficult or inaccessible locations can be eliminated. The advantages of the systematic system are that it affords low-bias sitewide coverage and the elimination of inaccessible areas. Disadvantages lies in the judgment used to select the grid frequency and the limited application for statistical analysis. In addition, it is easily challenged, and it can be expensive.

Combination sample programs are a blend of the random and systematic methods. Limited judgment is used to eliminate or at least minimize bias. These programs can be used to circumvent the biggest disadvantages of systematic or random programs, but the combined method is not completely random and the results can easily be challenged if the selections are not carefully documented.

Grab samples are individual samples taken from a specific location at a specific time. This is an appropriate method to use when the size, shape, location, and nature of a suspected release are defined.

Composite samples are made up of a combination of samples. They are used to make up a "representative" sample or to produce enough material quantity for analysis. Analysis represents an average condition—that is, sample locations could have higher results individually than are found in the composite. Composite samples are used for initial assessment purposes where the worst-case values can be extrapolated from the results.

Integrated samples are a composite of material taken from the same location over a period of time, and they are used in such situations as effluent discharges or waste streams.

The collection of the soil sample itself is a function of the homogeneity of the soil, the ease of collection, the ease of decontamination, and the construction of the sampling equipment. Shallow soil samples are typically collected with a *hand auger* that is manually rotated into the ground. Hand augers can be used up to 6 ft or so in depth in soft soils without interference. There are *power-driven hand augers* that can go a bit deeper. Sample material is collected into a tube (usually brass but sometimes plastic) that fits inside the tube. Other shallow soil sampling equipment could include a *shovel,* a *trier* (a hollow rod that is used to collect a core sample), a *thief* (a hollow tube within a hollow tube that is closed until a particular depth or material is reached at which time it is opened and inserted into the material to collect a sample), or a *trowel.*

Deeper soil samples are collected using *truck-mounted equipment* such as a *well-drilling rig* or *direct push technology.* Depths of from ten to several hundred feet are possible. *Hollow-stem augers* and *split-spoon augers* are commonly used to produce undisturbed samples. Soil sampling devices are inserted into the core at particular depths to gather samples. The core is evaluated for color, smell, moisture, and composition at the time the sample is taken.

Liquid samples can be collected from *drums, tanks, surface water using dippers, glass tubes,* and *composite liquid waste samplers* (COLIWASAs). Pumps can be used to assist in the collection, but they should be of a type and quality that do not introduce air into the sample.

Groundwater samples are collected by drilling a well and installing a *monitoring well. Temporary groundwater sampling stations* are those for which a well is not installed. These are often discouraged because the results are not repeatable. Monitoring wells are always purged before a sample is taken. A well is *purged* by removing from 3 to 5 well casing volumes (determined by multiplying the total depth of the well from the ground surface to the bottom of the water column by the cross-sectional area). The purpose of the purge is to obtain a representative sample of the groundwater rather than a ample that has been sitting in the well. Temperature, pH, and conductivity measures are taken of the sample. Purge water is collected into drums or another receptacle and handled according to the analysis results.

Quality assurance is a necessary part of the sampling process. There are established procedures and methods for meeting the quality assurance necessary for a responsible sampling process. These methods include *matrix spikes, laboratory blanks,* and *field duplicates.* Matrix spikes are duplicate field samples that are spiked in the laboratory with measured quantities of the contaminant. The volume of the added material can then be subtracted from an analyzed result to get an accurate analysis. Laboratory blanks are laboratory-grade samples that are analyzed the same way as the samples. Generally speaking, a lab blank is used for each 10 samples per round of samples. Field duplicates are extra samples of the same material used to check analysis accuracy. Duplicates are usually selected from areas suspected of having the highest contaminant concentrations.

Information may be discovered during the Phase II ESA that could indicate that the owner of a property is out of compliance with a federal or state environmental regulation, and the information might have significant economic repercussions for the owner, either in the form of additional work or cleanup or in the devaluation of the real estate. Reporting must therefore always be factual and limited to what is known.

Redevelopment Strategies

It is critical to understand the project objectives as well as the character of the site when evaluating redevelopment strategies. The strategies used for an urban redevelopment project that is strictly a real estate development undertaking may be very different from the strategies used for a project undertaken with additional economic or social development aspects. In either case, consideration must be given to the existing conditions and the degree of effort and expense necessary to mitigate site conditions for the proposed use.

Understanding the site character is the focus of the Phase I and II environmental site assessments. Site issues that influence redevelopment planning include the physical characteristics (Phase I and II, site evaluation issues such as storm water runoff, slope, floodplain); the real estate characteristics (location, utilities, site access, adjoining uses, zoning, and so on); and the market characteristics (location, proposed use, anticipated future use, bankability). Successful redevelopment and response action planning must consider all of the facets of redevelopment.

Based on these investigations, the site conditions are identified and data are analyzed to determine the risks associated with the redevelopment of the site. These risks are evaluated to identify the most effective and appropriate development response. In general, a redeveloper must choose from several standards for the cleanup of a site. While the standards differ from state to state, in general, the choices include a pollutant-specific, uniform risk-based standard, local background pollution levels, federal or state standards (water quality standards are often used), and site-specific risk-based standards. Once the standard is selected, a response method for meeting the criteria is selected. The possible responses include the following:

No action required

Avoidance

Administrative controls

Design controls

On-site remediation

Off-site remediation

In some cases the degree of contamination is so low that specific action is not necessary. In other cases small areas of contamination can be left in place, undisturbed in the redevelopment process so that no specific mitigation is required. By locating site features such as storm water management facilities away from areas of marginal contamination or by locating impermeable surfaces such as parking space over these same areas, more expensive controls or mitigation can be avoided.

Administrative controls are used to manage risks associated with contamination by limiting access through the use of deed restrictions. These controls generally limit the use of the site to a specific application or specific maintenance performance. Design controls, on the other hand, are site redevelopment features designed to reduce exposures to contamination or to provide passive treatment of contamination. Such features may include the coordination of impermeable surfaces to isolate contamination, the selection of contaminant-resistant vegetation, or the installation of fences or landscape barriers to isolate areas.

Remediation technologies are wide ranging. Since remediation is usually undertaken in the early phases of a project, typical brownfield sites will have relatively predictable remediation demands. As a general rule of thumb, the more short term and expensive the remediation, the greater the liability reduction it offers.

The purpose of the public policy shift is to encourage the investigation and to accelerate the cleanup of eligible properties. The public agenda is not concerned with the development program beyond the issues of public health and safety and the protection of the environment. To succeed, the redevelopment program must have a range of concerns wider than strictly the cleanup aspects of a site or building. The site must work for the proposed program. The redevelopment plan and the environmental action plan must be inclusive. In designing a response plan that accomplishes the site program, designers must consider the following:

Remediation costs (initial costs and life cycle costs)

Impact on current and future users

Future impacts (changes in regulations, potential for exacerbation, and so on)

"Bankability"

The developer or landowner, however, must also consider the business and risk management aspects of the plan, including the following:

Remediation costs (initial and life cycle costs)

Impact on users (effect on potential buyers or tenants)

Future impacts (changes in regulations, potential for exacerbation, and so on)

Impact of property restrictions on the marketability of the property

Reopeners

Nearly every state program includes *reopeners*—those conditions under which the state may revoke the letter of no further action or the certificate of completion if one or another reopener occurs. Reopeners usually include the following:

Imminent and substantial endangerment to the public health and environment

Certificate obtained through fraud or misrepresentation

Remedial response action failing to meet criteria

Use of land not in compliance with restrictions

Previously undiscovered contamination (reopens to RP)

New contamination or a change in condition that exacerbates contamination

Maintenance commitments not met

Environmental Insurance

Landowners and developers may also choose to purchase *environmental insurance* as part of the redevelopment and risk management strategy. Although there is a great deal of flexibility in the marketplace in terms of specific insurance, the environmental insurance market is composed of basically two types of insurance: stop-loss insurance and real estate insurance. These insurance cover specific types of insurable risks such as instances in which environmental restoration can exceed anticipated costs primarily due to preexisting but previously unknown contamination, excessive remediation costs due to changes in regulations, or contamination greater than anticipated. Insurance could also cover third-party or postremediation liability (migration off site, bodily harm, stigmatization, loss of market value, and so on).

Stop-loss insurance is purchased by the owner to provide a guarantee of the proposed cost of the remediation project. Generally it covers an exposure equivalent to 100 percent of the remediation cost estimate, although variations of this can be underwritten. Some insurers require approval of the response action plan, and others do not; however, some do not cover costs increased due to regulatory change. Stop-loss insurance is usually priced from 8 to 20 percent of the policy limit value and remains in effect until the work plan is completed or a specific policy limit is reached. There is no standardization of this type of insurance.

Environmental real estate insurance is written to protect current and future landowners. These policies may cover remediation costs that are incurred due to the discovery of previously unknown contamination, third-party claims for cleanup or injury, and the costs of cleanup required after a letter of no further action has been issued.

Environmental real estate insurance typically has a 5-year renewable term, and it may be renewed annually to insure 5 years of future coverage. Policy premiums range from \$25,000 to \$250,000 for a \$5 million policy limit with a \$50,000 deductible.

Appendix B

A Preparedness, Prevention, and Contingency Plan

As the National Pollution Discharge Elimination System (NPDES) rules on construction sites mature, it should be expected that a *preparedness, prevention, and contingency plan* (PPC plan) will be required on all construction sites. The following sample plan provides an example and an outline for developing such plans.

General Site Layout and Site Conditions

Property boundaries, loading and unloading areas, truck wash areas, storage areas for materials, location of drainage appurtenances, sediment removal devices, and site entrances and exits are to be shown on a site plan entitled "Preparedness, Prevention, and Contingency Plan" (PPC plan). Sources of additional information regarding the site, such as the *Erosion and Sediment Pollution Control Plan and Report,* can be included by references or attached to the plan.

The Emergency Response Plan

The emergency response coordinator for the site should be clearly identified, and an outline of the emergency response organization should be included with this report as Attachment A. Alternate emergency coordinators should be listed on the attachment as well. All listed personnel should be authorized to commit the necessary resources that might be required to carry out and fulfill the objectives of the PPC plan although it is the responsibility of the emergency response coordinator to complete the PPC plan for the site and to oversee and confirm the installation and maintenance of control and containment equipment or structures in accordance with the plan.

The coordinator shall arrange and provide training to all appropriate personnel on the site, including subcontractors reporting to the operator or responsible to the operator. Training will cover the elements of the waste disposal plans for the site, emergency response procedures for spills and releases, and material storage requirements, as well as the location of safety and emergency response equipment and material safety data sheets. The coordinator and the site superintendent shall oversee and confirm the location of emergency response and safety equipment, the installation of PPC plan elements such as waste disposal locations, material storage locations, and security devices or equipment to confirm compliance with the plan. If appropriate, local emergency response services (fire, police, and so on) should be contacted regarding the project and the PPC plan, and copies of the plan and site drawing should be provided to these agencies. Periodic site inspections should be conducted to confirm operational compliance with the elements of the plan, and the plan should be revised and adapted as changes in site conditions or circumstances require.

All appropriate personnel should be trained and authorized to initiate emergency response procedures to control and mitigate spills and releases on the site including but not limited to employing equipment and materials available on the site. A qualified contractor should be on retainer to contain, clean up and remove contaminated soils or water or other materials as required. The emergency coordinator shall contact the appropriate authorities in the event of a spill or release. The emergency response coordinator may delegate some of these responsibilities to other staff or contractors. Employees and subcontractors should know that in the event of an imminent or actual emergency situation, alarms should be activated and key personnel should be notified. The emergency response coordinator must assess the character of the spill or release as to the nature of the material, the source of the spill, and the amount and extent of the spill. This assessment must include an evaluation of the threat to human health and safety, or risk to the environment. This assessment must consider direct and indirect effects of emissions, discharge, fire, or explosion. If the emergency response coordinator determines that a threat to health and safety or the environment exists or will exist as a result of the event, the local authorities should be contacted. If appropriate, evacuation of the area should be discussed at this time. If a reportable quantity of a material has been released, it may be necessary to contact state authorities and the National Emergency Response Center at **(800) 424-8802.**

Contacts to the National Emergency Response Center will require the following information:

A. Name of the person reporting the incident

B. Name and location of the site

C. Phone number of the person reporting the incident

D. Date, time, and location of incident

E. Brief description of the incident—nature of materials and/or waste involved, nature of any injuries, and possible hazards to health, safety, and/or the environment

F. Estimated quantity of materials and/or waste released

G. Extent of contamination of land, water, or air

During an emergency, the emergency response coordinator must take all reasonable measures to ensure that fires, explosions, emissions, or discharges do not occur. These prevention measures may include stopping construction operations while a spill is controlled and cleaned up. Once control is established, the emergency response coordinator must provide for the collection, treatment, storage, and disposal of residues, contaminated soils, or materials that may have resulted from the emergency.

Employee Training Program

The employee training program should be given to all employees working on the site, and it should consist of all necessary instruction regarding the PPC plan and its elements from the emergency response coordinator at the startup of the project. This training meeting will be held at the earliest stages of the project, at about the same time the erosion and sediment control facilities are being installed and the earthwork is beginning. Employee training should include:

1. The purpose and legal basis for the PPC plan
2. The steps being taken on the site to comply with the requirements of the law and to protect the environment from spills
3. The location of waste disposal, material storage, and emergency response equipment and facilities, and the methods to be followed for waste disposal
4. Definitions of *spills* and *minor spills,* the steps to be taken in either case including notification of the emergency response coordinator or alternate, and the location of the call list
5. Training in the proper use of equipment, materials, and personal protective equipment required to respond to an emergency
6. Training in housekeeping procedures, techniques, and responsibilities, as well as routine maintenance of control and containment facilities

Note that specific housekeeping and maintenance assignments may be made at this time. A record should be kept of all training sessions. The record should include the date and time of the training as well as the names of those who attended, the name of the trainer, the subject matter covered, and copies of all materials distributed.

Emergency Equipment and Procedures

The project site must have the equipment required to respond to routine spills and to emergencies of a manageable nature as determined by the plan and the judgment of the emergency response coordinator. This equipment shall be maintained in an accessible and clearly marked location. This location should be kept locked to maintain the supply and integrity of the equipment and materials required by this plan. The location of the keys or the personnel with keys should be well known. The emergency response coordinator shall inspect the emergency response equipment at least monthly to confirm that supplies and equipment are accessible and in working order. A permanent inventory list shall be maintained and permanently mounted on the inside of the locker door. The inventory list shall include a complete compilation of the contents of the locker as required by this plan. The emergency response coordinator shall maintain a record of these inspections and any replacement materials required. These inspection records shall be maintained in the project file.

The emergency response equipment should consist of the following items:

- 50-lb bags of absorbent sweeping compound
- Two coarse push brooms
- Two flat long-handled shovels
- One box made of 8-ml plastic in a minimum size of 10 × 100 ft
- Two sealed beam flashlights, with one extra set of batteries for each (Batteries should be marked with the date at the time of installation.)
- Sets of disposable coveralls
- Gloves, including at least one pair impermeable acid-proof gloves
- One roll of duct tape (250 ft)
- Overboots
- A copy of the **emergency response procedures** from this plan
- A copy of the **call list** hung on site but removable from the locker
- A first-aid kit
- A chemical fire extinguisher
- A disposable flash camera
- A copy of the **incident report** form
- A set of nonsparking tools to include a hammer, adjustable wrench, pliers, pump pliers, and flat shovel

Procedures

Fuels and volatile petroleum products shall be stored in approved containers and stored out of doors in a secure, covered area or indoors in a properly ventilated area. Bulk storage of these materials (55-gal container or larger) shall be in a containment area sized to contain 110 percent of the largest single vessel within it.

In the event of a spill or a ruptured or leaking vessel, the material will be caught in the containment area. In the case of small spills, absorbent material will be used to collect the material. Contaminated absorbent material will be collected into an open barrel that is stored in a well-ventilated area. The material will be properly disposed of in accordance with DER regulations.

For spills that are too large to be collected by available absorbent material but that are within the containment area, the integrity of the containment area should be assessed. If the containment area shows signs of stress or imminent failure, steps shall be taken immediately to secure and support the containment. Where the containment is secure, the emergency response coordinator shall contact an oil recovery contractor to remove the spilled fuel. In the case of gasoline or other highly volatile materials, the local fire department shall be called to assess and secure the situation, including the use of foam to suppress the volatiles and reduce the risk of fire and/or explosion.

For spills from smaller vessels outside of the containment area, absorbent material will be used to collect the material. Steps should be taken immediately in the case of such spills to minimize the spread of materials over soils and to water. Such steps might include diverting the direction of flow or containing the spill with available materials.

In the event spills reach soils, the soils shall be removed and placed on plastic sheets and covered. Removal of contaminated soils shall proceed as soon as possible.

In the event spills reach a drainage facility, the situation should be assessed to determine if upstream drainage facilities can be diverted. In all cases the first step is contain the spill.

In the event a fuel oil or gasoline or kerosene spill that is not contained, that reaches the soil or water, and that is greater than a specified number of gallons, the incident should be reported to the **National Emergency Response Center** (800-424-8802), sewer authority, natural resources agency, fire department, police department, and local government.

Construction chemicals

All construction materials shall be stored, used, and disposed of in accordance with manufacturer's instructions, safety and health regulations, and solid waste disposal regulations. All on-site storage and disposal of materials shall conform to this plan. Construction materials shall include paint, acids, cleaning materials, solvents, fertilizers, soil conditioners, and building materials.

Bulk storage of these materials in a liquid form (55-gal container or larger) shall be in a containment area sized to contain 110 percent of the largest single vessel within it.

In the event of a spill or a ruptured or leaking vessel, the material will be caught in the containment area. In the case of small spills, absorbent material will be used to collect the material. Contaminated absorbent material will be collected into an open barrel that is stored in a well-ventilated area. The material will be properly disposed of in accordance with regulations.

For spills that are too large to be collected by available absorbent material but that are within the containment area, the integrity of the containment area should be assessed. If the containment area shows signs of stress or imminent failure, steps shall be taken immediately to secure and support the containment. Where the containment is secure, the emergency response coordinator shall contact an environmental cleanup contractor to remove the spilled fuel. In the case of solvents or other highly volatile materials, the local fire department shall be called to assess and secure the situation, and they will use foam to suppress the volatiles and reduce the risk of fire and/or explosion.

For spills from smaller vessels outside of the containment area, absorbent material will be used to collect the material. Steps should be taken immediately in the case of such spills to minimize the spread of materials over soils and to water. Such steps might include diverting the direction of flow or containing the spill with available materials.

In the event spills reach soils, the soils shall be removed and placed on plastic sheets and covered. Removal of contaminated soils shall proceed as soon as possible.

In the event spills reach a drainage facility, the situation should be assessed to determine if upstream drainage facilities can be diverted. In all cases the first step is contain the spill.

Dry, granular, powdered, or other types of materials shall be stored in a covered secure area. Releases and spills of these materials shall be cleaned up immediately. Care shall be taken to avoid the contamination of soils or surface water.

Cleanup methods may include the collection of these materials by sweeping, shoveling or other means. Contaminated materials unable to be used in the operation will be disposed of according to the manufacturer's instructions and in accordance with regulations. Spilled materials that are collected and deemed to be hazardous waste will be labeled and stored in a secure, properly identified, and protected area. Such materials are subject to hazardous waste storage, transport, and disposal requirements.

Personnel involved in the containment and cleanup of materials that pose a health risk or hazard are required to wear proper personal protective equipment. The emergency response coordinator shall evaluate each event to determine the presence and nature of risk as provided for in the material safety data sheet.

Solid waste

Solid waste shall include, wood, concrete, scrap metal, plastic, masonry products, roofing materials, and other building materials. These materials shall be collected into portable dumpsters located on the site and properly disposed of according to the solid waste regulations of DER. There shall be no burning or burying of waste materials on site.

Miscellaneous materials that may be encountered on the construction site include wash from concrete trucks, waste asphalt, pesticides and/or herbicides, tar, and demolition wastes as well as others.

In addition:

A. An area shall be designated for washing concrete trucks. The rinse water from these trucks can be extremely hazardous to fish. The wash area shall therefore be located well away from surface water or drainage paths. The area may be arranged in a sump condition to contain the washed material and allow it to dry.

B. Waste asphalt shall be collected away from drainage paths, consolidated, and disposed of as construction waste.

C. Pesticides and herbicides represent a significant risk to health and safety of the site personnel as well as to the environment. All such materials shall be used by trained and knowledgeable personnel wearing the proper safety and personal protective equipment, and in accordance with the manufacturer's instructions. At such time as the application is complete, the materials shall be removed from the site. There shall be no bulk storage of herbicides or pesticides on the site. In the event of a spill or release from smaller vessels outside of a containment area, absorbent material will be used to collect the material. Steps should be taken immediately in the case of such spills to minimize the spread of materials over soils and to water. Such steps might include diverting the direction of flow or containing the spill with available materials.

In the event spills reach soils, the soils shall be removed and placed on plastic sheets and covered. Removal of contaminated soils shall proceed as soon as possible.

In the event spills reach a drainage facility, the situation should be assessed to determine if upstream drainage facilities can be diverted. In all cases the first step is contain the spill.

Care should be taken to ensure that all cleanup personnel are wearing impermeable gloves and clothing, and wearing properly equipped respirators.

Inspections

The emergency response coordinator is responsible for conducting routine inspections of all facilities, equipment, and devices described by this plan as part of the prevention, preparedness, and contingency efforts. Attachment C includes an inspection report form. The emergency response coordinator or his or her trained delegate shall complete the inspection at least once per month commencing within the first 30 days of the start of work. The record of these inspections and actions taken shall be maintained in the project file.

The nature of the construction site makes denying access to the site impractical. Vehicular access is limited to the defined entrances and exits on the plan. Material storage and containment areas shall be secured by locked doors, and there shall be valve locks on tanks and drums. The emergency response equipment locker shall be locked. A copy of the call list and the emergency response coordinator's phone number shall be posted in a weather-resistant cover.

References

Chapter 1. Sustainability and Site Design

"Cement and Concrete: Environmental Considerations." *Environmental Building News* (*EBN*). 1993. Vol. 2, no. 2, March/April.

Chapter 2. Site Analysis

Borcherdt, G., and M. P. Kennedy. 1970. "Liquification Potential in Urban San Diego." *California Geology,* vol. 32.

Brick Institute of America (BIA). "Brick in Landscape Architecture, Pedestrian Applications." *Technical Notes on Brick Construction* 29. Brick Institute of America, Reston, Va. *www.bia.org/BIA/technotes/t29.htm.*

Griggs, Gary, and John A. Gilchrest. 1994. *Geologic Hazards, Resources and Environmental Planning,* 2d ed., Wadsworth. Belmont, Calif.

Chapter 3. Site Grading

Brown, D., R. Hallman, C. Lee, J. Skogerbee, K. Eskern, R. Price, N. Page, M. Claz, R. Kort, and H. Hopkins. 1986. *Reclamation and Vegetative Restoration of Problem Soils and Disturbed Lands.* Noyes Data Corp., Park Ridge, N.J.

Goldman, S. J., K. Jackson, and T. A. Bursetynsky. 1986. *Erosion and Sediment Control Handbook.* McGraw-Hill, New York.

Perry, Thomas O. 1991. "Conditions for Plant Growth." *A Reader for Natural Resource Experts. Global ReLeaf Reader.* American Forestry Association.

Chapter 4. Designing for People

Carstens, Diane Y., 1985. *Site Planning and Design for the Elderly,* Van Nostrand Reinhold, New York.

Construction Specifier, The. "Child's Play—Specifying Safety on the Playground." October 1985.

DeChiara, Joseph, and Lee E. Koppelman. 1984. *Time-Saver Standards for Site Planning.* McGraw-Hill, New York.

Gracar, Adam. 1998. "Safer Communities: Strategic Directions in Urban Planning." Paper presented at the Conference for Safer Communities, Melbourne, Australia, September 10–11.

Harris, Charles W., and Nicholas T. Dines. 1988. *Time-Saver Standards for Landscape Architecture.* McGraw-Hill, New York.

Martin, David, 1997. "Research Summary: Preventing Crime: What Works, What Doesn't, What's Promising." Report to the U.S. Congress.

Oakland County Road Commission. "Bike Route."

Pennsylvania Department of Community Affairs. Bureau of Recreation and Conservation. 1990. "Local Government Recreation Service Summary."

Pennsylvania Department of Environmental Resources. Office of Program Planning and Development. 1986. "Pennsylvania's Recreation Plan 1986–1990."

Pennsylvania Governor's Office of State Planning and Development. 1976. "Pennsylvania's Recreation Plan Summary."

U.S. Consumer Product Safety Commission (CPSC). 1980. *A Handbook for Public Playground Safety.*

Urban Land Institute. 1980. "Outdoor Recreation: Resources and Politics." *Environmental Comment 2.*

White, Rob. 1988. "Public Places and Community Crime Prevention." Paper presented at the Conference for Safer Communities, Melbourne, Australia. September 10–11.

Chapter 5. Street and Parking Lot Design

Appelyard, Donald, and Mark Lintell. 1972. "The Quality of City Streets: The Resident's Viewpoint." *Journal of the American Institute of Planners,* vol. 38, no. 2, March.

Burley, Jon, Dr. Ramey Rogness, and Cheryl Burley. 1993. "Edge Scour: Developing and Repairing a Grassed Road Way Shoulder." International Erosion Control Association, Proceedings of Conference XXIV.

Center for Watershed Protection. 1998. *Better Site Design: A Handbook for Changing Development Rules in Your Community.* Prepared for the Site Planning Roundtable.

Engstrom, Mark, and Marc Putnam. *Planning and Design of Townhouses and Condominiums.* Urban Land Institute, 1200 18th Street NW, Washington, D.C. 20036.

Harris, Charles W., and Nicholas T. Dines. 1988. *Time-Saver Standards for Landscape Architecture.* McGraw-Hill, New York.

Hornburger, W. S., E. A. Deakin, P. C. Bosselman, D. T. Smith, Jr., and B. Beukers. 1989. *Residential Street Design and Traffic Control.* Institute of Transportation Engineers. Prentice-Hall, Englewood-Cliffs, N.J.

Markowitz, F. 1995. *Shared Parking Planning Guidelines.* Institute of Transportation Engineers, August.

Land Development, National Association of Home Builders (NAHB). 1988. "Street Standards Survey Finds Narrower Streets Perform Well."

Chapter 6. Infrastructure

Burley, Jon, and Dr. Ramey Rogness, and Cheryl Burley. 1993. "Edge Scour: Developing and Repairing a Grassed Road Way Shoulder." International Erosion Control Association, Proceedings of Conference XXIV.

Cahill, Thomas H., P. E., Michele Adams, and Wesley Horner. 1988. "The Use of Porous Paving for Groundwater Recharge in Storm Water Management Systems." *Proceedings of the Floodplain/Storm Water Management Symposium.* State College, Pa.

Center for Watershed Protection. 1998. *Better Site Design: A Handbook for Changing Development Rules in Your Community.* Prepared for the Site Planning Roundtable.

"Effects of Urbanization on Water Quality," *Proceedings of the American Society of Civil Engineers, Transportation Division.* Environmental Impact.

"Impact of Land Use and NPS Loads on Lake Quality." *Journal of Environmental Engineering.*

Landpahir, Harlow C., and Fred Klatt, Jr. 1988. *Landscape Architecture Construction,* 2d ed. Elsevier Science Publishing, New York.

Maryland State Department of Education. 1999. *Conserving and Enhancing the Natural Environment: A Guide for Planning, Design, Construction and Maintenance on New & Existing School Sites.* Baltimore, Md.

Miller, Charlie, P.E. 1998. "Vegetated Roof Covers—A New Method for Controlling Runoff in Urbanized Areas." *Proceedings of the Pennsylvania Stormwater Management Symposium.* Villanova University, Villanova, Pa.

"Stormwater Management: Planning the Future!" 1995. *Proceedings of the Conference Stormwater Management, Planning for the Future,* Linthicum Heights, Md., November.

"Storm Water Recharge Beds Replace Detention Basins." 1988. *Highway and Heavy Construction,* July.

Chapter 7. Landscape Restoration

Brown, D., C. L. Hallman, J. Skogerbee, K. Eskern, and R. Price. 1986. *Reclamation and Vegetative Restoration of Problem Soils and Disturbed Lands.* Noyes Data Corp., Park Ridge, N.J.

Darmer, Gerhard. 1992. *Landscape and Surface Mining: Ecological Guidelines for Reclamation.* Edited by Norman L. Dietrich. Van Nostrand Reinhold, New York.
Goldman, S. J., K. Jackson, and T. A. Bursetynsky. 1986. *Erosion and Sediment Control Handbook.* McGraw-Hill, New York.
Hunter, Christopher J. 1991. *Better Trout Habitat.* Island Press, Washington, D.C.
Koenig, Bill. 1992. "Back to Nature, Lake Forest Gets Environmental Award." *The Indianapolis Star,* March 15.
Maryland Water Resources Administration. 1983. *Maryland Standards and Specifications, Soil Erosion and Sediment Control.* Baltimore, Md.
National Association of Home Builders (NAHB). 1990. "Sales Blossom on Wooded Sites:" *BUILDER,* July.
Perry, Thomas O. 1991. "Conditions for Plant Growth." *A Reader for Natural Resource Experts.* American Forestry Association.
Rogoshewski, P., H. Bryson, and K. Wagner. 1983. *Remediation Action Technology for Waste Disposal Sites.* Noyes Data Corp., Park Ridge, N.J.
Sobek, A., W. A. Sknuller, J. R. Freeman, and R. M. Smith. 1976. *Field and Laboratory Methods Applicate to Overburden and Minesites.* Industrial Environmental Research Laboratory.
Urban, James R. 1989. "Evaluation of Tree Planting Practices in the Urban Environment." *Proceedings of the Fourth Urban Forestry Conference,* St. Louis, Mo., October.
———. 1991. Presentation at the 1991 Annual Meeting of the American Society of Landscape Architects, Kansas City, Mo., November.
Vogel, Willis G. 1987. *A Manual for Training Reclamation Inspectors in the Fundamentals of Soils and Revegetation.* Prepared for the Office of Surface Mining and Enforcement by the U.S. Department of Agriculture. Berea, Ky., p. 67.
Zion, Robert L. 1968. *Trees for Architecture and the Landscape.* Von Nostrand Reinhold, New York.

Chapter 8. Site Layout

Arendt, Randall, 1991. "Cluster Development: A Profitable Way to Save Open Space." *Land Development,* fall.
Dramstad, Wenche E., James D. Olson, and Richard T. T. Forman. 1996. *Landscape Ecology Principles in Landscape Architecture and Land-Use Planning.* Island Press, Washington, D.C.
Fehr, Stephen S. 1991. "Reducing Land Use Barriers to Affordable Housing." *Planning Series No. 10." Planning Services Division of the Bureau of Community Planning, Pennsylvania Department of Community Affairs.*
Hylton, Thomas. 1995. *Save Our Lands, Save Our Towns: A Plan for Pennsylvania.* RB Books, Harrisburg, Pa.
Kreager, William. 1992. "Building Small Lot Homes in Your Community." *Land Development,* winter.
National Association of Home Builders (NAHB). 1986. *Cost Effective Site Planning: Single-family development.*
Naveh, Zeev, and Arthur S. Lieberman. 1993. *Landscape Ecology: Theory and Application,* 2d ed. Springer-Verlag, New York.
Sanders, Welford, and David Mosena. 1982. *Changing Development Standards for Affordable Housing,* Planning Advisory Service Report No. 371. The Joint Venture for Affordable Housing, American Planning Association.
Sanders, Welford, Judith Getzels, David Mosena, and JoAnn Butler. 1984. *Affordable Single Family Housing,* Planning Advisory Service Report No. 385. American Planning Association.
White, Rob. 1998. "Public Spaces and Community Crime Prevention." *Safer Communities: Strategic Directions in Urban Planning Conference,* Melbourne Australia.
Yaro, Robert D., Randall G. Arendt, Harry L. Dodson, ASLA, and Elizabeth A. Brabec, RLA. 1990. *Dealing with Change in the Connecticut River Valley: A Design Manual for Conservation and Development,* 4th printing. Lincoln Institute of Land Policy, Cambridge, Mass., and The Environmental Law Foundation, Oakland, Calif.

Chapter 10. Project Management Issues

Doherty, Tara, Melanie Murphy, and Riyaz Lalani. "Carbon Sequestration in City of Kitchener Parks." www.adm.uwaterloo.ca/infowas.

Kerzner, Harold. *Project Management,* 3rd ed. Van Nostrand Reinhold, New York.

Petroski, Henry. 2000. *Design Paradigms, Case Histories of Error and Judgment in Engineering.* Cambridge University Press, Cambridge, U.K.

Chapter 11. Historic Landscapes and Preserving Land

American Farmland Trust (AFT). 1992. *Does Farmland Protection Pay? The Cost of Community Services in Three Massachusetts Towns.* American Farmland Trust, Washington, D.C., June.

Babcock, Richard F., and Charles L. Siemon. 1985. *The Zoning Game Revisited.* Lincoln Institute of Land Policy, Cambridge, Mass., pp. 136–157.

Coughlin, Robert E. 1991. "Formulating and Evaluating Agricultural Zoning Programs." *Journal of the American Planning Association,* vol. 57, no. 2, spring, pp. 183–191.

Coughlin, Robert E., and John C. Keene. 1988. *National Agricultural Lands Study.* Regional Science Research Institute, Amherst, Mass.

Closser, Jay E., ASA. 1993. *Assessing Land Affected by Conservation Easements, A Resource Manual.* Lincoln Institute of Land Policy, Cambridge, Mass., pp. 4–8.

Czupryna, James J., ASA. 1983. *Newsletter of the American Society of Appraisers,* March.

Downing, Andrew Jackson. 1980. *Victorian Cottage Residences.* Dover Publications, New York.

Favretti, Rudy J., and Joy Putnam Favretti. 1995. *Landscapes and Gardens for Historic Buildings,* 2d ed. rev. *American Association for State and Local History Book Series.* Altamira Press, Walnut Creek, Calif.

Fischel, William A. 1990. "Do Growth Controls Matter? Working paper. A Review of Empirical Evidence on the Effectiveness and Efficiency of Local Land Use Regulations." Lincoln Institute of Land Policy, Cambridge, Mass.

Gambe, Hays B., and Roger H. Downs. "Effects of Nuclear Power Plants on Residential Property Values," *Journal of Regional Science.* pp. 457–478.

Garreau, Joel. 1991. *Edge City: Life on the New Frontier.* Doubleday, New York.

Gelman, Sheldon R., Donald J. Epp, Roger H. Downing, Richard D. Twark, and Raymond Eyerly. 1989. "Impact of Group Homes on the Values of Adjacent Residential Values." *Mental Retardation,* June, pp. 127–134.

Hennebery, David M., and Richard L. Barrows. 1990. "Capitalization of Exclusive Agricultural Zoning into Farmland Prices." *Land Economics,* vol. 66, no. 3, August, University of Wisconsin Press, pp. 250–258.

Kellogg, Robert L., Gale W. TeSelle, and J. Jeffery Goebel. 1994. "Highlights from the 1992 National Resources Inventory." *Journal of Soil and Water Conservation,* vol. 49, no. 6, November–December.

Knaap, Gerrit J. 1985. "The Price Effects of Urban Growth Boundaries in Metropolitan Portland, Oregon." *Land Economics,* vol. 61, February, pp. 28–35.

Lincoln Institute of Land Policy. 1993. "Is Land Conservation Bad for the Tax Base?" *Landlines,* the Lincoln Institute of Land Policy Newsletter, September.

Mill, David C. 1990. "Zoning Rights and Land Development Theory." *Land Economics,* vol, 66, no. 3, August, University of Wisconsin Press, p. 285.

Nelson, Jon P. "Three Mile Island and Residential Property Values: Empirical Analysis and Policy Implications." *Land Economics,* vol. 57, pp. 363–372. University of Wisconsin Press.

Pogodzinski, J. Michael, and Tim R. Sass. 1990. "A Review of Zoning Theory." *Land Economics,* vol. 66, no. 3, August, University of Wisconsin Press, pp. 294–313.

Pollakowski, Henry O., and Susan M. Wachter. 1990. "The Effects of Constraints on Housing Prices." *Land Economics,* vol. 66, no. 3, August, University of Wisconsin Press, pp. 315–324.

Shortle, Jane, Roger H. Downs, Donald Epp, and Hays B. Gamble. n.d. "Effects of Solid Waste Disposal Sites on Community Development and Residential Property Values." Unpublished paper prepared by the Institute for Research in Land and Water Resources.

Small, Stephen J. *The Federal Tax Law of Conservation Easements.* Land Trust Alliance, Washington, D.C.

Tishler, William H., ed. 1989. *American Landscape Architecture: Designers and Places.* National Trust for Historic Preservation: The Preservation Press, Washington, D.C.

U.S. Internal Revenue Service. *Federal Tax Code,* Reg. Sec. 14(h)(3)(i), example 10.
Vaillancourt, Francois, and Luc Monty. 1985. "The Effect of Agricultural Zoning on Land Prices, Quebec, 1975–1981." *Land Economics,* vol. 61, February, University of Wisconsin Press, pp. 36–42.
Williams, Howard L., and W. D. Davis. 1968. "Effect of Scenic Easement on the Market Value of Real Property." *The Appraiser Journal,* October, pp. 15–24.

Chapter 12. Landscape and Culture

Garreau, Joel. 1991. *Edge City: Life on the New Frontier.* Doubleday, New York.
Jackson, John B. 1986. *Discovering the Vernacular Landscape,* Yale University Press, New Haven, Conn.
Feldman, Elliot J., and Michael A. Goldeberg, eds. *Land Rites and Wrongs, The Management, Regulation and Use of Land in Canada and the United States.* Lincoln Institute of Land Policy, Cambridge, Mass.
Fischel, William A. "Do Growth Controls Matter? A Review of Empirical Evidence on the Effectiveness and Efficiency of Local Government Land Use Regulations." Working paper. Lincoln Institute of Land Policy, Cambridge, Mass.
Hendee, John C., and Randall C. Pitstick. 1992. "The Growth of Environmental and Conservation Related Organizations, 1980–1991." *Renewable Resources Journal,* summer.
Kempton, Willett, James S. Boster, and Jennifer A. Hartley. 1996. *Environmental Values in Americana Culture,* The MIT Press, Cambridge, Mass.
Kenyon, Daphne A. "The Economics of NIMBYs." Working paper. Lincoln Institute of Land Policy, Cambridge, Mass.
Lyle, John Tillman. 1994. *Regenerative Design for Sustainable Development.* Wiley, New York.
Pirsig, Robert. 1974. *Zen and the Art of Motorcycle Maintenance, An Inquiry into Values.* Bantam New Age, New York.
Reilly, William K., ed. 1973. *The Use of Land: A Policy Guide to Urban Growth.* A task force report sponsored by The Rockefeller Brothers Fund. Thomas Y. Crowell, New York.
Stilgoe, John R. 1982. *Common Landscape of America, 1580 to 1845.* Yale University Press, New Haven, Conn.
Times-Mirror Magazines. 1992. *Natural Resource Conservation: Where Environmentalism Is Headed in the 1990s.*
Wright, Frank Lloyd. n.d. "A Brief Biography." The Frank Lloyd Wright Foundation. www.frank lloydwright.org.

Index

P

Q

R

ABOUT THE AUTHOR

Thomas H. Russ is a licensed landscape architect with more than 20 years of professional experience in both environmental assessment and site design. He has worked as a project manager and principal in design firms and is currently Associate Professor of Environmental Technology at the College of Southern Maryland. Mr. Russ has written numerous papers on site and environmental design topics and is the author of *Redeveloping Brownfields for Landscape Architects, Designers, and Developers*.